# Lecture Notes in Physics

# The Lecture Notes in Physics

The series Lecture Notes in Physics (LNP), founded in 1969, reports new developments in physics research and teaching – quickly and informally, but with a high quality and the explicit aim to summarize and communicate current knowledge in an accessible way. Books published in this series are conceived as bridging material between advanced graduate textbooks and the forefront of research and to serve three purposes:

- to be a compact and modern up-to-date source of reference on a well-defined topic
- to serve as an accessible introduction to the field to postgraduate students and nonspecialist researchers from related areas
- to be a source of advanced teaching material for specialized seminars, courses and schools

Both monographs and multi-author volumes will be considered for publication. Edited volumes should, however, consist of a very limited number of contributions only. Proceedings will not be considered for LNP.

Volumes published in LNP are disseminated both in print and in electronic formats, the electronic archive being available at springerlink.com. The series content is indexed, abstracted and referenced by many abstracting and information services, bibliographic networks, subscription agencies, library networks, and consortia.

Proposals should be sent to a member of the Editorial Board, or directly to the managing editor at Springer:

Christian Caron
Springer Heidelberg
Physics Editorial Department I
Tiergartenstrasse 17
69121 Heidelberg / Germany
christian.caron@springer.com

For other titles published in this series, go to
www.springer.com/series/5304

Y.N. Grigoriev · N.H. Ibragimov · V.F. Kovalev ·
S.V. Meleshko

# Symmetries of Integro-Differential Equations

## With Applications in Mechanics and Plasma Physics

Springer

Prof. Yurii N. Grigoriev
Russian Academy of Sciences
Inst. Computational Technologies
Pr. Lavrentjeva 6
630090 Novosibirsk
Russia
grigor@ict.nsc.ru

Prof. Nail H. Ibragimov
Blekinge Institute of Technology
Dept. Mathematics & Science
371 79 Karlskrona
Sweden
nib@bth.se

Dr. Vladimir F. Kovalev
Russian Academy of Sciences
Inst. Mathematical Modelling
Miusskaya Square 4A
125047 Moscow
Russia
kovalev@imamod.ru

Sergey V. Meleshko
Suranaree University of Technology (SUT)
School of Mathematics
Institute of Science
30000 Nakhon Ratchasima
Thailand
sergey@math.sut.ac.th

Grigoriev, Y.N. et al., *Symmetries of Integro-Differential Equations: With Applications in Mechanics and Plasma Physics*, Lect. Notes Phys. 806 (Springer, Dordrecht 2010), DOI 10.1007/978-90-481-3797-8

Lecture Notes in Physics ISSN 0075-8450   e-ISSN 1616-6361
ISBN 978-90-481-3796-1                     e-ISBN 978-90-481-3797-8
DOI 10.1007/978-90-481-3797-8
Springer Dordrecht Heidelberg London New York

Library of Congress Control Number: 2010930651

Springer is part of Springer Science+Business Media (www.springer.com)

# Preface

The present book is an introduction to a new field in applied group analysis. The book deals with symmetries of integro-differential, stochastic and delay equations that form the basis of a large variety of mathematical models, used to describe various phenomena in fluid mechanics and plasma physics and other fields of nonlinear science.

Because of its baffling complexity the mathematical study of nonlocal equations is far from completion, although the equations have been intensively studied in numerous applications over more than fifty last years using both numerical and analytical methods. The principal aim of analytical approaches is to obtain exact solutions, admitted symmetries, conservation laws and other mathematical properties, which allow one to make sound decisions in more detailed applied investigations.

Classical Lie group theory provides a universal tool for calculating symmetry groups for systems of differential equations. Consequently, group theoretical methods appear efficient in analyzing different phenomena using mathematical models that employ differential equations. However Lie's methods cannot be directly applied to integro-differential equations, infinite systems of differential equations, delay equations, etc. Hence it is natural to extend the ideas of modern group analysis to these mathematical objects that up to recently were not in mainstream of classical group theoretical approaches.

The book is designed for specialists in nonlinear physics interested in methods of applied group analysis for investigating nonlinear problems in physical, engineering and natural sciences. It is based on our research results and various courses and lectures given to undergraduate and graduate students as well as professional audiences over the past thirty years. The book can also serve as a textbook on symmetries of integro-differential, stochastic and delay equations for graduate students in applied mathematics, physics and engineering.

In the preparation of this monograph the roles were distributed in the following way. The first chapter was written by N.H. Ibragimov. The second and third chapters are the result of collaboration between Y.N. Grigoriev and S.V. Meleshko. The fourth chapter was prepared by V.F. Kovalev. Chapters five and six are the work of S.V. Meleshko.

# Organization of the Book

The contents of this book have been assembled from results scattered across many different articles and books published over the last thirty years.

The monograph includes six chapters. The first chapter contains an introduction to the methods of Lie group analysis of ordinary and partial differential equations. The basic notions of this mathematical area: continuous transformation groups, algebras of their generators, determining equations and methods of finding invariant solutions of differential equations are presented and illustrated by numerous examples. New trends in modern group analysis are also reflected. The intention of the chapter is to give the basic ideas of classical and modern group analysis to beginner readers and provide useful materials for advanced specialists.

The second chapter presents a survey of different methods for constructing symmetries and finding invariant solutions of integro-differential equations. An introduction to these methods is carried out using simple model equations, allowing the reader to follow the calculations in detail. The chapter includes substantial generalization of the original scheme of the group analysis method to equations with nonlocal operators. In the concluding sections of the chapter this regular method of obtaining admitted Lie groups is illustrated by applications to different integro-differential equations.

The results of group analysis of the Boltzmann kinetic equation and some similar equations with squarely nonlinear integral operators are described in the third chapter. These equations form the foundation of the kinetic theory of rarefied gas and coagulation. The main point of interest here is the isomorphism of the Lie group of point transformations admitted by the full Boltzmann equation and the Euler inviscid gas dynamic system. This remarkable fact allows us to obtain representations of all invariant solutions with one and two independent variables of the Boltzmann equation. For equations with few number of independent variables the proposed method allows us to derive constructive proofs of the completeness of admitted Lie groups. The representations of all invariant solutions are also presented.

The fourth chapter is entirely devoted to a group analysis of the Vlasov–Maxwell and related type equations. The equations form the basis of the collisionless plasma kinetic theory, and are also applied in gravitational astrophysics, in shallow-water theory, in the theory of pulverulent suspensions, etc. Nonlocal operators in these equations appear in the form of the functionals defined by integrals of the distribution functions over momenta of particles. Much of the importance of the approach used in this chapter for calculating symmetries stems from the procedure of solving determining equations using variational differentiation. The set of symmetries obtained comprises symmetries for the Vlasov–Maxwell equations of the non-relativistic and relativistic electron and electron–ion plasmas in both one- and three-dimensional cases, and symmetries for Benney equations. In the concluding sections of this chapter the procedure for symmetry calculation and the renormalization group algorithm go hand in hand to present illustrations from plasma kinetic theory, plasma dynamics, and nonlinear optics, which demonstrate the potentialities of the method in construction of analytic solutions to nonlocal problems of nonlinear physics.

The fifth and sixth chapters present new fields of application of group analysis to stochastic and delay differential equations. In the fifth chapter a definition of determining equations for calculation of the Lie algebras admitted by stochastic dynamical systems is formulated. This gives an opportunity to derive determining equations for symmetries of Itô and Stratonovich dynamical systems.

The sixth chapter deals with symmetries of delay differential equations. In recent years these equations have been intensively studied in biology, in population dynamics and bioscience problems, in control problems, etc. The equations have a nonlocal character because their solutions demand a knowledge of not only current conditions, but also of conditions at certain previous moments. The concept of determining equations is also introduced here, and is then used for classification of invariant solutions of the second-order ordinary delay differential equations.

## Acknowledgements

We express our gratitude to Alan Jeffrey who persuaded us to undertake this venture and was unfailing in his assistance. We owe sincere thanks to our friends, colleagues and students for their support. Finally, we thank the Editors at Springer for their cooperation and courtesy during the preparation of the monograph.

| | |
|---|---|
| Novosibirsk, Russia | Y.N. Grigoriev |
| Karlskrona, Sweden | N.H. Ibragimov |
| Moscow, Russia | V.F. Kovalev |
| Nakhon Ratchasima, Thailand | S.V. Meleshko |

# Contents

# Chapter 1
# Introduction to Group Analysis of Differential Equations

In this chapter we introduce the basic concepts from Lie group analysis: continuous transformation groups, their generators, Lie equations, groups admitted by differential equations, integration of ordinary differential equations using their symmetries, group classification and invariant solutions of partial differential equations. It contains also an introduction to the theory of Lie–Bäcklund transformations groups and approximate groups. The reader interested in studying more about Lie group methods of integration of differential equations is referred to [8] and to the recent textbook [10].

## 1.1 One-Parameter Groups

### 1.1.1 Definition of a Transformation Group

We will consider here only one-parameter groups. Let $T_a$ be an invertible transformation depending on a real parameter $a$ and acting in the $(x, y)$-plane:

$$\bar{x} = f(x, y, a), \quad \bar{y} = g(x, y, a), \tag{1.1.1}$$

where the functions $f$ and $g$ satisfy the conditions

$$f\big|_{a=0} = x, \quad g\big|_{a=0} = y. \tag{1.1.2}$$

The invertibility is guaranteed if one requires that the Jacobian of $f, g$ with respect to $x, y$ is not zero in a neighborhood of $a = 0$. Further, it is assumed that the functions $f$ and $g$ as well as their derivatives that appear in the subsequent discussion are continuous in $x, y, a$.

**Definition 1.1.1** A set $G$ of transformations (1.1.1) is a *one-parameter transformation group* if it contains the identical transformation $I = T_0$ and includes the inverse $T_a^{-1}$ as well as the composition $T_b T_a$ of all its elements $T_a, T_b \in G$. By a suitable choice of the group parameter $a$, the main group property $T_b T_a \in G$ can be written

$$T_b T_a = T_{a+b},$$

Y.N. Grigoriev et al., *Symmetries of Integro-Differential Equations*,
Lecture Notes in Physics 806,
DOI 10.1007/978-90-481-3797-8_1, © Springer Science+Business Media B.V. 2010

that is

$$f\big(f(x, y, a), g(x, y, a), b\big) = f(x, y, a + b),$$
$$g\big(f(x, y, a), g(x, y, a), b\big) = g(x, y, a + b). \tag{1.1.3}$$

In practical applications, the conditions (1.1.3) hold only for sufficiently small values of $a$ and $b$. Then one arrives at what is called a *local one-parameter group* $G$. For brevity, local groups are also termed groups.

## 1.1.2 Generator of a One-Parameter Group

The expansion of the functions $f, g$ into the Taylor series in $a$ near $a = 0$, taking into account the initial condition (1.1.2), yields the *infinitesimal transformation* of the group $G$ (1.1.1):

$$\bar{x} \approx x + \xi(x, y)a, \quad \bar{y} \approx y + \eta(x, y)a, \tag{1.1.4}$$

where

$$\xi(x, y) = \frac{\partial f(x, y, a)}{\partial a}\bigg|_{a=0}, \quad \eta(x, y) = \frac{\partial g(x, y, a)}{\partial a}\bigg|_{a=0}. \tag{1.1.5}$$

The vector $(\xi, \eta)$ with components (1.1.5) is the tangent vector (at the point $(x, y)$) to the curve described by the transformed points $(\bar{x}, \bar{y})$, and is therefore called the *tangent vector field* of the group $G$.

**Example 1.1.1** The group of rotations

$$\bar{x} = x \cos a + y \sin a, \quad \bar{y} = y \cos a - x \sin a$$

has the following infinitesimal transformation:

$$\bar{x} \approx x + ya, \quad \bar{y} \approx y - xa.$$

The tangent vector field (1.1.5) is sometimes also written as a first-order differential operator

$$X = \xi(x, y)\frac{\partial}{\partial x} + \eta(x, y)\frac{\partial}{\partial y}, \tag{1.1.6}$$

which behaves as a *scalar* under an arbitrary change of variables, unlike the *vector* $(\xi, \eta)$. Lie called the operator (1.1.6) the *symbol* of the infinitesimal transformation (1.1.4) or of the corresponding group $G$. In the current literature, the operator $X$ (1.1.6) is called the *generator* of the group $G$ of transformations (1.1.1).

**Example 1.1.2** The generator of the group of rotations from Example 1.1.1 has the form

$$X = y\frac{\partial}{\partial x} - x\frac{\partial}{\partial y}. \tag{1.1.7}$$

### 1.1.3 Construction of a Group with a Given Generator

Given an infinitesimal transformation (1.1.4), or the generator (1.1.6), the transformations (1.1.1) of the corresponding one-parameter group $G$ are defined by solving the following equations known as the *Lie equations*:

$$\frac{df}{da} = \xi(f, g), \quad f\big|_{a=0} = x,$$

$$\frac{dg}{da} = \eta(f, g), \quad g\big|_{a=0} = y. \tag{1.1.8}$$

We will write (1.1.8) also in the following equivalent form:

$$\frac{d\bar{x}}{da} = \xi(\bar{x}, \bar{y}), \quad \bar{x}\big|_{a=0} = x,$$

$$\frac{d\bar{y}}{da} = \eta(\bar{x}, \bar{y}), \quad \bar{y}\big|_{a=0} = y. \tag{1.1.9}$$

**Example 1.1.3** Consider the infinitesimal transformation

$$\bar{x} \approx x + ax^2, \quad \bar{y} \approx y + axy.$$

The corresponding generator has the form

$$X = x^2 \frac{\partial}{\partial x} + xy \frac{\partial}{\partial y}. \tag{1.1.10}$$

The Lie equations (1.1.9) are written as follows:

$$\frac{d\bar{x}}{da} = \bar{x}^2, \quad \bar{x}\big|_{a=0} = x,$$

$$\frac{d\bar{y}}{da} = \bar{x}\bar{y}, \quad \bar{y}\big|_{a=0} = y.$$

The differential equations of this system are easily solved and yield

$$\bar{x} = -\frac{1}{a + C_1}, \quad \bar{y} = \frac{C_2}{a + C_1}.$$

The initial conditions imply that $C_1 = -1/x$, $C_2 = -y/x$. Consequently we arrive at the following one-parameter group of *projective transformations*:

$$\bar{x} = \frac{x}{1 - ax}, \quad \bar{y} = \frac{y}{1 - ax}. \tag{1.1.11}$$

One can represent the solution to the Lie equations (1.1.9) by means of infinite power series (Taylor series). Then the group transformation (1.1.1) for a generator $X$ (1.1.6) is given by the so-called *exponential map*:

$$\bar{x} = e^{aX}(x), \quad \bar{y} = e^{aX}(y), \tag{1.1.12}$$

where

$$e^{aX} = 1 + \frac{a}{1!}X + \frac{a^2}{2!}X^2 + \cdots + \frac{a^s}{s!}X^s + \cdots. \tag{1.1.13}$$

**Example 1.1.4** Consider again the generator (1.1.10) discussed in Example 1.1.3:

$$X = x^2 \frac{\partial}{\partial x} + xy \frac{\partial}{\partial y}.$$

According to (1.1.12)–(1.1.13), one has to find $X^s(x)$ and $X^s(y)$ for all $s = 1, 2, \ldots$. We calculate several terms, e.g.

$$X(x) = x^2, \quad X^2(x) = X(X(x)) = X(x^2) = 2!x^3, \quad X^3(x) = X(2!x^3) = 3!x^4,$$

and then make a guess:

$$X^s(x) = s!x^{s+1}.$$

The proof of the latter equation is given by induction:

$$X^{s+1}(x) = X(s!x^{s+1}) = (s+1)!x^2 x^s = (s+1)!x^{s+2}.$$

Furthermore, one obtains

$$X(y) = xy, \quad X^2(y) = X(xy) = yX(x) + xX(y) = yx^2 + xxy = 2!yx^2,$$
$$X^3(y) = 2![yX(x^2) + x^2 X(y)] = 2![y(2x^3) + x^2 xy] = 3!yx^3,$$

then makes a guess

$$X^s(y) = s!yx^s$$

and proves it by induction:

$$X^{s+1}(y) = s!X(yx^s) = s![syx^{s+1} + x^s(xy)] = (s+1)!yx^{s+1}.$$

Substitution of the above expressions in the exponential map yields:

$$e^{aX}(x) = x + ax^2 + \cdots + a^s x^{s+1} + \cdots.$$

One can rewrite the right-hand side as $x(1 + ax + \cdots + a^s x^s + \cdots)$. The series in brackets is manifestly the Taylor expansion of the function $1/(1 - ax)$ provided that $|ax| < 1$. Consequently,

$$\bar{x} = e^{aX}(x) = \frac{x}{1 - ax}.$$

Likewise, one obtains

$$e^{aX}(y) = y + ayx + a^2 yx^2 + \cdots + a^s yx^s + \cdots$$
$$= y(1 + ax + \cdots + a^s x^s + \cdots).$$

Hence,

$$\bar{y} = e^{aX}(y) = \frac{y}{1 - ax}.$$

Thus, we have arrived at the transformations (1.1.11):

$$\bar{x} = \frac{x}{1 - ax}, \quad \bar{y} = \frac{y}{1 - ax}.$$

### *1.1.4 Introduction of Canonical Variables*

**Theorem 1.1.1** *Every one-parameter group of transformations* (1.1.1) *reduces to the group of translations* $\bar{t} = t + a$, $\bar{u} = u$ *with the generator* $X = \frac{\partial}{\partial t}$ *by a suitable change of variables*

$$t = t(x, y), \quad u = u(x, y).$$

*The variables* $t, u$ *are called canonical variables.*

*Proof* Under a change of variables the differential operator (1.1.6) transforms according to the formula

$$X = X(t)\frac{\partial}{\partial t} + X(u)\frac{\partial}{\partial u}. \tag{1.1.14}$$

Therefore canonical variables are found from the linear partial differential equations of the first order:

$$
\begin{aligned}
X(t) &\equiv \xi(x, y)\frac{\partial t(x, y)}{\partial x} + \eta(x, y)\frac{\partial t(x, y)}{\partial y} = 1, \\
X(u) &\equiv \xi(x, y)\frac{\partial u(x, y)}{\partial x} + \eta(x, y)\frac{\partial u(x, y)}{\partial y} = 0.
\end{aligned}
\tag{1.1.15}
$$

$\square$

### *1.1.5 Invariants (Invariant Functions)*

**Definition 1.1.2** A function $F(x, y)$ is an invariant of the group $G$ of transformations (1.1.1) if $F(\bar{x}, \bar{y}) = F(x, y)$, i.e.

$$F\big(f(x, y, a), g(x, y, a)\big) = F(x, y) \tag{1.1.16}$$

identically in the variables $x$, $y$ and the group parameter $a$.

**Theorem 1.1.2** *A function* $F(x, y)$ *is an invariant of the group* $G$ *if and only if it solves the following first-order linear partial differential equation*

$$XF \equiv \xi(x, y)\frac{\partial F}{\partial x} + \eta(x, y)\frac{\partial F}{\partial y} = 0. \tag{1.1.17}$$

*Proof* Let $F(x, y)$ be an invariant. Let us take the Taylor expansion of $F(f(x, y, a), g(x, y, a))$ with respect to $a$:

$$F\big(f(x, y, a), g(x, y, a)\big) \approx F(x + a\xi, y + a\eta) \approx F(x, y) + a\left(\xi\frac{\partial F}{\partial x} + \eta\frac{\partial F}{\partial y}\right),$$

or

$$F(\bar{x}, \bar{y}) = F(x, y) + aX(F) + o(a),$$

and substitute it in (1.1.16):

$$F(x, y) + aX(F) + o(a) = F(x, y).$$

It follows that $aX(F) + o(a) = 0$, whence $X(F) = 0$, i.e. (1.1.17).

Conversely, let $F(x, y)$ be a solution of (1.1.17). Assuming that the function $F(x, y)$ is analytic and using its Taylor expansion, one can extend the exponential map (1.1.12) to the function $F(x, y)$ as follows:

$$F(\bar{x}, \bar{y}) = e^{aX} F(x, y) \overset{\text{def}}{=} \left( 1 + \frac{a}{1!}X + \frac{a^2}{2!}X^2 + \cdots + \frac{a^s}{s!}X^s + \cdots \right) F(x, y).$$

Since $XF(x, y) = 0$, one has $X^2 F = X(XF) = 0, \ldots, X^s F = 0$. We conclude that $F(\bar{x}, \bar{y}) = F(x, y)$, i.e. (1.1.16) thus proving the theorem.

It follows from Theorem 1.1.2 that every one-parameter group of transformations in the plane has one independent invariant, which can be taken to be the left-hand side of any first integral $\psi(x, y) = C$ of the characteristic equation for (1.1.17):

$$\frac{dx}{\xi(x, y)} = \frac{dy}{\eta(x, y)}. \tag{1.1.18}$$

Any other invariant $F$ is then a function of $\psi$, i.e. $F(x, y) = \Phi(\psi(x, y))$.  □

**Example 1.1.5** Consider the group with the generator

$$X = x\frac{\partial}{\partial x} + 2y\frac{\partial}{\partial y}.$$

The characteristic equation (1.1.18) is written

$$\frac{dx}{x} = \frac{dy}{2y}$$

and yields the first integral $\psi = y/x^2$. Hence, the general invariant is given by $F(x, y) = \Phi(y/x^2)$ with an arbitrary function $\Phi$ of one variable.

The concepts introduced above can be generalized in an obvious way to the multi-dimensional case by considering groups of transformations

$$\bar{x}^i = f^i(x, a), \quad i = 1, \ldots, n, \tag{1.1.19}$$

in the $n$-dimensional space $R^n$ of points $x = (x^1, \ldots, x^n)$ instead of transformations (1.1.1) in the $(x, y)$-plane. The generator of the group of transformations (1.1.19) is written

$$X = \xi^i(x)\frac{\partial}{\partial x^i}, \tag{1.1.20}$$

where

$$\xi^i(x) = \left. \frac{\partial f^i(x, a)}{\partial a} \right|_{a=0}.$$

The Lie equations (1.1.9) become

$$\frac{d\bar{x}^i}{da} = \xi^i(\bar{x}), \quad \bar{x}^i\big|_{a=0} = x^i. \tag{1.1.21}$$

The exponential map (1.1.12) is written:

$$\bar{x}^i = e^{aX}(x^i), \quad i = 1, \ldots, n, \tag{1.1.22}$$

where

$$e^{aX} = 1 + \frac{a}{1!}X + \frac{a^2}{2!}X^2 + \cdots + \frac{a^s}{s!}X^s + \cdots. \tag{1.1.23}$$

The extension of the exponential map to a function $F(x)$ is written

$$F(\bar{x}) = e^{aX}F(x) \equiv F(x) + aX(F(x)) + \frac{a^2}{2!}X^2(F(x)) + \cdots. \tag{1.1.24}$$

Definition 1.1.2 of invariant functions of several variables remains the same, namely an invariant is defined by the equation $F(\bar{x}) = F(x)$. The invariant test given by Theorem 1.1.2 has the same formulation with the evident replacement of (1.1.17) by its $n$-dimensional version:

$$\sum_{i=1}^{n} \xi^i(x)\frac{\partial F}{\partial x^i} = 0. \tag{1.1.25}$$

Then $n - 1$ functionally independent first integrals $\psi_1(x), \ldots, \psi_{n-1}(x)$ of the characteristic system for (1.1.25):

$$\frac{dx^1}{\xi^1(x)} = \frac{dx^2}{\xi^2(x)} = \cdots = \frac{dx^n}{\xi^n(x)} \tag{1.1.26}$$

provides a basis of invariants. Namely, any invariant $F(x)$ is given by

$$F(x) = \Phi\big(\psi_1(x), \ldots, \psi_{n-1}(x)\big). \tag{1.1.27}$$

### 1.1.6 Invariant Equations (Manifolds)

Let $x = (x^1, \ldots, x^n) \in R^n$. Consider an $(n - s)$-dimensional manifold $M \subset R^n$ defined by a system of equations[1]

$$F_1(x) = 0, \ldots, F_s(x) = 0, \quad s < n. \tag{1.1.28}$$

It is assumed that

$$\text{rank}\left\|\frac{\partial F_k}{\partial x^i}\right\|_M = s. \tag{1.1.29}$$

---

[1] Manifolds are treated locally and all functions under consideration are supposed to be continuous and differentiable sufficiently many times.

**Definition 1.1.3** The system of equations (1.1.28) is said to be invariant with respect to the group $G$ of transformations (1.1.19),

$$\bar{x}^i = f^i(x, a), \quad i = 1, \ldots, n,$$

if each solution $x = (x^1, \ldots, x^n)$ of the system (1.1.28) is mapped to a solution $\bar{x} = (\bar{x}^1, \ldots, \bar{x}^n)$ of the same system, i.e.

$$F_1(\bar{x}) = 0, \ldots, F_s(\bar{x}) = 0. \tag{1.1.30}$$

We also say that (1.1.28) admit the group $G$. The invariance of (1.1.28) means that the manifold $M \subset R^n$ defined by (1.1.28) is also invariant in the sense that each point $x$ on the surface $M$ is moved by $G$ along the surface $M$, i.e. $x \in M$ implies that $\bar{x} \in M$.

**Theorem 1.1.3** *The system of equations* (1.1.28) *admits the group $G$ of transformations* (1.1.19) *with the generator $X$* (1.1.20) *if and only if*

$$X F_k \big|_M = 0, \quad k = 1, \ldots, s. \tag{1.1.31}$$

*Proof* (See also [8], Sect. 7.2.) Let the system (1.1.28) be invariant under the group $G$, i.e. let (1.1.30) hold for every point $x \in M$ and every admissible value of the group parameter $a$. Taking into account that

$$F_k(\bar{x}) = F_k(x) + a X F_k + o(a), \quad k = 1, \ldots, s,$$

and that $F_k(x) = 0$ whenever $x \in M$, one arrives at (1.1.31).

Let us prove now that (1.1.31) imply the invariance of the system (1.1.28), i.e. that (1.1.30) hold for any point $x \in M$. We assume in what follows that the functions $F_k(z)$ and $X F_k(z)$ are analytic in a neighborhood of the manifold $M$. Then (1.1.31) can be written in the form

$$X F_k(z) = \lambda_k^l(z) F_l(z), \quad k = 1, \ldots, s, \tag{1.1.32}$$

where the coefficients $\lambda_k^l(z)$ are bounded in a neighborhood of $M$. Equations (1.1.32), together with (1.1.24), provide the proof. Indeed, it follows from (1.1.32) that

$$X^2 F_k = X(\lambda_k^l) F_l + \lambda_k^l X(F_l) = \left[ X(\lambda_k^p) + \lambda_k^l \lambda_l^p \right] F_p.$$

Iteration and substitution into (1.1.24) yields $F_k(\bar{x}) = \Lambda_k^l(x) F_l(x)$. It follows that (1.1.30) hold, thus completing the proof. $\square$

**Remark 1.1.1** The condition (1.1.29) is used for reducing (1.1.28) to the form (1.1.32).

### *1.1.7 Representation of Regular Invariant Manifolds via Invariants*

**Definition 1.1.4** Let $G$ be a one-parameter group of transformations (1.1.19) with the generator (1.1.20),

$$X = \xi^i(x)\frac{\partial}{\partial x^i}.$$

An invariant manifold $M$ of the group $G$ is said to be *regular* with respect to $G$ if at least one of the coefficients $\xi^i(x)$ does not vanish on $M$, and it is *singular* if all coefficients $\xi^i(x)$ of the generator $X$ vanish on $M$.

Invariant manifolds of a given group $G$ can be equivalently represented by different systems of equations (1.1.28). A general procedure for constructing invariant manifolds is provided by the following theorem on representation of *regular* invariant manifolds by invariant functions (for the proof, see [16], §8.7, or [8], Sect. 7.2.2).

**Theorem 1.1.4** *Let $G$ be a group of transformations (1.1.19). Any regular $(n-s)$-dimensional manifold $M \subset R^n$ can be represented by a system of equations (1.1.28) with invariant functions $F_k$, i.e. (see (1.1.27))*

$$F_k(x) = \Phi_k\big(\psi_1(x), \ldots, \psi_{n-1}(x)\big), \quad k = 1, \ldots, s, \tag{1.1.33}$$

*where $\psi_1(x), \ldots, \psi_{n-1}(x)$ is a basis of invariants of the group $G$.*

**Example 1.1.6** Let $G$ be the group of dilations

$$\bar{x} = xe^a, \quad \bar{y} = ye^a, \quad \bar{z} = ze^{2a}$$

in the three-dimensional space $R^3$. The generator of this group is

$$X = x\frac{\partial}{\partial x} + y\frac{\partial}{\partial y} + 2z\frac{\partial}{\partial z}.$$

The characteristic equations (1.1.26) are written

$$\frac{dx}{x} = \frac{dy}{y} = \frac{dz}{z}$$

and yield the following basis of invariants for the group $< G$:

$$\psi_1 = \frac{x^2}{z}, \quad \psi_2 = \frac{y^2}{z}.$$

According to Theorem 1.1.4, any regular two-dimensional invariant manifold (a surface in $R^3$) is given by $\Phi(\psi_1, \psi_2) = 0$:

$$\Phi\left(\frac{x^2}{z}, \frac{y^2}{z}\right) = 0.$$

In particular, taking $\Phi(\psi_1, \psi_2) = \psi_1 + \psi_2 - C$ with any constant $C$ we obtain a paraboloid

$$\frac{x^2 + y^2}{z} - C = 0.$$

The left-hand side of this equation is an invariant function with respect to the group $G$. But if multiply the above equation by $z$, we represent the same invariant paraboloid by the equation

$$x^2 + y^2 - Cz = 0$$

whose left-hand side is not an invariant function.

## 1.2 Symmetries and Integration of Ordinary Differential Equations

### 1.2.1 The Frame of Differential Equations

Any differential equation has two components, namely, the *frame* and the *class of solutions* (see [8]). For example, the frame of a first-order ordinary differential equation

$$F(x, y, y') = 0$$

is the surface $F(x, y, p) = 0$ in the space of three *independent variables* $x, y, p$. It is obtained by replacing the first derivative $y'$ in the differential equation $F(x, y, y') = 0$ by the variable $p$.

The class of solutions is defined in accordance with certain "natural" mathematical assumptions or from a physical significance of the differential equations under discussion.

The crucial step in integrating differential equations is a "simplification" of the frame by a suitable change of the variables $x, y$. The Lie group analysis suggests methods for simplification of the frame by using *symmetry groups* (or *admissible groups*) of differential equations.

Consider, as an example, the following Riccati equation:

$$y' + y^2 - \frac{2}{x^2} = 0. \tag{1.2.1}$$

Its frame is defined by the algebraic equation

$$p + y^2 - \frac{2}{x^2} = 0 \tag{1.2.2}$$

and is a "hyperbolic paraboloid". For the Riccati equation (1.2.1), a one-parameter symmetry group is provided by the following scaling transformations (non-homogeneous dilations) obtained in Sect. 1.2.7:

$$\bar{x} = xe^a, \quad \bar{y} = ye^{-a}. \tag{1.2.3}$$

Indeed, the transformations (1.2.3) after the extension to the first derivative $y'$ and the substitution $y' = p$ are written

$$\bar{x} = xe^a, \quad \bar{y} = ye^{-a},$$
$$\bar{p} = pe^{-2a}. \tag{1.2.4}$$

One can readily verify that the frame of (1.2.2) is invariant with respect to the transformations (1.2.4). Let us check the infinitesimal invariance condition (1.1.31). The generator (1.1.20) of the group of transformations (1.2.4) has the form

$$X = x\frac{\partial}{\partial x} - y\frac{\partial}{\partial y} - 2p\frac{\partial}{\partial p}.$$

One can easily check that the invariance condition is satisfied. Indeed:

$$X\left(p + y^2 - \frac{2}{x^2}\right) = -2p - 2y^2 + \frac{4}{x^2} = -2\left(p + y^2 - \frac{2}{x^2}\right),$$

and hence $X(p + y^2 - \frac{2}{x^2})|_{(1.2.2)} = 0$. For the transformations (1.2.3), the canonical variables are

$$t = \ln x, \quad u = xy. \tag{1.2.5}$$

In the canonical variables (1.2.5), the Riccati equation (1.2.1) becomes:

$$u' + u^2 - u - 2 = 0 \quad (u' = du/dt). \tag{1.2.6}$$

Its frame is obtained by substituting $u' = q$ in (1.2.6) and is given by the following algebraic equation:

$$q + u^2 - u - 2 = 0. \tag{1.2.7}$$

The left-hand side of (1.2.7) does not involve the variable $t$. Thus the curved frame (1.2.2) has been reduced to a cylindrical surface protracted along the $t$-axis. Namely it is a "parabolic cylinder". We see that, in integrating differential equations, the decisive step is that of simplifying the frame by converting it into a cylinder. For such purpose, it is sufficient to simplify the symmetry group by introducing canonical variables. In consequence, any first-order ordinary differential equation with a known symmetry reduces to the integrable form $u' = f(u)$ similar to (1.2.6).

Of course, in certain particular examples the equation in question may be solved by other means. For example, it is well-known that the substitution $y = (\ln |u|)'$ reduces (1.2.1) to Euler's equation

$$x^2 u'' - 2u = 0$$

having the general solution $u = C_1 x^{-1} + C_2 x^2$. Hence, the general solution of (1.2.1) has the form

$$y = \frac{d}{dx}\ln\left|\frac{C_1}{x} + C_2 x^2\right| = \frac{2C_2 x^3 - C_1}{x(C_2 x^3 + C_1)}.$$

If $C_2 \neq 0$ one has the solution

$$y = \frac{2x^3 - C}{x(x^3 + C)}$$

depending on one arbitrary constant $C = C_1/C_2$. The case $C_2 = 0$ yields the singular solution $y = -1/x$.

## 1.2.2 Prolongation of Group Transformations and Their Generators

The transformation of derivatives $y', y'', \ldots$ under the action of the point transformations (1.1.1), regarded as a change of variables, is well-known from Calculus. It is convenient to write these transformation formulae by using the operator of *total differentiation*:

$$D = \frac{\partial}{\partial x} + y'\frac{\partial}{\partial y} + y''\frac{\partial}{\partial y'} + \cdots .$$

Then the transformation formulae, e.g. for the first and second derivatives are written

$$\bar{y}' \equiv \frac{d\bar{y}}{d\bar{x}} = \frac{Dg}{Df} = \frac{g_x + y'g_y}{f_x + y'f_y} \equiv P(x, y, y', a), \tag{1.2.8}$$

$$\bar{y}'' \equiv \frac{d\bar{y}'}{d\bar{x}} = \frac{DP}{Df} = \frac{P_x + y'P_y + y''P_{y'}}{f_x + y'f_y}. \tag{1.2.9}$$

Starting from the group $G$ of point transformations (1.1.1) and then adding the transformation (1.2.8), one obtains the group $G_{(1)}$, which acts in the space of the three variables $(x, y, y')$. Further, by adding the transformation (1.2.9) one obtains the group $G_{(2)}$ acting in the space $(x, y, y', y'')$.

**Definition 1.2.1** The groups $G_{(1)}$ and $G_{(2)}$ are termed the first and second *prolongations* of $G$, respectively. The higher prolongations are determined similarly.

Substituting into (1.2.8), (1.2.9) the infinitesimal transformation (1.1.4),

$$\bar{x} \approx x + a\xi, \quad \bar{y} \approx y + a\eta,$$

and neglecting all terms of higher order in $a$, one obtains the following infinitesimal transformations of derivatives:

$$\bar{y}' = \frac{y' + aD(\eta)}{1 + aD(\xi)} \approx [y' + aD(\eta)][1 - aD(\xi)]$$

$$\approx y' + [D(\eta) - y'D(\xi)]a \equiv y' + a\zeta_1,$$

$$\bar{y}'' = \frac{y'' + aD(\zeta_1)}{1 + aD(\xi)} \approx [y'' + aD(\zeta_1)][1 - aD(\xi)]$$

$$\approx y'' + [D(\zeta_1) - y''D(\xi)]a \equiv y'' + a\zeta_2.$$

Therefore the generators of the prolonged groups $G_{(1)}$, $G_{(2)}$ are

$$X_{(1)} = \xi \frac{\partial}{\partial x} + \eta \frac{\partial}{\partial y} + \zeta_1 \frac{\partial}{\partial y'}, \quad \zeta_1 = D(\eta) - y'D(\xi), \tag{1.2.10}$$

$$X_{(2)} = X_{(1)} + \zeta_2 \frac{\partial}{\partial y''}, \quad \zeta_2 = D(\zeta_1) - y''D(\xi). \tag{1.2.11}$$

These are called the *first* and *second* prolongations of the *infinitesimal operator* (1.1.9). The term *prolongation formulae* is frequently used to denote the expressions for the additional coordinates:

$$\zeta_1 = D(\eta) - y'D(\xi) = \eta_x + (\eta_y - \xi_x)y' - y'^2\xi_y, \tag{1.2.12}$$

$$\zeta_2 = D(\zeta_1) - y''D(\xi) = \eta_{xx} + (2\eta_{xy} - \xi_{xx})y'$$
$$+ (\eta_{yy} - 2\xi_{xy})y'^2 - y'^3\xi_{yy} + (\eta_y - 2\xi_x - 3y'\xi_y)y''. \tag{1.2.13}$$

### 1.2.3 Group Admitted by Differential Equations

Let $G$ be a group of point transformations and let $G_{(1)}$, $G_{(2)}$ be its first and second prolongations, defined in the previous section.

**Definition 1.2.2** We say that a group $G$ of point transformations (1.1.1) is a symmetry group of a first-order ordinary differential equation

$$F(x, y, y') = 0, \tag{1.2.14}$$

or that (1.2.14) admits the group $G$ if (1.2.14) is form invariant under the transformations (1.1.1), or, in other words, if the frame of (1.2.14) is invariant (in the sense of Definition 1.1.3) with respect to the first prolongation $G_{(1)}$ of the group $G$.

Likewise, an $n$th order differential equation

$$F(x, y, y', \ldots, y^{(n)}) = 0 \tag{1.2.15}$$

admits a group $G$ if the frame (the surface in the space $x, y, y', \ldots, y^{(n)}$) is invariant with respect to the $n$th prolongation $G_{(n)}$ of $G$.

Consider (1.2.15) written in the form solved with respect to the $y^{(n)}$:

$$y^{(n)} = f(x, y, y', \ldots, y^{(n-1)}) \tag{1.2.16}$$

with a smooth function $f$. The main property of a symmetry group first proved by S. Lie (the proof for first-order equations is given, e.g. in [13], Chap. 16, Sect. 1, Theorem 1) is the following.

**Theorem 1.2.1** *A group $G$ is a symmetry group for (1.2.16) if and only if $G$ converts any classical solution (i.e. $n$ times continuously differentiable) of (1.2.16) into a classical solution of the same equation.*

### 1.2.4 Determining Equation for Infinitesimal Symmetries

According to Sect. 1.1.3, it is sufficient to find *infinitesimal symmetries*, i.e. generators (1.1.6) of symmetry groups.

Here, the algorithm of construction of infinitesimal symmetries is discussed for second-order equations

$$F(x, y, y', y'') = 0. \tag{1.2.17}$$

The infinitesimal invariance criterion has the form:

$$X_{(2)}F\big|_{F=0} \equiv (\xi F_x + \eta F_y + \zeta_1 F_{y'} + \zeta_2 F_{y''})\big|_{F=0} = 0, \tag{1.2.18}$$

where $\zeta_1$ and $\zeta_2$ are computed from the prolongation formulae (1.2.12) and (1.2.13). Equation (1.2.18) is called the *determining equation* for the group admitted by the ordinary differential equation (1.2.17).

If the differential equation is written in the explicit form

$$y'' = f(x, y, y'), \tag{1.2.19}$$

the determining equation (1.2.18), after substituting the values of $\zeta_1$, $\zeta_2$ from (1.2.12), (1.2.13) with $y''$ given by the right-hand side of (1.2.19), assumes the form

$$\eta_{xx} + (2\eta_{xy} - \xi_{xx})y' + (\eta_{yy} - 2\xi_{xy})y'^2$$
$$- y'^3\xi_{yy} + (\eta_y - 2\xi_x - 3y'\xi_y)f$$
$$- [\eta_x + (\eta_y - \xi_x)y' - y'^2\xi_y]f_{y'} - \xi f_x - \eta f_y = 0. \tag{1.2.20}$$

Here $f(x, y, y')$ is a known function (we are dealing with a *given* differential equation (1.2.19) while the coordinates $\xi$ and $\eta$ of the generator (1.1.6)),

$$X = \xi(x, y)\frac{\partial}{\partial x} + \eta(x, y)\frac{\partial}{\partial y},$$

are unknown functions of $x, y$. Since the left-hand side of (1.2.20) contains the quantity $y'$ considered as an *independent variable* along with $x, y$, the determining equation splits into several independent equations, thus becoming an overdetermined system of differential equations for $\xi(x, y)$, $\eta(x, y)$. Solving this system, we find all the infinitesimal symmetries of (1.2.19).

### 1.2.5 An Example on Calculation of Symmetries

Let us find the operators

$$X = \xi(x, y)\frac{\partial}{\partial x} + \eta(x, y)\frac{\partial}{\partial y}$$

admitted by the second-order equation

$$y'' + \frac{1}{x}y' - e^y = 0. \tag{1.2.21}$$

Here $f = e^y - \frac{1}{x}y'$ and the determining equation (1.2.20) has the form

$$\eta_{xx} + (2\eta_{xy} - \xi_{xx})y' + (\eta_{yy} - 2\xi_{xy})y'^2$$
$$- y'^3\xi_{yy} + (\eta_y - 2\xi_x - 3y'\xi_y)\left(e^y - \frac{y'}{x}\right)$$
$$+ \frac{1}{x}[\eta_x + (\eta_y - \xi_x)y' - y'^2\xi_y] - \xi\frac{y'}{x^2} - \eta e^y = 0.$$

The left-hand side of this equation is a third-degree polynomial in the variable $y'$. Therefore the determining equation decomposes into the following four equations, obtained by setting the coefficients of the various powers of $y'$ equal to zero:

$$(y')^3: \quad \xi_{yy} = 0, \tag{1.2.22}$$

$$(y')^2: \quad \eta_{yy} - 2\xi_{xy} + \frac{2}{x}\xi_y = 0, \tag{1.2.23}$$

$$y': \quad 2\eta_{xy} - \xi_{xx} + \left(\frac{\xi}{x}\right)_x - 3\xi_y e^y = 0, \tag{1.2.24}$$

$$(y')^0: \quad \eta_{xx} + \frac{1}{x}\eta_x + (\eta_y - 2\xi_x - \eta)e^y = 0. \tag{1.2.25}$$

Integration of (1.2.22) and (1.2.23) with respect to $y$ yields:

$$\xi = p(x)y + a(x), \quad \eta = \left(p' - \frac{p}{x}\right)y^2 + q(x)y + b(x).$$

Let us substitute these expressions for $\xi$, $\eta$ into (1.2.24), (1.2.25). As the dependence of $\xi$ and $\eta$ on $y$ is polynomial, while the left-hand sides of (1.2.24), (1.2.25) contain $e^y$, we must have

$$\xi_y = 0, \quad \eta_y - 2\xi_x - \eta = 0.$$

The first of these gives us $p = 0$, that is, the equality $\xi = a(x)$; taking this into account, the second condition can be written in the form

$$q(x) - 2a'(x) - b(x) - q(x)y = 0.$$

Hence $q = 0$, $2a' + b = 0$. Therefore

$$\xi = a(x), \quad \eta = -2a'(x).$$

Substituting these expressions into (1.2.24), we have

$$\left(a' - \frac{a}{x}\right)' = 0,$$

from which $a = C_1 x \ln x + C_2 x$; here (1.2.25) is satisfied identically.

As a result, we have obtained the general solution of the determining equations (1.2.22)–(1.2.25) in the form

$$\xi = C_1 x \ln x + C_2 x, \quad \eta = -2[C_1(1 + \ln x) + C_2 x]$$

with constant coefficients $C_1$, $C_2$. In view of the linearity of the determining equations, the general solution can be represented as a linear combination of two independent solutions

$$\xi_1 = x \ln x, \quad \eta_1 = -2(1 + \ln x);$$
$$\xi_2 = x, \quad \eta_2 = -2.$$

This means that (1.2.21) admits two linearly independent operators

$$X_1 = x \ln x \frac{\partial}{\partial x} - 2(1 + \ln x) \frac{\partial}{\partial y}, \quad X_2 = x \frac{\partial}{\partial x} - 2 \frac{\partial}{\partial y}, \tag{1.2.26}$$

and that the set of all admissible operators is a two-dimensional vector space with basis (1.2.26).

### 1.2.6 Lie Algebras. Specific Property of Determining Equations

**Definition 1.2.3** Let $X$ and $X'$ be first-order linear differential operators of the form (1.1.6):

$$X = \xi(x, y) \frac{\partial}{\partial x} + \eta(x, y) \frac{\partial}{\partial y}, \quad X' = \xi'(x, y) \frac{\partial}{\partial x} + \eta'(x, y) \frac{\partial}{\partial y}. \tag{1.2.27}$$

Their commutator $[X, X']$ is defined by $[X, X'] = XX' - X'X$. It is a first-order linear differential operator and has the form:

$$[X, X'] = \left(X(\xi') - X'(\xi)\right) \frac{\partial}{\partial x} + \left(X(\eta') - X'(\eta)\right) \frac{\partial}{\partial y}. \tag{1.2.28}$$

**Definition 1.2.4** A vector space $L$ of operators (1.1.6) is called a Lie algebra if it is closed under the commutator, i.e. if $[X, X'] \in L$ for any $X, X' \in L$. The Lie algebra is denoted by the same letter $L$, and its dimension is the dimension of the vector space $L$.

If a Lie algebra $L$ has the dimension $r < \infty$ it is denoted by $L_r$. If the vector space $L_r$ is spanned by linearly independent operators $X_1, \ldots, X_r$, then the operators $X_1, \ldots, X_r$ provide a basis of the Lie algebra $L_r$. The condition that $[X, X'] \in L$ for any $X, X' \in L$ is equivalent to the following:

$$[X_i, X_j] = c_{ij}^k X_k, \quad c_{ij}^k = \text{const.} \quad (i, j, k = 1, \ldots, r). \tag{1.2.29}$$

**Definition 1.2.5** Let $L_r$ be a Lie algebra spanned by $X_1, \ldots, X_r$. A subspace $K_s$ $(s < r)$ of the vector space $L_r$ spanned by linearly independent operators $Y_1, \ldots, Y_s \in L_r$ is called a *subalgebra* of $L_r$ if

$$[Y, Y'] \in K_s \quad \text{for any } Y, Y' \in K_s.$$

This condition is equivalent to the following:

$$[Y_i, Y_j] \in K_s, \quad i, j = 1, \ldots, s.$$

Let us return to general properties of determining equations. As can be seen from (1.2.20), a determining equation is a linear partial differential equation with the unknown functions $\xi$ and $\eta$ of the variables $x$ and $y$. Therefore the set of its solutions forms a vector space, which was already noted in the previous example. However, a specific property of determining equations is given by the following statement due to S. Lie.

**Theorem 1.2.2** *The set of all solutions of any determining equation forms a Lie algebra.*

Investigation of the determining equations for symmetries of second-order ordinary differential equations lead Lie to the following significant result [13] (see also [8]).

**Theorem 1.2.3** *For a second-order equation* (1.2.19), *the symmetry Lie algebra L has the dimension $r \leq 8$. The maximal dimension $r = 8$ is attained if and only if* (1.2.19) *either is linear or can be linearized by a change of variables.*

We will discuss below two methods of integration of first-order ordinary differential equations with a known infinitesimal symmetry.

### 1.2.7 Integration of First-Order Equations: Lie's Integrating Factor

We begin with the method of Lie's integrating factor. Consider a first-order ordinary differential equation written in the form

$$Q(x, y)dx + P(x, y)dy = 0. \tag{1.2.30}$$

Lie [13] showed that if (1.2.30) admits a one-parameter group with the generator (1.1.6)

$$X = \xi(x, y)\frac{\partial}{\partial x} + \eta(x, y)\frac{\partial}{\partial y}$$

and if $\xi Q + \eta P \neq 0$, then the function

$$\mu = \frac{1}{\xi Q + \eta P} \tag{1.2.31}$$

is an integrating factor for (1.2.30).

**Example 1.2.1** Consider the Riccati equation (1.2.1):

$$y' + y^2 - \frac{2}{x^2} = 0. \tag{1.2.32}$$

Its symmetry group can be readily found by considering dilations $\bar{x} = ax$, $\bar{y} = by$. Substitution in (1.2.32) yields:

$$\bar{y}' + \bar{y}^2 - \frac{2}{\bar{x}^2} = \frac{b}{a}y' + b^2 y^2 - \frac{2}{a^2 x^2}.$$

The invariance of (1.2.32) requires $b/a = b^2 = 1/a^2$. Hence $b = 1/a$. Therefore the equation admits a one-parameter group of dilations (which can be written in the form $\bar{x} = xe^a$, $\bar{y} = ye^{-a}$) with the generator

$$X = x\frac{\partial}{\partial x} - y\frac{\partial}{\partial y}. \tag{1.2.33}$$

Writing (1.2.32) in the form (1.2.30),

$$dy + (y^2 - 2/x^2)dx = 0 \tag{1.2.34}$$

and applying the formula (1.2.31), one obtains the integrating factor

$$\mu = \frac{x}{x^2 y^2 - xy - 2}.$$

After multiplication by this factor, (1.2.34) is brought to the following form:

$$\frac{xdy + (xy^2 - 2/x)dx}{x^2 y^2 - xy - 2} = \frac{xdy + ydx}{x^2 y^2 - xy - 2} + \frac{dx}{x} = d\left(\ln x + \frac{1}{3}\ln\frac{xy - 2}{xy + 1}\right) = 0,$$

whence

$$\frac{xy - 2}{xy + 1} = \frac{C}{x^3} \quad \text{or} \quad y = \frac{2x^3 + C}{x(x^3 - C)}.$$

## 1.2.8  Integration of First-Order Equations: Method of Canonical Variables

Given a one-parameter symmetry group, one can use the canonical variables introduced in Sect. 1.1.4 for integrating first-order equations. Since the property of invariance of an equation with respect to a group is independent of the choice of variables, introduction of canonical variables reduces the equation in question to an equation which does not depend on one of the variables, and hence can be integrated by quadrature. Consider examples.

**Example 1.2.2** Let us solve the Riccati equation (1.2.32),

$$y' + y^2 - \frac{2}{x^2} = 0,$$

by the method of canonical variables using the symmetry (1.2.33):

$$X = x\frac{\partial}{\partial x} - y\frac{\partial}{\partial y}.$$

The partial differential equations

$$X(t) = x\frac{\partial t}{\partial x} - y\frac{\partial t}{\partial y} = 1, \quad X(u) = x\frac{\partial u}{\partial x} - y\frac{\partial u}{\partial y} = 0$$

yield the following canonical variables:

$$t = \ln|x|, \quad u = xy.$$

Let us rewrite (1.2.32) in the canonical variables. We have:

$$\frac{dy}{dx} = \frac{d}{dx}\left(\frac{u}{x}\right) = -\frac{u}{x^2} + \frac{1}{x}\frac{du}{dx} = -\frac{u}{x^2} + \frac{1}{x}\frac{du}{dt}\frac{dt}{dx} = -\frac{u}{x^2} + \frac{u'}{x^2}.$$

Therefore, the left-hand side of the equation in question is written as follows:

$$\frac{dy}{dx} + y^2 - \frac{2}{x^2} = \frac{u'}{x^2} - \frac{u}{x^2} + \frac{u^2}{x^2} - \frac{2}{x^2} = \frac{1}{x^2}(u' + u^2 - u - 2) = 0.$$

Thus, the Riccati equation is rewritten in the canonical variables in the following integrable form:

$$\frac{du}{dt} + u^2 - u - 2 = 0.$$

It is integrated by separation of variables:

$$\frac{du}{u^2 - u - 2} = -dt.$$

Decomposing the integrand into elementary fractions:

$$\frac{1}{u^2 - u - 2} = \frac{1}{3}\left[\frac{1}{u-2} - \frac{1}{u+1}\right],$$

we evaluate the integral in elementary functions and obtain:

$$\ln\left(\frac{u-2}{u+1}\right) = -3t + \ln C.$$

Now we solve this equation with respect to $u$,

$$u = \frac{C + 2e^{3t}}{e^{3t} - C},$$

substitute $t = \ln|x|$, $u = xy$ and arrive at the solution of the Riccati equation (cf. Example 1.2.1):

$$y = \frac{2x^3 + C}{x(x^3 - C)}.$$

**Example 1.2.3** The equation

$$y' = \frac{y}{x} + \frac{y^2}{x^2} \tag{1.2.35}$$

is homogeneous, i.e. it admits the group of dilations (scaling transformations) $\bar{x} = xe^a$, $\bar{y} = ye^a$ with the generator

$$X = x\frac{\partial}{\partial x} + y\frac{\partial}{\partial y}. \tag{1.2.36}$$

Canonical variables for the operator (1.2.36) are

$$t = \ln|x|, \quad u = \frac{y}{x}. \tag{1.2.37}$$

In these variables, (1.2.35) is written

$$\frac{du}{dt} = u^2.$$

Whence, upon integration:

$$\frac{1}{u} = C - t.$$

Substituting here $t = \ln|x|$ and $y = xu$, we obtain the solution of the original equation:

$$y = \frac{x}{C - \ln|x|}.$$

**Example 1.2.4** The equation

$$y' = \frac{y}{x} + \frac{y^3}{x^4} \tag{1.2.38}$$

admits the group of projective transformations

$$\bar{x} = \frac{x}{1 - ax}, \quad \bar{y} = \frac{y}{1 - ax},$$

with the generator

$$X = x^2\frac{\partial}{\partial x} + xy\frac{\partial}{\partial y}. \tag{1.2.39}$$

Introducing the canonical variables

$$t = -\frac{1}{x}, \quad u = \frac{y}{x}, \tag{1.2.40}$$

we rewrite (1.2.38) in the form

$$\frac{du}{dt} = u^3.$$

Integration yields

$$u = \pm\frac{1}{\sqrt{C - 2t}},$$

whence, substituting the expressions for $t$ and $u$, we obtain the following general solution to our equation:

$$y = \pm x\sqrt{\frac{x}{2 + Cx}}.$$

**Table 1.2.9.1** Structure and standard forms of $L_2$

| Type | Structure of $L_2$ | Standard form of $L_2$ |
|------|-------------------|------------------------|
| I | $[X_1, X_2] = 0, \xi_1 \eta_2 - \eta_1 \xi_2 \neq 0$ | $X_1 = \frac{\partial}{\partial t}, X_2 = \frac{\partial}{\partial u}$ |
| II | $[X_1, X_2] = 0, \xi_1 \eta_2 - \eta_1 \xi_2 = 0$ | $X_1 = \frac{\partial}{\partial u}, X_2 = t \frac{\partial}{\partial u}$ |
| III | $[X_1, X_2] = X_1, \xi_1 \eta_2 - \eta_1 \xi_2 \neq 0$ | $X_1 = \frac{\partial}{\partial u}, X_2 = t \frac{\partial}{\partial t} + u \frac{\partial}{\partial u}$ |
| IV | $[X_1, X_2] = X_1, \xi_1 \eta_2 - \eta_1 \xi_2 = 0$ | $X_1 = \frac{\partial}{\partial u}, X_2 = u \frac{\partial}{\partial u}$ |

**Table 1.2.9.2** Four types of second-order equations admitting $L_2$

| Type | Standard form of $L_2$ | Canonical form of the equation |
|------|------------------------|-------------------------------|
| I | $X_1 = \frac{\partial}{\partial t}, X_2 = \frac{\partial}{\partial u}$ | $u'' = f(u')$ |
| II | $X_1 = \frac{\partial}{\partial u}, X_2 = t \frac{\partial}{\partial u}$ | $u'' = f(t)$ |
| III | $X_1 = \frac{\partial}{\partial u}, X_2 = t \frac{\partial}{\partial t} + u \frac{\partial}{\partial u}$ | $u'' = \frac{1}{t} f(u')$ |
| IV | $X_1 = \frac{\partial}{\partial u}, X_2 = u \frac{\partial}{\partial u}$ | $u'' = f(t)u'$ |

## 1.2.9 Standard Forms of Two-Dimensional Lie Algebras

Lie's method of integration of second-order ordinary differential equations employs *canonical variables* in two-dimensional Lie algebras. Introduction of canonical variables reduces any second-order differential equation admitting a two-dimensional Lie algebra $L_2$ into an integrable form.

*Canonical variables* reduce a basis of every two-dimensional Lie algebra $L_2$ to the simplest form and provide four *standard forms* of second-order equations with two symmetries. The basic statements are as follows.

**Theorem 1.2.4** *Any two-dimensional Lie algebra can be transformed, by a proper choice of its basis and suitable variables $t, u$, called canonical variables, to one of the four non-similar standard forms presented in Table* 1.2.9.1.

**Remark 1.2.1** In types III and IV, the condition $[X_1, X_2] = X_1$ can be satisfied by a proper change of the basis in $L_2$ provided that $[X_1, X_2] \neq 0$.

Let a second-order equation

$$y'' = f(x, y, y')  \tag{1.2.41}$$

admit two or more symmetries. Let us single out from these symmetries a two-dimensional Lie algebra $L_2$, determine its type according to Table 1.2.9.1, find canonical variables $t, u$ for $L_2$, and rewrite (1.2.41) in the variables $t, u$:

$$u'' = g(t, u, u').  \tag{1.2.42}$$

Theorem 1.2.4 guarantees that (1.2.42) belongs to one of four integrable equations given in Table 1.2.9.2.

## *1.2.10  Lie's Method of Integration for Second-Order Equations*

The method of integration of second-order non-linear differential equations (1.2.41) requires the following calculations. First of all, one needs to find the symmetries of the equation in question. Let the equation have two or more symmetries. We single out from these symmetries a two-dimensional Lie algebra $L_2$ and determine its type according to the *Structure* column of Table 1.2.9.1. Then we find canonical variables by solving the following equations in accordance with the type:

$$
\begin{aligned}
\textbf{Type I:} \quad & X_1(t) = 1, \ X_2(t) = 0; \quad X_1(u) = 0, \ X_2(u) = 1. \\
\textbf{Type II:} \quad & X_1(t) = 0, \ X_2(t) = 0; \quad X_1(u) = 1, \ X_2(u) = t. \\
\textbf{Type III:} \quad & X_1(t) = 0, \ X_2(t) = t; \quad X_1(u) = 1, \ X_2(u) = u. \\
\textbf{Type IV:} \quad & X_1(t) = 0, \ X_2(t) = 0; \quad X_1(u) = 1, \ X_2(u) = u.
\end{aligned}
\tag{1.2.43}
$$

Now we rewrite the differential equation in the canonical variables choosing $t$ as a new independent variable and $u$ as a dependent one. It will have one of the integrable forms given in Table 1.2.9.2. It remains to integrate the resulting equation and rewrite the solution in the original variables $x$, $y$. This completes the integration procedure.

**Example 1.2.5** Let us apply the integration method to the following non-linear second-order equation:

$$
y'' + e^{3y} y'^4 + y'^2 = 0.
\tag{1.2.44}
$$

First, we have to find the symmetries of (1.2.44). Here

$$
f = -(e^{3y} y'^4 + y'^2)
$$

and the determining equation (1.2.20) is written as follows:

$$
\begin{aligned}
& \eta_{xx} + (2\eta_{xy} - \xi_{xx})y' + (\eta_{yy} - 2\xi_{xy})y'^2 - y'^3 \xi_{yy} \\
& + 3e^{3y} y'^4 \eta - (\eta_y - 2\xi_x - 3y'\xi_y)(e^{3y} y'^4 + y'^2) \\
& + [\eta_x + (\eta_y - \xi_x)y' - y'^2 \xi_y](4e^{3y} y'^3 + 2y') = 0.
\end{aligned}
$$

The left-hand side of this equation is a polynomial of fifth degree in $y'$. Since it should vanish identically in $y'$, we equate to zero the coefficients of $y'^5$, $y'^4$, ... and obtain the following four independent equations:

$$
\begin{aligned}
& (y')^5 : \xi_y = 0, \\
& (y')^4 : 3(\eta_y + \eta) - 2\xi_x = 0, \\
& (y')^3 : \eta_x = 0, \\
& (y')^1 : \xi_{xx} = 0.
\end{aligned}
$$

The coefficients for $(y')^2$ and $(y')^0$ vanish together with the coefficients of $(y')^4$ and $(y')^1$, respectively. The above four differential equations for two unknown functions $\xi(x, y)$ and $\eta(x, y)$ are readily solved and yield:

$$\xi = C_1 + 3C_3 x, \quad \eta = 2C_3 + C_2 e^{-y}, \quad C_1, C_2, C_3 = \text{const.}$$

Hence, the general form of the operator

$$X = \xi(x, y)\frac{\partial}{\partial x} + \eta(x, y)\frac{\partial}{\partial y}$$

admitted by (1.2.44) is

$$X = C_1 X_1 + C_2 X_2 + C_3 X_3,$$

where

$$X_1 = \frac{\partial}{\partial x}, \quad X_2 = e^{-y}\frac{\partial}{\partial y}, \quad X_3 = 3x\frac{\partial}{\partial x} + 2\frac{\partial}{\partial y}. \tag{1.2.45}$$

In other words, (1.2.44) admits the three-dimensional Lie algebra $L_3$ spanned by the operators (1.2.45).

The operators $X_1$ and $X_2$ span a two-dimensional subalgebra $L_2 \subset L_3$ and has the type I. Canonical variables $t$ and $u$ are obtained by solving (1.2.43) for type I, i.e. the following equations:

$$\frac{\partial t}{\partial x} = 1, \quad e^{-y}\frac{\partial t}{\partial y} = 0; \quad \frac{\partial u}{\partial x} = 0, \quad e^{-y}\frac{\partial u}{\partial y} = 1.$$

We take the following solutions to this system:

$$t = x, \quad u = e^y.$$

Thus, we set $u = u(t)$ and rewrite the equation in question in the new variables to obtain

$$u'' + u'^4 = 0.$$

The standard substitution $u' = v$ reduces it to the first-order equation $v' + v^4 = 0$, whence

$$v = \frac{1}{\sqrt[3]{3x + C_1}}.$$

Now we integrate the equation

$$\frac{du}{dx} = \frac{1}{\sqrt[3]{3x + C_1}}$$

and obtain:

$$u = \frac{1}{2}\left[\sqrt[3]{(3x + C_1)^2} + C_2\right].$$

Substitution of the expressions for $t, u$ yields the solution to (1.2.44):

$$y = \ln\left|\sqrt[3]{(3x + C_1)^2} + C_2\right| - \ln 2.$$

**Example 1.2.6** Integrate the non-linear equation

$$y'' + 2\left(y' - \frac{y}{x}\right)^3 = 0 \tag{1.2.46}$$

which admits the algebra $L_2$ of type II spanned by

$$X_1 = x^2 \frac{\partial}{\partial x} + xy \frac{\partial}{\partial y}, \quad X_2 = xy \frac{\partial}{\partial x} + y^2 \frac{\partial}{\partial y}. \tag{1.2.47}$$

*Solution.* The equations $X_1(t) = 0$, $X_1(u) = 1$; $X_2(t) = 0$, $X_2(u) = t$ provide the canonical variables

$$t = \frac{y}{x}, \quad u = -\frac{1}{x}. \tag{1.2.48}$$

Since the variable $t$ involves the dependent variable $y$, $t$ can be a new independent variable only if one excludes the *singular* solutions of (1.2.46) along which $t$ is identically constant. These singular solutions are the straight lines:

$$y = Kx, \quad K = \text{const.}$$

In the variables (1.2.48) the equation (1.2.46) becomes

$$u'' = 2$$

and yields $u = t^2 + C_1 t + C_2$. Substituting the expressions for $t$ and $u$, we obtain:

$$y^2 + C_1 xy + C_2 x^2 + x = 0.$$

Solving this equation with respect to $y$ and introducing the new constants $A = -C_1/2$, $B = A^2 - C_2$, we obtain the solution to (1.2.46):

$$y = Kx, \quad y = Ax \pm \sqrt{Bx^2 - x}. \tag{1.2.49}$$

## 1.3  Symmetries and Invariant Solutions of Partial Differential Equations

### 1.3.1  Discussion of Symmetries for Evolution Equations

Consider evolutionary partial differential equations of the second order with one spatial variable $x$:

$$u_t = F(t, x, u, u_x, u_{xx}), \quad \partial F/\partial u_{xx} \neq 0. \tag{1.3.1}$$

**Definition 1.3.1** A one-parameter group $G$ of transformations (1.1.19) of the variables $t, x, u$:

$$\bar{t} = f(t, x, u, a), \quad \bar{x} = g(t, x, u, a), \quad \bar{u} = h(t, x, u, a) \tag{1.3.2}$$

is called a *group admitted* by (1.3.1), or a *symmetry group* of (1.3.1), if (1.3.1) has the same form in the new variables $\bar{t}, \bar{x}, \bar{u}$:

$$\bar{u}_{\bar{t}} = F(\bar{t}, \bar{x}, \bar{u}, \bar{u}_{\bar{x}}, \bar{u}_{\bar{x}\bar{x}}). \tag{1.3.3}$$

The function $F$ has the same form in both (1.3.1) and (1.3.3).

According to this definition, the transformations (1.3.2) of the group $G$ map every solution $u = u(t, x)$ of (1.3.1) into a solution $\bar{u} = \bar{u}(\bar{t}, \bar{x})$ of (1.3.3). Since (1.3.3) is identical with (1.3.1), the definition of an admitted group can be formulated as follows.

**Definition 1.3.2** A one-parameter group $G$ of transformations (1.3.2) is called a *group admitted* by (1.3.1) if the transformations (1.3.2) map any solution of (1.3.1) into a solution of the same equation.

The *infinitesimal transformations* of the group $G$ of transformations (1.3.2) are written

$$\bar{t} \approx t + a\tau(t, x, u), \quad \bar{x} \approx x + a\xi(t, x, u), \quad \bar{u} \approx u + a\eta(t, x, u) \quad (1.3.4)$$

and provide the following generator of the group $G$:

$$X = \tau(t, x, u)\frac{\partial}{\partial t} + \xi(t, x, u)\frac{\partial}{\partial x} + \eta(t, x, u)\frac{\partial}{\partial u} \quad (1.3.5)$$

acting on any differentiable function $J(t, x, u)$ as follows:

$$X(J) = \tau(t, x, u)\frac{\partial J}{\partial t} + \xi(t, x, u)\frac{\partial J}{\partial x} + \eta(t, x, u)\frac{\partial J}{\partial u}.$$

The generator (1.3.5) of a group $G$ admitted by (1.3.1) is known as an *infinitesimal symmetry* of (1.3.1).

The transformations (1.3.2) of the group with the generator (1.3.5) are found by solving the *Lie equations*

$$\frac{d\bar{t}}{da} = \tau(\bar{t}, \bar{x}, \bar{u}), \quad \frac{d\bar{x}}{da} = \xi(\bar{t}, \bar{x}, \bar{u}), \quad \frac{d\bar{u}}{da} = \eta(\bar{t}, \bar{x}, \bar{u}), \quad (1.3.6)$$

with the initial conditions:

$$\bar{t}\big|_{a=0} = t, \quad \bar{x}\big|_{a=0} = x, \quad \bar{u}\big|_{a=0} = u. \quad (1.3.7)$$

Let us turn now to (1.3.3). The quantities $\bar{u}_{\bar{t}}$, $\bar{u}_{\bar{x}}$ and $\bar{u}_{\bar{x}\bar{x}}$ involved in (1.3.3) are obtained via the usual rule of change of derivatives by treating (1.3.2) as a change of variables. Then, expanding the resulting expressions for $\bar{u}_{\bar{t}}$, $\bar{u}_{\bar{x}}$, $\bar{u}_{\bar{x}\bar{x}}$ into Taylor series with respect to the parameter $a$ and keeping only the terms linear in $a$, one obtains the infinitesimal form of these expressions:

$$\bar{u}_{\bar{t}} \approx u_t + a\zeta_0(t, x, u, u_t, u_x),$$
$$\bar{u}_{\bar{x}} \approx u_x + a\zeta_1(t, x, u, u_t, u_x), \quad (1.3.8)$$
$$\bar{u}_{\bar{x}\bar{x}} \approx u_{xx} + a\,\zeta_2(t, x, u, u_t, u_x, u_{tx}, u_{xx}),$$

where $\zeta_0, \zeta_1, \zeta_2$ are given by the following *prolongation formulae*:

$$\zeta_0 = D_t(\eta) - u_t D_t(\tau) - u_x D_t(\xi),$$
$$\zeta_1 = D_x(\eta) - u_t D_x(\tau) - u_x D_x(\xi), \quad (1.3.9)$$
$$\zeta_2 = D_x(\zeta_1) - u_{tx} D_x(\tau) - u_{xx} D_x(\xi).$$

Here $D_t$ and $D_x$ denote the *total differentiations* with respect to $t$ and $x$:

$$D_t = \frac{\partial}{\partial t} + u_t \frac{\partial}{\partial u} + u_{tt} \frac{\partial}{\partial u_t} + u_{tx} \frac{\partial}{\partial u_x},$$

$$D_x = \frac{\partial}{\partial x} + u_x \frac{\partial}{\partial u} + u_{tx} \frac{\partial}{\partial u_t} + u_{xx} \frac{\partial}{\partial u_x}.$$

Substitution of (1.3.4) and (1.3.8) in (1.3.3) yields:

$$\bar{u}_{\bar{t}} - F(\bar{t}, \bar{x}, \bar{u}, \bar{u}_{\bar{x}}, \bar{u}_{\bar{x}\bar{x}})$$

$$\approx u_t - F(t, x, u, u_x, u_{xx})$$

$$+ a\left(\zeta_0 - \frac{\partial F}{\partial u_{xx}}\zeta_2 - \frac{\partial F}{\partial u_x}\zeta_1 - \frac{\partial F}{\partial u}\eta - \frac{\partial F}{\partial x}\xi - \frac{\partial F}{\partial t}\tau\right).$$

Therefore, by virtue of (1.3.1), the equation (1.3.3) yields

$$\zeta_0 - \frac{\partial F}{\partial u_{xx}}\zeta_2 - \frac{\partial F}{\partial u_x}\zeta_1 - \frac{\partial F}{\partial u}\eta - \frac{\partial F}{\partial x}\xi - \frac{\partial F}{\partial t}\tau = 0, \qquad (1.3.10)$$

where $u_t$ is replaced by $F(t, x, u, u_x, u_{xx})$ in $\zeta_0, \zeta_1, \zeta_2$.

Equation (1.3.10) determines all infinitesimal symmetries of (1.3.1) and therefore it is called the *determining equation*. Conventionally, it is written in the compact form

$$X\big(u_t - F(t, x, u, u_x, u_{xx})\big)\big|_{u_t=F} = 0, \qquad (1.3.11)$$

where the *prolongation* of the operator $X$ (1.3.5) to the first and second order derivatives is understood:

$$X = \tau \frac{\partial}{\partial t} + \xi \frac{\partial}{\partial x} + \eta \frac{\partial}{\partial u} + \zeta_0 \frac{\partial}{\partial u_t} + \zeta_1 \frac{\partial}{\partial u_x} + \zeta_2 \frac{\partial}{\partial u_{xx}},$$

and the symbol $|_{u_t=F}$ means that $u_t$ is replaced by $F(t, x, u, u_x, u_{xx})$.

The determining equation (1.3.10) (or its equivalent (1.3.11)) is a linear homogeneous partial differential equation of the second order for unknown functions $\tau(t, x, u)$, $\xi(t, x, u)$, $\eta(t, x, u)$. In consequence, the set of all solutions to the determining equation is a vector space $L$. Furthermore, the determining equation possesses the following significant and less evident property. The vector space $L$ is a *Lie algebra*, i.e. it is closed with respect to the *commutator*. In other words, $L$ contains, together with any operators $X_1, X_2$, their commutator $[X_1, X_2]$ defined by

$$[X_1, X_2] = X_1 X_2 - X_2 X_1.$$

In particular, if $L = L_r$ is finite-dimensional and has a basis $X_1, \ldots, X_r$, then the Lie algebra condition is written in the form

$$[X_\alpha, X_\beta] = c_{\alpha\beta}^\gamma X_\gamma$$

with constant coefficients $c_{\alpha\beta}^\gamma$ known as the *structure constants* of $L_r$.

Note that (1.3.10) should be satisfied identically with respect to all the variables involved, the variables $t, x, u, u_x, u_{xx}, u_{tx}$ are treated as five independent variables. Consequently, the determining equation decomposes into a system of several equations. As a rule, this is an over-determined system since it contains more equations than three unknown functions $\tau, \xi$ and $\eta$. Therefore, in practical applications, the determining equation can be readily solved. The following statement due to Lie [12] simplifies the calculation of the symmetries of evolution equations.[2]

**Lemma 1.3.1** *The symmetry transformations* (1.3.2) *of* (1.3.1) *have the form*

$$\bar{t} = f(t, a), \quad \bar{x} = g(t, x, u, a), \quad \bar{u} = h(t, x, u, a). \tag{1.3.12}$$

*It means that one can search the infinitesimal symmetries in the form*

$$X = \tau(t)\frac{\partial}{\partial t} + \xi(t, x, u)\frac{\partial}{\partial x} + \eta(t, x, u)\frac{\partial}{\partial u}. \tag{1.3.13}$$

For the operators (1.3.13), the prolongation formulae (1.3.9) are written as follows:

$$\zeta_0 = D_t(\eta) - u_x D_t(\xi) - \tau'(t)u_t, \quad \zeta_1 = D_x(\eta) - u_x D_x(\xi),$$
$$\zeta_2 = D_x(\zeta_1) - u_{xx} D_x(\xi) = D_x^2(\eta) - u_x D_x^2(\xi) - 2u_{xx} D_x(\xi). \tag{1.3.14}$$

## 1.3.2 Calculation of Symmetries for Burgers' Equation

Let us find the symmetries of the Burgers equation

$$u_t = u_{xx} + uu_x. \tag{1.3.15}$$

According to Lemma 1.3.1, the infinitesimal symmetries have the form (1.3.13). For the Burgers equation, the determining equation (1.3.10) has the form

$$\zeta_0 - \zeta_2 - u\zeta_1 - \eta u_x = 0, \tag{1.3.16}$$

where $\zeta_0, \zeta_1$ and $\zeta_2$ are given by (1.3.14). Let us single out and annul the terms with $u_{xx}$. Bearing in mind that $u_t$ has to be replaced by $u_{xx} + uu_x$ and substituting in $\zeta_2$ the expressions

$$D_x^2(\xi) = D_x(\xi_x + \xi_u u_x) = \xi_u u_{xx} + \xi_{uu} u_x^2 + 2\xi_{xu} u_x + \xi_{xx},$$
$$D_x^2(\eta) = D_x(\eta_x + \eta_u u_x) = \eta_u u_{xx} + \eta_{uu} u_x^2 + 2\eta_{xu} u_x + \eta_{xx} \tag{1.3.17}$$

we arrive at the following equation:

$$2\xi_u u_x + 2\xi_x - \tau'(t) = 0.$$

It splits into two equations, namely $\xi_u = 0$ and $2\xi_x - \tau'(t) = 0$. The first equation shows that $\xi$ depends only on $t, x$, and integration of the second equation yields

$$\xi = \frac{1}{2}\tau'(t)x + p(t). \tag{1.3.18}$$

---

[2]In [12], Sect. III, Lie proves a more general statement about contact transformations of parabolic equations.

It follows from (1.3.18) that $D_x^2(\xi) = 0$. Now the determining equation (1.3.16) reduces to the form

$$u_x^2 \eta_{uu} + \left[ \frac{1}{2}\tau'(t)u + \frac{1}{2}\tau''(t)x + p'(t) + 2\eta_{xu} + \eta \right] u_x + u\eta_x + \eta_{xx} - \eta_t = 0$$

and splits into three equations:

$$\eta_{uu} = 0,$$

$$\frac{1}{2}\left(\tau'(t)u + \tau''(t)x\right) + p'(t) + 2\eta_{xu} + \eta = 0, \tag{1.3.19}$$

$$u\eta_x + \eta_{xx} - \eta_t = 0.$$

The first equation (1.3.19) yields $\eta = \sigma(t, x)u + \mu(t, x)$, and the second equation (1.3.19) becomes:

$$\left( \frac{1}{2}\tau'(t) + \sigma \right) u + \frac{1}{2}\tau''(t)x + p'(t) + 2\sigma_x + \mu = 0,$$

whence

$$\sigma = -\frac{1}{2}\tau'(t), \quad \mu = -\frac{1}{2}\tau''(t)x - p'(t).$$

Thus, we have

$$\eta = -\frac{1}{2}\tau'(t)u - \frac{1}{2}\tau''(t)x - p'(t). \tag{1.3.20}$$

Finally, substitution of (1.3.20) in the third equation (1.3.19) yields

$$\frac{1}{2}\tau'''(t)x + p''(t) = 0,$$

whence $\tau'''(t) = 0$, $p''(t) = 0$, and hence

$$\tau(t) = C_1 t^2 + 2C_2 t + C_3, \quad p(t) = C_4 t + C_5.$$

Invoking (1.3.18) and (1.3.20), we ultimately arrive at the following general solution of the determining equation (1.3.16):

$$\tau(t) = C_1 t^2 + 2C_2 t + C_3,$$

$$\xi = C_1 tx + C_2 x + C_4 t + C_5,$$

$$\eta = -(C_1 t + C_2)u - C_1 x - C_4.$$

It contains five arbitrary constants $C_i$. Hence, the infinitesimal symmetries of the Burgers equation (1.3.15) form the five-dimensional Lie algebra $L_5$ spanned by the following linearly independent operators:

$$X_1 = \frac{\partial}{\partial t}, \quad X_2 = \frac{\partial}{\partial x}, \quad X_3 = 2t\frac{\partial}{\partial t} + x\frac{\partial}{\partial x} - u\frac{\partial}{\partial u},$$

$$X_4 = t\frac{\partial}{\partial x} - \frac{\partial}{\partial u}, \quad X_5 = t^2\frac{\partial}{\partial t} + tx\frac{\partial}{\partial x} - (x + tu)\frac{\partial}{\partial u}. \tag{1.3.21}$$

Let $G$ be a group admitted by (1.3.1). Then every transformation (1.3.2) belonging to the group $G$ carries over any solution of the differential equation (1.3.1) into a solution of the same equation. It means that the solutions of a partial differential equation are permuted among themselves under the action of a symmetry group. The solutions may also be individually unaltered, then they are called *invariant solutions*. Accordingly, group analysis provides two basic ways for construction of exact solutions: *group transformations* of known solutions and construction of *invariant solutions*.

### 1.3.3 Invariant Solutions and Their Calculation

If a group transformation maps a solution into itself, we arrive at what is called a *self-similar* or *group invariant solution*. According to Theorem 1.1.4 on invariant representation of invariant manifolds, the invariant solutions under a one-parameter group with a generator $X$ are obtained as follows.

Let $X$ be a given infinitesimal symmetry (1.3.5) of (1.3.1). One calculates two independent *invariants* $J_1 = \lambda(t, x)$ and $J_2 = \mu(t, x, u)$ by solving the first-order linear partial differential equation

$$X(J) \equiv \tau(t, x, u)\frac{\partial J}{\partial t} + \xi(t, x, u)\frac{\partial J}{\partial x} + \eta(t, x, u)\frac{\partial J}{\partial u} = 0,$$

or its characteristic system:

$$\frac{dt}{\tau(t, x, u)} = \frac{dx}{\xi(t, x, u)} = \frac{du}{\eta(t, x, u)}. \tag{1.3.22}$$

Then one designates one of the invariants as a function of the other, e.g.

$$\mu = \phi(\lambda), \tag{1.3.23}$$

and solves (1.3.23) with respect to $u$. Finally, one substitutes the expression for $u$ in (1.3.1) and obtains an ordinary differential equation for the unknown function $\phi(\lambda)$ of one variable. This procedure reduces the number of independent variables by one.

**Example 1.3.1** Let us find the solutions of the Burgers equation that are invariant under the time translations generated by the operator $X_1$ from (1.3.21). The invariance condition leads to the stationary solutions

$$u = \Phi(x)$$

for which the Burgers equation is written

$$\Phi'' + \Phi\Phi' = 0. \tag{1.3.24}$$

Integrating once, we obtains

$$\Phi' + \frac{\Phi^2}{2} = C_1.$$

We integrate now this first-order equation by setting $C_1 = 0$, $C_1 = v^2 > 0$, and $C_1 = -\omega^2 < 0$ and obtain:

$$\Phi(x) = \frac{2}{x+C},$$

$$\Phi(x) = v\text{th}\left(C + \frac{v}{2}x\right),$$  (1.3.25)

$$\Phi(x) = \omega\text{tg}\left(C - \frac{\omega}{2}x\right).$$

### 1.3.4  Group Transformations of Solutions

Let (1.3.2) be an admitted group for (1.3.1), and let a function

$$u = \Phi(t, x)$$

solve (1.3.1). Since (1.3.2) is a symmetry transformation, the above solution can be also written in the new variables:

$$\bar{u} = \Phi(\bar{t}, \bar{x}).$$

Replacing here $\bar{u}, \bar{t}, \bar{x}$ from (1.3.2), we get

$$h(t, x, u, a) = \Phi\big(f(t, x, u, a), g(t, x, u, a)\big).$$  (1.3.26)

Solving (1.3.26) with respect to $u$ one obtains a one-parameter family (with the parameter $a$) of new solutions to (1.3.1).

**Example 1.3.2** Consider the Burgers equation (1.3.15),

$$u_t = u_{xx} + uu_x,$$

and apply the above procedure to the admitted one-parameter group generated by the operator $X_5$ from (1.3.21):

$$X_5 = t^2\frac{\partial}{\partial t} + tx\frac{\partial}{\partial x} - (x + tu)\frac{\partial}{\partial u}.$$

The one-parameter group generated by $X_5$ has the form

$$\bar{t} = \frac{t}{1 - at}, \quad \bar{x} = \frac{C_2}{1 - at}, \quad \bar{u} = (1 - at)u - ax.$$  (1.3.27)

Using the transformations (1.3.27) and applying (1.3.26) to any known solution $u = \Phi(t, x)$ of the Burgers equation, one obtains the following one-parameter set of new solutions:

$$u = \frac{ax}{1 - at} + \frac{1}{1 - at}\Phi\left(\frac{t}{1 - at}, \frac{x}{1 - at}\right).$$  (1.3.28)

Let us apply the transformation (1.3.28), e.g. to the first stationary solution (1.3.25):

$$\Phi(x) = \frac{2}{x + C}, \quad C = \text{const.},$$

one obtains the new non-stationary solutions

$$u = \frac{ax}{1 - at} + \frac{2}{x + C(1 - at)}$$

depending on the parameter $a$.

## 1.3.5 Optimal Systems of Subalgebras

The concept of optimal systems of subalgebras of a given Lie algebra was used by Ovsyannikov [16] for describing essentially different invariant solutions. This concept is useful in dealing with nonlinear mathematical models. A simple method of construction of an optimal system is illustrated in this section by of means of the five-dimensional Lie algebra $L_5$ spanned by the symmetries (1.3.21) of the Burgers equation. The result is used in the next section for describing all invariant solutions of the Burgers equation.

The symmetry Lie algebra $L_5$ spanned by the operators (1.3.21) allows one to construct invariant solutions of the Burgers equation (1.3.15),

$$u_t = u_{xx} + uu_x,$$

by using any one-dimensional subalgebra of the algebra $L_5$, i.e. on any operator $X \in L_5$. However, there are infinite number of one-dimensional subalgebras of $L_5$ since an arbitrary operator from $L_5$ is written

$$X = l^1 X_1 + \cdots + l^5 X_5, \tag{1.3.29}$$

and hence depends on five arbitrary constants $l^1, \ldots, l^5$. Ovsyannikov [16] has noticed, however, that if two subalgebras are *similar*, i.e. connected with each other by a transformation of the symmetry group, then their corresponding invariant solutions are connected with each other by the same transformation. Consequently, it is sufficient to deal with an optimal system of subalgebras obtained in our case as follows. We put into one class all similar operators $X \in L_5$ and select a representative of each class. The set of the representatives of all these classes is an *optimal system of one-dimensional subalgebras*.

An optimal system of one-dimensional subalgebras of the Lie algebra $L_5$ is constructed as follows [11]. The transformations of the symmetry group with the Lie algebra $L_5$ provide the 5-parameter group of linear transformations of the operators $X \in L_5$ or, equivalently, linear transformations of the vector

$$l = (l^1, \ldots, l^5), \tag{1.3.30}$$

where $l^1, \ldots, l^5$ are taken from (1.3.29). To find these linear transformations, we use their generators (see, e.g. Sect. 1.4 in [6])

$$E_\mu = c_{\mu\nu}^\lambda l^\nu \frac{\partial}{\partial l^\lambda}, \quad \mu = 1, \ldots, 5, \tag{1.3.31}$$

where $c_{\mu\nu}^\lambda$ are the structure constants of the Lie algebra $L_5$ defined by

$$[X_\mu, X_\nu] = c_{\mu\nu}^\lambda X_\lambda.$$

For computing the operators (1.3.31) it is convenient to use the following commutator table of the operators (1.3.21):

|       | $X_1$    | $X_2$  | $X_3$    | $X_4$  | $X_5$   |        |
|-------|----------|--------|----------|--------|---------|--------|
| $X_1$ | 0        | 0      | $2X_1$   | $X_2$  | $X_3$   |        |
| $X_2$ | 0        | 0      | $X_2$    | 0      | $X_4$   | (1.3.32) |
| $X_3$ | $-2X_1$  | $-X_2$ | 0        | $X_4$  | $2X_5$  |        |
| $X_4$ | $-X_2$   | 0      | $-X_4$   | 0      | 0       |        |
| $X_5$ | $-X_3$   | $-X_4$ | $-2X_5$  | 0      | 0       |        |

Let us find, e.g. the operator $E_1$. According to (1.3.31), it is written

$$E_1 = c_{1\nu}^\lambda l^\nu \frac{\partial}{\partial l^\lambda},$$

where $c_{1\nu}^\lambda$ are defined by the commutators $[X_1, X_\nu] = c_{\mu\nu}^\lambda X_\lambda$, i.e. by the first raw in table (1.3.32). Namely, the non-vanishing $c_{\mu\nu}^\lambda$ are

$$c_{13}^1 = 2, \quad c_{14}^2 = 1, \quad c_{15}^2 = 1.$$

Therefore we have:

$$E_1 = 2l^3 \frac{\partial}{\partial l^1} + l^4 \frac{\partial}{\partial l^2} + l^5 \frac{\partial}{\partial l^3}.$$

Substituting in (1.3.31) all structure constants given by table (1.3.32) we obtain:

$$E_1 = 2l^3 \frac{\partial}{\partial l^1} + l^4 \frac{\partial}{\partial l^2} + l^5 \frac{\partial}{\partial l^3}, \quad E_2 = l^3 \frac{\partial}{\partial l^2} + l^5 \frac{\partial}{\partial l^4},$$

$$E_3 = -2l^1 \frac{\partial}{\partial l^1} - l^2 \frac{\partial}{\partial l^2} + l^4 \frac{\partial}{\partial l^4} + 2l^5 \frac{\partial}{\partial l^5}, \tag{1.3.33}$$

$$E_4 = -l^1 \frac{\partial}{\partial l^2} - l^3 \frac{\partial}{\partial l^4}, \quad E_5 = -l^1 \frac{\partial}{\partial l^3} - l^2 \frac{\partial}{\partial l^4} - 2l^3 \frac{\partial}{\partial l^5}.$$

Let us find the transformations provided by the generators (1.3.33). For the generator $E_1$, the Lie equations with the parameter $a_1$ are written

$$\frac{d\tilde{l}^1}{da_1} = 2\tilde{l}^3, \quad \frac{d\tilde{l}^2}{da_1} = \tilde{l}^4, \quad \frac{d\tilde{l}^3}{da_1} = \tilde{l}^5, \quad \frac{d\tilde{l}^4}{da_1} = 0, \quad \frac{d\tilde{l}^5}{da_1} = 0.$$

Integrating these equations and using the initial condition $\tilde{l}|_{a_1=0} = l$, we obtain:

$$E_1: \quad \tilde{l}^1 = l^1 + 2a_1 l^3 + a_1^2 l^5, \quad \tilde{l}^2 = l^2 + a_1 l^4,$$

$$\tilde{l}^3 = l^3 + a_1 l^5, \quad \tilde{l}^4 = l^4, \quad \tilde{l}^5 = l^5. \tag{1.3.34}$$

Taking the other operators (1.3.33) we obtain the following transformations:

$$E_2: \quad \tilde{l}^1 = l^1, \; \tilde{l}^2 = l^2 + a_2 l^3, \; \tilde{l}^3 = l^3, \; \tilde{l}^4 = l^4 + a_2 l^5, \; \tilde{l}^5 = l^5, \quad (1.3.35)$$

$$E_3: \quad \tilde{l}^1 = a_3^{-2} l^1, \; l^2 = a_3^{-1} l^2, \; \tilde{l}^3 = l^3, \; \tilde{l}^4 = a_3 l^4, \; \tilde{l}^5 = a_3^2 l^5, \quad (1.3.36)$$

where $a_3 > 0$ since the integration of the Lie equations yields, e.g.

$$l^4 = l^4 e^{\tilde{a}_3} = a_3 l^4.$$

$$E_4: \quad \tilde{l}^1 = l^1, \; \tilde{l}^2 = l^2 - a_4 l^1, \; \tilde{l}^3 = l^3, \; \tilde{l}^4 = l^4 - a_4 l^3, \; \tilde{l}^5 = l^5. \quad (1.3.37)$$

$$E_5: \quad \tilde{l}^1 = l^1, \quad \tilde{l}^2 = l^2, \quad \tilde{l}^3 = l^3 - a_5 l^1,$$

$$\tilde{l}^4 = l^4 - a_5 l^2, \quad \tilde{l}^5 = l^5 - 2a_5 l^3 + a_5^2 l^1. \quad (1.3.38)$$

Note that the transformations (1.3.34)–(1.3.38) map the vector $X \in L_5$ given by (1.3.29) to the vector $\tilde{X} \in L_5$ given by the following formula:

$$\tilde{X} = \tilde{l}^1 X_1 + \cdots + \tilde{l}^5 X_5. \quad (1.3.39)$$

Now we can prove the following statement on an optimal system of one-dimensional subalgebras of symmetry algebra for the Burgers equation.

**Theorem 1.3.1** *The following operators provide an optimal system of one-dimensional subalgebras of the Lie algebra $L_5$ with the basis (1.3.21):*

$$X_1, \quad X_2, \quad X_3, \quad X_4, \quad X_1 + X_4, \quad X_1 - X_4,$$

$$X_5, \quad X_1 + X_5, \quad X_2 + X_5, \quad X_2 - X_5, \quad (1.3.40)$$

*where $k$ is an arbitrary parameter.*

*Proof* We first clarify if the transformations (1.3.34)–(1.3.38) have invariants $J(l^1, \ldots, l^5)$. The reckoning shows that the $5 \times 5$ matrix $\|c_{\mu\nu}^\lambda l^\nu\|$ of the coefficients of the operators (1.3.33) has the rank four. It means that the transformations (1.3.34)–(1.3.38) have precisely one functionally independent invariant. The integration of the equations

$$E_\mu(J) = 0, \quad \mu = 1, \ldots, 5,$$

shows that the invariant is

$$J = (l^3)^2 - l^1 l^5. \quad (1.3.41)$$

Knowledge of the invariant (1.3.41) simplifies further calculations significantly.

The last equation in (1.3.38) shows that if $l^1 \neq 0$, we get $\tilde{l}_5 = 0$ by solving the quadratic equation $l^5 - 2a_5 l^3 + a_5^2 l^1 = 0$ for $a_5$, i.e. by taking

$$a_5 = \frac{l^3 \pm \sqrt{J}}{l^1}, \quad (1.3.42)$$

where $J$ is the invariant (1.3.41). We can use (1.3.42) only if $J \geq 0$.

Now we begin the construction of the optimal system. The method requires a simplification of the general vector (1.3.30) by means of the transformations (1.3.34)–(1.3.38). As a result, we will find the simplest representatives of each class of similar vectors (1.3.30). Substituting these representatives in (1.3.29), we will obtain the optimal system of one-dimensional subalgebras of $L_5$. We will divide the construction to several cases. □

### 1.3.5.1 The Case $l^1 = 0$

I will divide this case into the following two subcases.

$1°$. $l^3 \neq 0$. In other words, we consider the vectors (1.3.30) of the form

$$(0, l^2, l^3, l^4, l^5), \quad l^3 \neq 0.$$

First we take $a_5 = l^5/(2l^3)$ in (1.3.38) and reduce the above vector to the form

$$(0, l^2, l^3, l^4, 0).$$

Then we subject the latter vector to the transformation (1.3.37) with $a_4 = l^4/l^3$ and obtain the vector

$$(0, l^2, l^3, 0, 0).$$

Since the operator $X$ is defined up to a constant factor and $l^3 \neq 0$, we divide the above vector by $l^3$ and transform it using (1.3.35) to the form

$$(0, 0, 1, 0, 0).$$

Substituting it in (1.3.29), we obtain the operator

$$X_3. \tag{1.3.43}$$

$2°$. $l^3 = 0$. Thus, we consider the vectors (1.3.30) of the form

$$(0, l^2, 0, l^4, l^5).$$

$2°(1)$. If $l^2 \neq 0$, we can assume $l^2 = 1$ (see above), use the transformation (1.3.38) with $a_5 = Al^4$ and get the vector

$$(0, 1, 0, 0, l^5).$$

If $l^5 \neq 0$ we can make $l^5 = \pm 1$ by the transformation (1.3.36). Thus, taking into account the possibility $l^5 = 0$, we obtain the following representatives for the optimal system:

$$X_2, \quad X_2 + X_5, \quad X_2 - X_5. \tag{1.3.44}$$

$2°(2)$. Let $l^2 = 0$. If $l^5 \neq 0$ we can set $l^5 = 1$. Now we apply the transformation (1.3.35) with $a_2 = -l^4$ and obtain the vector $(0, 0, 0, 0, 1)$. If $l^5 = 0$ we get the vector $(0, 0, 0, 1, 0)$. Thus, the case $l^2 = 0$ provides the operators

$$X_4, \quad X_5. \tag{1.3.45}$$

### 1.3.5.2 The Case $l^1 \neq 0$, $J > 0$

Now we can define $a_5$ by (1.3.42) and annul $\bar{l}^5$ by the transformation (1.3.38). Thus, we will deal with the vector

$$(l^1, l^2, l^3, l^4, 0), \quad l^1 \neq 0.$$

Since $J$ is invariant under the transformations (1.3.34)–(1.3.38), the condition $J > 0$ yields that in the above vector we have $l^3 \neq 0$. Therefore we can use the transformation (1.3.37) with $a_4 = l^4/l^3$ and get $\bar{l}^4 = 0$. Then we apply the transformation (1.3.34) with $a_1 = -l^1/(2l^3)$ and obtain $\bar{l}^1 = 0$, thus arriving at the vector $(0, l^2, l^3, 0, 0)$, and hence at the previous operator (1.3.43). Hence, this case contributes no additional subalgebras to the optimal system.

### 1.3.5.3 The Case $l^1 \neq 0$, $J = 0$

In this case (1.3.42) reduces to $a_5 = l^3/l^1$.

If $l^3 \neq 0$, we use the transformation (1.3.38) with $a_5 = l^3/l^1$ and obtain $\bar{l}^5 = 0$. Due to the invariance of $J$ we conclude that the equation $J = 0$ yields $(\bar{l}^3)^2 - \bar{l}^1\bar{l}^5 = 0$. Since $\bar{l}^5 = 0$, it follows that $\bar{l}^3 = 0$. Thus we can deal with the vectors of the form

$$(l^1, l^2, 0, l^4, 0), \quad l^1 \neq 0. \tag{1.3.46}$$

Furthermore, if $l^3 = 0$, we have $J = -l^1 l^5$, and the equation $J = 0$ yields $l^5 = 0$ since $l^1 \neq 0$. Therefore we again have the vectors of the form (1.3.46) where we can assume $l^1 = 1$. Subjecting the vector (1.3.46) with $l^1 = 1$ to the transformation (1.3.37) with $a_4 = l^2$ we obtain $\bar{l}^2 = 0$, and hence map the vector (1.3.46) to the form

$$(1, 0, 0, l^4, 0).$$

If $l^4 \neq 0$, we use the transformation (1.3.36) with an appropriately chosen $a_3$ and obtain $l^4 = \pm 1$. taking into account the possibility $l^4 = 0$, we see that this case contributes the following operators:

$$X_1, \quad X_1 + X_4, \quad X_1 - X_4. \tag{1.3.47}$$

### 1.3.5.4 The Case $l^1 \neq 0$, $J < 0$

It is obvious from the condition $J = (l^3)^2 - l^1 l^5 < 0$ that $l^5 \neq 0$. Therefore we successively apply the transformations (1.3.38), (1.3.37) and (1.3.35) with $a_5 = l^3/l^1$, $a_4 = l^2/l^1$ and $a_2 = -l^4/l^5$, respectively and obtain $\bar{l}^3 = \bar{l}^2 = \bar{l}^4 = 0$. The components $l^1$ and $l^5$ of the resulting vector

$$(l^1, 0, 0, 0, l^5)$$

have the common sign since the condition $J < 0$ yields $l^1 l^5 > 0$. Therefore using the transformation (1.3.36) with an appropriate value of the parameter $a_3$ and invoking that we can multiply the vector $l$ by any constant, we obtain $l^1 = l^5 = 1$, i.e. the operator

$$X_1 + X_5. \tag{1.3.48}$$

Finally, collecting the operators (1.3.43), (1.3.44), (1.3.45), (1.3.47) and (1.3.48), we arrive at the optimal system (1.3.40), thus completing the proof.

### 1.3.6  All Invariant Solutions of the Burgers Equation

Constructing the invariant solution for each operator from the optimal system of sub-algebras (1.3.40), we obtain the following *optimal system of invariant solutions* [11].

**Theorem 1.3.2** *An optimal system of invariant solutions for the Burgers equation is provided by the following solutions, where $\sigma$, $\gamma$ and $K$ are arbitrary constants.*

$X_1$:

$$\text{(i)}\quad u = \frac{2}{x+\gamma};$$

$$\text{(ii)}\quad u = \sigma\frac{\gamma e^{\sigma x}+1}{\gamma e^{\sigma x}-1} \equiv \tilde{\sigma}\tanh\left(\tilde{\gamma}+\frac{\tilde{\sigma}}{2}x\right); \qquad (1.3.49)$$

$$\text{(iii)}\quad u = \sigma\tan\left(\gamma-\frac{\sigma}{2}x\right).$$

$X_3$:

$$u = \frac{\varphi(\lambda)}{\sqrt{t}}, \quad \lambda = \frac{x}{\sqrt{t}},$$

$$\text{where}\quad \varphi' + \frac{1}{2}\varphi^2 + \frac{1}{2}\lambda\varphi = K. \qquad (1.3.50)$$

$$X_4: \qquad u = \frac{K-x}{t}. \qquad (1.3.51)$$

$$X_1 + X_4: \qquad u = \varphi(\lambda) - t, \quad \lambda = x - \frac{t^2}{2},$$

$$\text{where}\quad \varphi' + \frac{1}{2}\varphi^2 + \lambda = K. \qquad (1.3.52)$$

$$X_1 - X_4: \qquad u = t + \varphi(\lambda), \quad \lambda = x + \frac{t^2}{2},$$

$$\text{where}\quad \varphi' + \frac{1}{2}\varphi^2 - \lambda = K. \qquad (1.3.53)$$

$X_5$:

$$u = -\lambda + \frac{\varphi(\lambda)}{t}, \quad \lambda = \frac{x}{t}, \quad \text{where}$$

$$\text{(i)}\quad \varphi(\lambda) = \sigma\frac{\gamma e^{\sigma\lambda}-1}{\gamma e^{\sigma\lambda}+1}, \quad |\varphi| < \sigma;$$

$$\text{(ii)}\quad \varphi(\lambda) = \sigma\frac{\gamma e^{\sigma\lambda}+1}{\gamma e^{\sigma\lambda}-1}, \quad |\varphi| > \sigma; \qquad (1.3.54)$$

$$\text{(iii)}\quad \varphi(\lambda) = \sigma\tan\left(\gamma-\frac{\sigma}{2}\lambda\right).$$

$$X_1 + X_5: \qquad u = -\frac{tx}{1+t^2} + \frac{\varphi(\lambda)}{\sqrt{1+t^2}}, \quad \lambda = \frac{x}{\sqrt{1+t^2}},$$

$$\text{where}\quad \varphi' + \frac{1}{2}\varphi^2 + \frac{1}{2}\lambda^2 = K. \qquad (1.3.55)$$

$$X_2 + X_5: \qquad u = -\frac{x}{t} - \frac{1}{t^2} + \frac{\varphi(\lambda)}{t}, \quad \lambda = \frac{x}{t} + \frac{1}{2t^2},$$

$$\text{where}\quad \varphi' + \frac{1}{2}\varphi^2 - \lambda = K. \qquad (1.3.56)$$

$$X_2 - X_5: \qquad u = -\frac{x}{t} + \frac{1}{t^2} + \frac{\varphi(\lambda)}{t}, \qquad \lambda = \frac{x}{t} - \frac{1}{2t^2},$$

$$\text{where} \quad \varphi' + \frac{1}{2}\varphi^2 + \lambda = K. \tag{1.3.57}$$

If one subjects each solution from the optimal system of invariant solutions (1.3.49)–(1.3.57) to all transformations of the group admitted by the Burgers equation, one obtains *all invariant solutions* of the Burgers equation. We see that the invariant solutions of the Burgers equation are given either by elementary functions or by solving a Riccati equation. Furthermore, one can verify that the set of all invariant solutions involves 76 parameters.

We see that the invariant solutions of the Burgers equation are given either by elementary functions or by solving a Riccati equation.

Furthermore, the solutions to the Riccati equations describing the solutions (1.3.52), (1.3.53), (1.3.56) and (1.3.57) can be represented by special functions. Namely, setting $\varphi = \sqrt{2}\psi$, $\mu = \lambda + K$ and using the substitution

$$\psi = \frac{d \ln |z|}{d\mu} \equiv \frac{z'}{z}$$

we reduce the Riccati equation in (1.3.53) and (1.3.56) to the *Airy equation*

$$\frac{d^2 z}{d\mu^2} - \mu z = 0. \tag{1.3.58}$$

The general solution to (1.3.58) is the linear combination

$$z = C_1 \mathrm{Ai}(\mu) + C_2 \mathrm{Bi}(\mu), \qquad C_1, C_2 = \text{const.}, \tag{1.3.59}$$

of the *Airy functions* (see, e.g. [14], [17])

$$\mathrm{Ai}(\mu) = \frac{1}{\pi} \int_0^\infty \cos\left(\mu\tau + \frac{1}{3}\tau^3\right) d\tau,$$

$$\mathrm{Bi}(\mu) = \frac{1}{\pi} \int_0^\infty \left[\exp\left(\mu\tau - \frac{1}{3}\tau^3\right) + \sin\left(\mu\tau + \frac{1}{3}\tau^3\right)\right] d\tau. \tag{1.3.60}$$

Hence, the function $\varphi(\lambda)$ in the solutions (1.3.53) and (1.3.56) is given by

$$\varphi(\lambda) = \sqrt{2}\frac{d}{d\lambda} \ln\left|C_1 \mathrm{Ai}(\lambda + K) + C_2 \mathrm{Bi}(\lambda + K)\right|. \tag{1.3.61}$$

One can obtain likewise that $\varphi(\lambda)$ in (1.3.52) and (1.3.57) is given by

$$\varphi(\lambda) = \sqrt{2}\frac{d}{d\lambda} \ln\left|C_1 \mathrm{Ai}(K - \lambda) + C_2 \mathrm{Bi}(K - \lambda)\right|. \tag{1.3.62}$$

Finally, it is worth noting that the optimal system of subalgebras is not unique, it depends on the choice of a representative in each class of similar operators. Consequently, the form of the solutions included in an optimal system of invariant solutions depends on the choice of representatives. However, this choice does not affect the amount of the optimal system of invariant solutions since the number of the classes of similar operators does not depend on the choice of representatives. Moreover, this choice does not affect the final form of the 76-parameter set of all invariant solutions obtained from an optimal system of invariant solutions by the transformations of the general group admitted by the Burgers equation.

## 1.4  General Definitions of Symmetry Groups

### 1.4.1  Differential Variables and Function

We will use the following notation. Consider the algebraically independent variables

$$x = \{x^i\}, \quad u = \{u^\alpha\}, \quad u_{(1)} = \{u_i^\alpha\}, \quad u_{(2)} = \{u_{ij}^\alpha\}, \dots, \qquad (1.4.1)$$

where $\alpha = 1, \dots, m$, and $i, j = 1, \dots, n$. The variables $u_{ij}^\alpha, \dots$ are assumed to be symmetric in subscripts, i.e. $u_{ij}^\alpha = u_{ji}^\alpha$. The operator

$$D_i = \frac{\partial}{\partial x^i} + u_i^\alpha \frac{\partial}{\partial u^\alpha} + u_{ij}^\alpha \frac{\partial}{\partial u_j^\alpha} + \cdots \quad (i = 1, \dots, n), \qquad (1.4.2)$$

is called the *total differentiation* with respect to $x^i$. The operator $D_i$ is a formal sum of an infinite number of terms. However, it truncates when acting on any function of a finite number of the variables $x, u, u_{(1)}, \dots$. In consequence, the total differentiations $D_i$ are well defined on the set of all functions depending on a finite number of $x, u, u_{(1)}, \dots$.

Though the variables (1.4.1) are assumed to be *algebraically independent*, they are connected by the following *differential relations*:

$$u_i^\alpha = D_i(u^\alpha), \quad u_{ij}^\alpha = D_j(u_i^\alpha) = D_j D_i(u^\alpha). \qquad (1.4.3)$$

The variables $x^i$ are called independent variables, and the variables $u^\alpha$ are known as *differential* (or dependent) *variables* with the successive derivatives $u_{(1)}, u_{(2)}$, etc. The universal space of modern group analysis is the space $\mathscr{A}$ of differential functions introduced by Ibragimov [3] (see also [4], Sect. 19) as a generalization of differential polynomials considered by J.F. Ritt in the 1950s.

**Definition 1.4.1** A locally analytic function (i.e., locally expandable in a Taylor series with respect to all arguments) of a finite number of variables (1.4.1) is called a *differential function*. The highest order of derivatives appearing in the differential function is called the order of this function. The set of all differential functions of all finite orders is denoted by $\mathscr{A}$. This set is a vector space with respect to the usual addition of functions and becomes an associative algebra if multiplication is defined by the usual multiplication of functions. A significant property of the space $\mathscr{A}$ is that it is closed under the action of total derivatives (1.4.2).

**Definition 1.4.2** A group $G$ of transformations of the form

$$\bar{x}^i = f^i(x, u, a), \quad f^i|_{a=0} = x^i, \tag{1.4.4}$$

$$\bar{u}^\alpha = \varphi^\alpha(x, u, a), \quad \varphi^\alpha|_{a=0} = u^\alpha, \tag{1.4.5}$$

is called a group of point transformations in the space of dependent and independent variables. The generator of the group $G$ is

$$X = \xi^i(x, u)\frac{\partial}{\partial x^i} + \eta^\alpha(x, u)\frac{\partial}{\partial u^\alpha}, \tag{1.4.6}$$

where

$$\xi^i = \frac{\partial f^i}{\partial a}\bigg|_{a=0}, \quad \eta^\alpha = \frac{\partial \varphi^\alpha}{\partial a}\bigg|_{a=0}. \tag{1.4.7}$$

Let $F_k \in \mathscr{A}$ be any differential functions and let $p$ be the maximum of orders of the differential functions $F_k, k = 1, \ldots, s$. Consider the system of equations

$$F_k(x, u, u_{(1)}, \ldots, u_{(p)}) = 0, \quad k = 1, \ldots, s. \tag{1.4.8}$$

If one treats the variables $u^\alpha$ as functions of $x$ so that

$$u^\alpha = u^\alpha(x), \quad u_i^\alpha = \frac{\partial u^\alpha(x)}{\partial x^i}, \ldots,$$

then one arrives at the usual concept of a *system of differential equations* (1.4.8) of order $p$.

## 1.4.2 Frame and Extended Frame

Recall the definitions of the *frame* and *extended frame* of differential equations given in [5] (see also [6], Chap. 1).

**Definition 1.4.3** Let us treat $x, u, u_{(1)}, \ldots$ as functionally independent variables connected only by the differential relations (1.4.3). Then (1.4.8) determine a surface in the space of the independent variables $x, u, u_{(1)}, \ldots, u_{(p)}$. This surface is called the *frame* (or *skeleton*) of the system of differential equations (1.4.8).

**Definition 1.4.4** Consider the frame equation (1.4.8) together with its differential consequences,

$$F_k = 0, \quad D_i F_k = 0, \quad D_i D_j F_k = 0, \ldots. \tag{1.4.9}$$

The totality of points $(x, u, u_{(1)}, \ldots)$ satisfying (1.4.9) is called the *extended frame* of the system of differential equations (1.4.8) and is denoted by $[F]$.

We will assume that

$$\text{rank}\left\|\frac{\partial F_k}{\partial x^i}, \frac{\partial F_k}{\partial u^\alpha}, \frac{\partial F_k}{\partial u_i^\alpha}, \ldots\right\| = s$$

on the frame of the differential equations under consideration.

### 1.4.3 Definition Using Solutions

The first definition of a symmetry group of an arbitrary system of differential equations coincides with Definition 1.3.2 for a single evolution equation.

**Definition 1.4.5** The system of differential equations (1.4.8) is said to be invariant under the group $G$ of transformations (1.4.4), (1.4.5) if the transformations (1.4.4), (1.4.5) convert every solution of the system (1.4.8) into a solution of the same system. Here the solutions of differential equations are considered as classical ones, i.e., are assumed to be smooth functions $u^\alpha = u^\alpha(x)$. If the system of equations (1.4.8) is invariant under the group $G$ then $G$ is also known as a symmetry group for the system (1.4.8) or a group admitted by this system.

### 1.4.4 Definition Using the Frame

Though the first definition is conceptually simple, it depends upon knowledge of solutions. Therefore, in practical calculation of symmetries the following second, geometric definition is more efficient.

**Definition 1.4.6** The system of differential equations (1.4.8) is said to be invariant under the group $G$ if the frame of the system is an invariant surface with respect to the prolongation of the transformations (1.4.4), (1.4.5) of the group $G$ to the derivatives $u_{(1)}, \ldots, u_{(p)}$.

According to this definition and the invariance test of equations given by Theorem 1.1.3, one obtains the following infinitesimal test for obtaining symmetries of differential equations.

**Theorem 1.4.1** *The group $G$ with the generator $X$ is admitted by the system of differential equations* (1.4.8) *if and only if*

$$X_{(p)} F_k \big|_{(1.4.8)} = 0, \quad k = 1, \ldots, s, \tag{1.4.10}$$

*where $X_{(p)}$ is the $p$-th prolongation of $X$ and $|_{(1.4.8)}$ means evaluated on the frame the system of differential equations* (1.4.8). *Equations* (1.4.10) *are the determining equations.*

Let $z_0 = (x_0, u_0, \ldots, u_{0(p)})$ be a point on the frame of the system (1.4.8), i.e., $F_k(x_0, u_0, \ldots, u_{0(p)}) = 0$ $(k = 1, \ldots, s)$. The system of differential equations (1.4.8) is said to be *locally solvable* at $z_0$ if there is a solution passing through this point, i.e., there exist a solution $u = h(x)$ of differential equations (1.4.8) defined in a neighborhood of the point $x_0$ such that $u_0 = h(x_0), \ldots, u_{0(p)} = \partial p_h / \partial x_p(x_0)$. The system (1.4.8) is said to be *locally solvable* if it has this property at every generic point of the frame.

It can be shown that for locally solvable systems the first and the second definitions of the symmetry group are equivalent, i.e. that Definition 1.4.5 and Definition 1.4.6 provide exactly the same symmetry group. A discussion of this equivalence is to be found in Lie [13], Chap. 6, Sect. 1, and Ovsyannikov [16], Sect. 15.1. See also Olver [15], Sect. 2.6, for a modern treatment of this subject.

### 1.4.5 Definition Using the Extended Frame

If the system (1.4.8) is not locally solvable, e.g. if the system (1.4.8) is over-determined, it may happen (see further Example 1.4.1) that Definition 1.4.6 provides only a subgroup of the symmetry group given by Definition 1.4.5. Therefore, Ibragimov proposed ([4], Sect. 17.1, see also [6], Chap. 1) the following third definition and proved the appropriate infinitesimal test for the invariance of over-determined systems of differential equations.

**Definition 1.4.7** The system of differential equations (1.4.8) is said to be invariant under the group $G$ if the extended frame $[F]$ is invariant with respect to the infinite-order prolongation of $G$.

The infinitesimal test for this invariance is written as follows (see [4], Theorem 17.1).

**Theorem 1.4.2** *Let $X$ be the generator of a group $G$. The system of differential equations (1.4.8) are invariant under the group $G$ in the sense of Definition 1.4.7 if and only if the following equations are satisfied:*

$$X_{(p)} F_k\big|_{[F]} = 0, \quad k = 1, \ldots, s. \tag{1.4.11}$$

*Equations (1.4.11) are also called determining equations.*

**Remark 1.4.1** According to Theorem 1.4.2, the invariance test does not involve all the differential consequences (1.4.9) of the differential equations (1.4.8). In fact, it can be easily shown that it suffices to consider only a finite number of the differential consequences (1.4.9) such that they form a system in involution. It is also worth noting that we do not need to take into account the additional equations such as $X_{(p)}(D_i F_k) = 0$ since they are satisfied identically due to (1.4.11).

For locally solvable systems, all three definitions of symmetry groups are equivalent. For over-determined systems, the first and third definitions are equivalent, whereas the second definition provides, in general, only a subgroup of the symmetry group given by the third definition.

**Example 1.4.1** Consider the over-determined system ([6], Sect. 1.3.10)

$$u_t = (u_x)^{-4/3} u_{xx}, \quad v_t = -3(u_x)^{-1/3}, \quad v_x = u. \tag{1.4.12}$$

This is a system of three equations for two dependent variables $u$ and $v$. The maximal order of equations involved in the system is $p = 2$. Let us first solve the determining equations (1.4.10). The left-hand side of (1.4.10) depends upon the variables $x, t, u, v, u_x, u_{xx}, u_{xt}$, and $v_{xx}$ in accordance with the prolongation formulae. The solution of the determining equations yields the 6-dimensional Lie algebra spanned by

$$X_1 = \frac{\partial}{\partial x}, \quad X_2 = \frac{\partial}{\partial t}, \quad X_3 = \frac{\partial}{\partial v}, \quad X_4 = \frac{\partial}{\partial u} + x \frac{\partial}{\partial v},$$
$$X_5 = 4t \frac{\partial}{\partial t} + 3u \frac{\partial}{\partial u} + 3v \frac{\partial}{\partial v}, \quad X_6 = 2x \frac{\partial}{\partial x} - u \frac{\partial}{\partial u} + v \frac{\partial}{\partial v}. \tag{1.4.13}$$

This is the Lie algebra of the maximal symmetry group for (1.4.12) obtained by the second definition (Definition 1.4.6).

Consider now the determining equations (1.4.11). Differentiation of the third equation (1.4.12) yields $v_{xx} = u_x$. Therefore, we replace $v_{xx}$ in the determining equation by $u_x$. Then the left-hand side of (1.4.11) involves only the variables $x, t, u, v, u_x, u_{xx}$ and $u_{xt}$. Solving the determining equations (1.4.11), one obtains the 7-dimensional Lie algebra spanned by the operators (1.4.13) and by

$$X_7 = x^2 \frac{\partial}{\partial x} + xv \frac{\partial}{\partial v} + (v - xu) \frac{\partial}{\partial u}. \tag{1.4.14}$$

Thus, the third definition (Definition 1.4.7) provides a more general symmetry group than the second definition.

## 1.5  Lie–Bäcklund Transformation Groups

This is section provides an introduction to the theory of Lie–Bäcklund transformation groups and contains the basic definitions, theorems and algorithms used for computation of Lie–Bäcklund symmetries of differential equations. The space $\mathscr{A}$ of differential functions introduced in Sect. 1.4.1 play a central role in this theory.

### 1.5.1  Lie–Bäcklund Operators

Geometrically, Lie–Bäcklund transformations appear in attempting to find a higher-order generalization of the classical contact (first-order tangent) transformations (see Bäcklund's paper [1], its English translation is available in [9]) and are identified with infinite-order tangent transformations. A historical survey of the development of this branch of group analysis and a detailed discussion of the modern theory with many applications are to be found in [4] (see also [7], Chap. 1). We will use here a shortcut to the theory of Lie–Bäcklund transformation groups by using a generalization of infinitesimal generators of point and contact transformation groups. The generalization is known as a Lie–Bäcklund operator and is defined as follows.

**Definition 1.5.1** Let $\xi^i, \eta^\alpha \in \mathscr{A}$ be differential functions depending on any finite number of variables $x, u, u_{(1)}, u_{(2)}, \ldots$. A differential operator

$$X = \xi^i \frac{\partial}{\partial x^i} + \eta^\alpha \frac{\partial}{\partial u^\alpha} + \zeta_i^\alpha \frac{\partial}{\partial u_i^\alpha} + \zeta_{i_1 i_2}^\alpha \frac{\partial}{\partial u_{i_1 i_2}^\alpha} + \cdots, \qquad (1.5.1)$$

where

$$\zeta_i^\alpha = D_i(\eta^\alpha - \xi^j u_j^\alpha) + \xi^j u_{ij}^\alpha,$$
$$\zeta_{i_1 i_2}^\alpha = D_{i_1} D_{i_2}(\eta^\alpha - \xi^j u_j^\alpha) + \xi^j u_{j i_1 i_2}^\alpha, \ldots \qquad (1.5.2)$$

is called a *Lie–Bäcklund operator*. The Lie–Bäcklund operator (1.5.1) is often written in the abbreviated form

$$X = \xi^i \frac{\partial}{\partial x^i} + \eta^\alpha \frac{\partial}{\partial u^\alpha} + \cdots, \qquad (1.5.3)$$

where the prolongation given by (1.5.1)–(1.5.2) is understood.

The operator (1.5.1) is formally an infinite sum. However, it truncates when acting on any differential function. Hence, *the action of Lie–Bäcklund operators is well defined on the space $\mathscr{A}$.*

Consider two Lie–Bäcklund operators

$$X_\nu = \xi_\nu^i \frac{\partial}{\partial x^i} + \eta_\nu^\alpha \frac{\partial}{\partial u^\alpha} + \cdots, \qquad \nu = 1, 2,$$

and define their commutator by the usual formula:

$$[X_1, X_2] = X_1 X_2 - X_2 X_1.$$

**Theorem 1.5.1** *The commutator $[X_1, X_2]$ is identical with the Lie–Bäcklund operator given by*

$$[X_1, X_2] = \big(X_1(\xi_2^i) - X_2(\xi_1^i)\big) \frac{\partial}{\partial x^i} + \big(X_1(\eta_2^\alpha) - X_2(\eta_1^\alpha)\big) \frac{\partial}{\partial u^\alpha} + \cdots, \quad (1.5.4)$$

*where the terms denoted by dots are obtained by prolonging the coefficients of $\partial/\partial x^i$ and $\partial/\partial u^\alpha$ in accordance with (1.5.1) and (1.5.2).*

According to Theorem 1.5.1, the set of all Lie–Bäcklund operators is an infinite dimensional Lie algebra with respect to the commutator (1.5.4). It is called the *Lie–Bäcklund algebra* and denoted by $L_\mathscr{B}$. The Lie–Bäcklund algebra is endowed with the following properties (see [4]).

**I.** $D_i \in L_\mathscr{B}$. In other words, the total differentiation (1.4.2) is a Lie–Bäcklund operator. Furthermore,

$$X_* = \xi_*^i D_i \in L_\mathscr{B} \qquad (1.5.5)$$

for any $\xi_*^i \in \mathscr{A}$.

**II.** Let $L_*$ be the set of all Lie–Bäcklund operators of the form (1.5.5). Then $L_*$ is an ideal of $L_\mathscr{B}$, i.e., $[X, X_*] \in L_*$ for any $X \in L_\mathscr{B}$. Indeed,

$$[X, X_*] = \big(X(\xi_*^i) - X_*(\xi^i)\big) D_i \in L_*.$$

**III.** In accordance with property II, two operators $X_1, X_2 \in L_{\mathscr{B}}$ are said to be *equivalent* (i.e., $X_1 \sim X_2$) if $X_1 - X_2 \in L_*$. In particular, every operator $X \in L_{\mathscr{B}}$ is equivalent to an operator (1.5.1) with $\xi^i = 0$, $i = 1, \ldots, n$. Namely, , $X \sim \tilde{X}$ where

$$\tilde{X} = X - \xi^i D_i = (\eta^\alpha - \xi^i u_i^\alpha)\frac{\partial}{\partial u^\alpha} + \cdots. \tag{1.5.6}$$

**Definition 1.5.2** The operators of the form

$$X = \eta^\alpha \frac{\partial}{\partial u^\alpha} + \cdots, \quad \eta^\alpha \in \mathscr{A}, \tag{1.5.7}$$

are called *canonical Lie–Bäcklund operators.*

Using this definition, we can formulate the property III as follows.

**Theorem 1.5.2** *Any operator $X \in L_{\mathscr{B}}$ is equivalent to a canonical Lie–Bäcklund operator.*

**Example 1.5.1** Let us take $n = m = 1$ and denote $u_1 = u_x$. The generator of the group of translations along the $x$-axis and its canonical Lie–Bäcklund form (1.5.6) are written as follows:

$$X = \frac{\partial}{\partial x} \sim \tilde{X} = u_x \frac{\partial}{\partial u} + \cdots.$$

**Example 1.5.2** Let $x, y$ be the independent variables, and $k, c = $ const. The generator of non-homogeneous dilations and its canonical Lie–Bäcklund form (1.5.6) are written:

$$X = x\frac{\partial}{\partial x} + ky\frac{\partial}{\partial y} + cu\frac{\partial}{\partial u} \sim \tilde{X} = (cu - xu_x - kyu_y)\frac{\partial}{\partial u} + \cdots.$$

**Example 1.5.3** Let $t, x$ be the independent variables. The generator of the Galilean boost and its canonical Lie–Bäcklund form (1.5.6) are written:

$$X = t\frac{\partial}{\partial x} + \frac{\partial}{\partial u} \sim \tilde{X} = (1 - tu_x)\frac{\partial}{\partial u} + \cdots.$$

The canonical operators leave invariant the independent variables $x^i$. Therefore, the use of the canonical form is convenient, e.g., for investigating symmetries of integro-differential equations.

**IV.** The following statements describe all Lie–Bäcklund operators equivalent to generators of Lie point and Lie contact transformation groups.

**Theorem 1.5.3** *The Lie–Bäcklund operator* (1.5.1) *is equivalent to the infinitesimal operator of a one-parameter point transformation group if and only if its coordinates assume the form*

$$\xi^i = \xi_1^i(x, u) + \xi_*^i, \quad \eta^\alpha = \eta_1^\alpha(x, u) + \left(\xi_2^i(x, u) + \xi_*^i\right)u_i^\alpha,$$

where $\xi_*^i \in \mathscr{A}$ is an arbitrary differential function, and $\xi_1^i, \xi_2^i, \eta_1^\alpha$ are arbitrary functions of $x$ and $u$.

**Theorem 1.5.4** *Let $m = 1$. Then the operator (1.5.1) is equivalent to the infinitesimal operator of a one-parameter contact transformation group if and only if its coordinates assume the form*

$$\xi^i = \xi_1^i(x, u, u_{(1)}) + \xi_*^i, \quad \eta = \eta_1(x, u, u_{(1)}) + \xi_*^i u_i,$$

*where $\xi_*^i \in \mathscr{A}$ is an arbitrary differential function, and $\xi_1^i, \eta_1$ are arbitrary first-order differential functions, i.e. depend upon $x, u$ and $u_{(1)}$.*

## 1.5.2 Integration of Lie–Bäcklund Equations

Consider the sequence

$$z = (x, u, u_{(1)}, u_{(2)}, \ldots) \tag{1.5.8}$$

with the elements $z^\nu, \nu \geq 1$, were

$$z^i = x^i, 1 \leq i \leq n, \quad z^{n+\alpha} = u^\alpha, 1 \leq \alpha \leq m.$$

Denote by $[z]$ any finite subsequence of $z$. Then elements of the space $\mathscr{A}$ of differential functions are written as $f([z])$.

**Definition 1.5.3** Given an operator (1.5.1), the following infinite system is called *Lie–Bäcklund equations*:

$$\frac{d}{da} \bar{x}^i = \xi^i([\bar{z}]), \quad \frac{d}{da} \bar{u}^\alpha = \eta^\alpha([\bar{z}]),$$
$$\frac{d}{da} \bar{u}_i^\alpha = \zeta_i^\alpha([\bar{z}]), \quad \frac{d}{da} \bar{u}_{ij}^\alpha = \zeta_{ij}^\alpha([\bar{z}]), \ldots, \tag{1.5.9}$$

where $\alpha = 1, \ldots, m$ and $i, j, \ldots = 1, \ldots, n$.

In the case of canonical operators (1.5.7), the infinite system of equations (1.5.9) can be replaced by the finite system

$$\frac{d}{da} \bar{u}^\alpha = \eta^\alpha([\bar{z}]), \quad \alpha = 1, \ldots, m. \tag{1.5.10}$$

Indeed, upon solving the system (1.5.10), the transformations of the successive derivatives are obtained by the total differentiation:

$$\bar{u}_i^\alpha = D_i(\bar{u}^\alpha), \quad \bar{u}_{ij}^\alpha = D_i D_j(\bar{u}^\alpha), \ldots. \tag{1.5.11}$$

We will use the abbreviated form (1.5.3) of Lie–Bäcklund operators and write the system (1.5.9), together with the initial conditions, as follows:

$$\frac{d}{da}\bar{x}^i = \xi^i([\bar{z}]), \quad \bar{x}^i\big|_{a=0} = x^i,$$

$$\frac{d}{da}\bar{u}^\alpha = \eta^\alpha([\bar{z}]), \quad \bar{u}^\alpha\big|_{a=0} = u^\alpha, \tag{1.5.12}$$

The formal integrability of the infinite system (1.5.12) has been proved by Ibragimov (see, e.g. [4], Sect. 15.1; it is also discussed in [6]). For the convenience of the reader, we formulate here the existence theorem. The following notation is convenient for formulating and proving the theorem.

Let $f$ and $g$ be formal power series in one symbol $a$ with coefficients from the space $\mathscr{A}$, i.e. let

$$f(z,a) = \sum_{k=0}^{\infty} f_k([z])a^k, \quad f_k([z]) \in \mathscr{A}, \tag{1.5.13}$$

and

$$g(z,a) = \sum_{k=0}^{\infty} g_k([z])a^k, \quad g_k([z]) \in \mathscr{A}.$$

Their linear combination $\lambda f([z]) + \mu g([z])$ with constant coefficients $\lambda, \mu$ and product $f([z]) \cdot g([z])$ are defined by

$$\lambda \sum_{k=0}^{\infty} f_k([z])a^k + \mu \sum_{k=0}^{\infty} g_k([z])a^k = \sum_{k=0}^{\infty}\big(\lambda f_k([z]) + \mu g_k([z])\big)a^k, \tag{1.5.14}$$

and

$$\left(\sum_{p=0}^{\infty} f_p([z])a^p\right) \cdot \left(\sum_{q=0}^{\infty} g_q([z])a^q\right) = \sum_{k=0}^{\infty}\left(\sum_{p+q=k} f_p([z])g_q([z])\right)a^k, \tag{1.5.15}$$

respectively. The space of all formal power series (1.5.13) endowed with the addition (1.5.14) and the multiplication (1.5.15) is denoted by $[[\mathscr{A}]]$.

Lie point and Lie contact transformations, together with their prolongations of all orders, are represented by elements of the space $[[\mathscr{A}]]$. Moreover, the utilization of this space is necessary in the theory of Lie–Bäcklund transformation groups. Therefore, $[[\mathscr{A}]]$ is called *the representation space of modern group analysis* ([6], Sect. 1.2).

The existence theorem is formulated as follows.

**Theorem 1.5.5** *The Lie–Bäcklund equations* (1.4.5) *have a solution in the space* $[[\mathscr{A}]]$. *The solution is unique. It is given by formal power series*

$$\bar{x}^i = x^i + \sum_{k=0}^{\infty} A_k^i([z])a^k, \quad A_k^i([z]) \in \mathscr{A},$$

$$\bar{u}^\alpha = u^\alpha + \sum_{k=0}^{\infty} B_k^\alpha([z])a^k, \quad B_k^\alpha([z]) \in \mathscr{A}, \tag{1.5.16}$$

*and satisfies the group property.*

**Definition 1.5.4** The group of formal transformations (1.5.16) is called a one-parameter Lie–Bäcklund transformation group.

Recall that a point transformation group acting in the finite dimensional space of variables $x = (x^1, \ldots, x^n)$ and generated by an operator $X$ can be represented by the exponential map (1.1.22):

$$\bar{x}^i = \exp(aX)(x^i), \quad i = 1, \ldots, n, \tag{1.5.17}$$

where

$$\exp(aX) = 1 + aX + \frac{a^2}{2!} X^2 + \frac{a^3}{3!} X^3 + \cdots. \tag{1.5.18}$$

Likewise, the solution (1.5.16) to the Lie–Bäcklund equations (1.5.12) can be represented by the exponential map

$$\bar{x}^i = \exp(aX)(x^i), \ \bar{u}^\alpha = \exp(aX)(u^\alpha), \ \bar{u}_i^\alpha = \exp(aX)(u_i^\alpha), \ldots, \tag{1.5.19}$$

where $X$ is a Lie–Bäcklund operator (1.5.1) and $\exp(aX)$ is given by (1.5.18).

If we consider canonical operators (1.5.7) then (1.5.12) reduce to the finite system of equations (1.5.10) supplemented by the initial conditions, i.e. by the system

$$\frac{d}{da} \bar{u}^\alpha = \eta^\alpha([\bar{z}]), \quad \bar{u}^\alpha \big|_{a=0} = u^\alpha. \tag{1.5.20}$$

Consequently, Lie–Bäcklund transformation groups can be constructed by virtue of the following theorem.

**Theorem 1.5.6** *Given a canonical Lie–Bäcklund operator,*

$$X = \eta^\alpha \frac{\partial}{\partial u^\alpha} + \cdots,$$

*the corresponding formal one-parameter group is represented by the series*

$$\bar{u}^\alpha = u^\alpha + a\eta^\alpha + \frac{a^2}{2!} X(\eta^\alpha) + \cdots + \frac{a^n}{n!} X^{n-1}(\eta^\alpha) + \cdots \tag{1.5.21}$$

*together with its differential consequences:*

$$\bar{u}_i^\alpha = u_i^\alpha + aD_i(\eta^\alpha) + \frac{a^2}{2!} X(D_i(\eta^\alpha)) + \cdots + \frac{a^n}{n!} X^{n-1}(D_i(\eta^\alpha)) + \cdots,$$

$$\bar{u}_{i_1\cdots i_s}^\alpha = u_{i_1\cdots i_s}^\alpha + aD_{i_1}\cdots D_{i_s}(\eta^\alpha) + \cdots + \frac{a^n}{n!} X^{n-1}(D_{i_1}\cdots D_{i_s}(\eta^\alpha)) + \cdots.$$

**Example 1.5.4** Let

$$X = u_1 \frac{\partial}{\partial u} + u_2 \frac{\partial}{\partial u_1} + \cdots.$$

Here $\eta = u_1$ and therefore

$$X(\eta) = u_2, \ X^2(\eta) = u_3, \ \ldots, \ X^{n-1}(\eta) = u_n.$$

Hence, the transformation (1.5.21) has the form

$$\bar{u} = u + \sum_{n=1}^{\infty} \frac{a^n}{n!} u_n.$$

**Example 1.5.5** Let

$$X = u_2 \frac{\partial}{\partial u} + u_3 \frac{\partial}{\partial u_1} + u_4 \frac{\partial}{\partial u_2} + \cdots.$$

Here, $\eta = u_2$ and

$$X(\eta) = u_4, \quad X^2(\eta) = u_6, \quad \dots, \quad X^{n-1}(\eta) = u_{2n}.$$

Hence, the transformation (1.5.21) is given by the power series

$$\bar{u} = u + \sum_{n=1}^{\infty} \frac{a^n}{n!} u_{2n}.$$

### 1.5.3 Lie–Bäcklund Symmetries

Lie–Bäcklund symmetries of differential equations are given by Definition 1.4.7 from Sect. 1.4.5. Thus, we use the following definition.

**Definition 1.5.5** Let $G$ be a Lie–Bäcklund transformation group generated by a Lie–Bäcklund operator (1.5.1),

$$X = \xi^i \frac{\partial}{\partial x^i} + \eta^\alpha \frac{\partial}{\partial u^\alpha} + \zeta_i^\alpha \frac{\partial}{\partial u_i^\alpha} + \zeta_{i_1 i_2}^\alpha \frac{\partial}{\partial u_{i_1 i_2}^\alpha} + \cdots. \qquad (1.5.1)$$

The group $G$ is called a group of *Lie–Bäcklund symmetries* of a system of differential equations

$$F_k(x, u, u_{(1)}, \dots, u_{(p)}) = 0, \quad k = 1, \dots, s, \qquad (1.5.22)$$

if the extended frame of (1.5.22) defined by (see Definition 1.4.4)

$$[F]: \qquad F_k = 0, \quad D_i F_k = 0, \quad D_i D_j F_k = 0, \dots \qquad (1.5.23)$$

is invariant under $G$. The operator $X$ (1.5.1) is called an *infinitesimal Lie–Bäcklund symmetry* for (1.5.22).

The infinitesimal invariance criteria proved in [4] is formulated in the following statements.

**Theorem 1.5.7** *The operator* (1.5.1) *is an infinitesimal Lie–Bäcklund symmetry for* (1.5.22) *if and only if*

$$X F_k \big|_{[F]} = 0, \quad X D_i (F_k) \big|_{[F]} = 0, \quad X D_i D_j (F_k) \big|_{[F]} = 0, \dots \quad (k = 1, \dots, s).$$

Theorem 1.5.7 contains an infinite number of equations. However, it can be simplified and reduced to a finite number of equations by means of the following result.

**Lemma 1.5.1** *The equations*

$$XF_k\big|_{[F]} = 0$$

*yield the infinite series of equations*

$$XD_i(F_k)\big|_{[F]} = 0, \quad XD_iD_j(F_k)\big|_{[F]} = 0, \ldots.$$

Thus, one arrives at the following finite test for calculating Lie–Bäcklund symmetries of differential equations.

**Theorem 1.5.8** *The operator* (1.5.1) *is an infinitesimal Lie–Bäcklund symmetry for* (1.5.22) *if and only if the following equations hold*:

$$XF_k\big|_{[F]} = 0, \quad k = 1, \ldots, s. \tag{1.5.24}$$

*Equations* (1.5.24) *are the determining equations for Lie–Bäcklund symmetries.*

**Remark 1.5.1** Every operator of the form (1.5.5), i.e. $X_* = \xi_*^i D_i \in L_{\mathscr{B}}$ is an infinitesimal Lie–Bäcklund symmetry for any system of differential equations. Furthermore all operators (1.5.1) satisfying the conditions

$$\xi^i\big|_{[F]} = 0, \quad \eta^\alpha\big|_{[F]} = 0 \tag{1.5.25}$$

solve the determining equations (1.5.24). All operators $X_* \in L_{\mathscr{B}}$ and the operators obeying the conditions (1.5.25) are termed *trivial Lie–Bäcklund symmetries* ([7], Sect. 1.3.2).

**Example 1.5.6** The equations of motion of a planet (Kepler's problem):

$$m\frac{d^2x^k}{dt^2} = \mu\frac{x^k}{r^3}, \quad k = 1, 2, 3,$$

have the following three nontrivial infinitesimal Lie–Bäcklund symmetries different from Lie point and contact symmetries (see [4]):

$$X_i = \left(2x^i v^k - x^k v^i - (x \cdot v)\delta_i^k\right)\frac{\partial}{\partial x^k}, \quad i = 1, 2, 3.$$

Here the independent variable is time $t$, the dependent variables are the coordinates of the position vector $x = (x^1, x^2, x^3)$ of the planet. The vector $v = (v^1, v^2, v^3)$ is the velocity of the planet, i.e. $v = dx/dt$.

# 1.6 Approximate Transformation Groups

A detailed discussion of the material presented here as well as of the theory of multi-parameter approximate groups can be found in [2].

## 1.6.1 Approximate Transformations and Generators

In what follows, functions $f(x, \varepsilon)$ of $n$ variables $x = (x^1, \ldots, x^n)$ and a parameter $\varepsilon$ are considered locally in a neighborhood of $\varepsilon = 0$. These functions are continuous in the $x$'s and $\varepsilon$, as are also their derivatives to as high an order as enters in the subsequent discussion.

If a function $f(x, \varepsilon)$ satisfies the condition

$$\lim_{\varepsilon \to 0} \frac{f(x, \varepsilon)}{\varepsilon^p} = 0,$$

it is written $f(x, \varepsilon) = o(\varepsilon^p)$ and $f$ is said to be *of order less than* $\varepsilon^p$. If

$$f(x, \varepsilon) - g(x, \varepsilon) = o(\varepsilon^p),$$

the functions $f$ and $g$ are said to be *approximately equal* (with an error $o(\varepsilon^p)$) and written

$$f(x, \varepsilon) = g(x, \varepsilon) + o(\varepsilon^p),$$

or, briefly $f \approx g$ when there is no ambiguity.

The approximate equality defines an equivalence relation, and we join functions into equivalence classes by letting $f(x, \varepsilon)$ and $g(x, \varepsilon)$ to be members of the same class if and only if $f \approx g$.

Given a function $f(x, \varepsilon)$, let

$$f_0(x) + \varepsilon f_1(x) + \cdots + \varepsilon^p f_p(x)$$

be the approximating polynomial of degree $p$ in $\varepsilon$ obtained via the Taylor series expansion of $f(x, \varepsilon)$ in powers of $\varepsilon$ about $\varepsilon = 0$. Then any function $g \approx f$ (in particular, the function $f$ itself) has the form

$$g(x, \varepsilon) \approx f_0(x) + \varepsilon f_1(x) + \cdots + \varepsilon^p f_p(x) + o(\varepsilon^p).$$

Consequently the function

$$f_0(x) + \varepsilon f_1(x) + \cdots + \varepsilon^p f_p(x)$$

is called a *canonical representative* of the equivalence class of functions containing $f$.

Thus, the equivalence class of functions $g(x, \varepsilon) \approx f(x, \varepsilon)$ is determined by the ordered set of $p + 1$ functions

$$f_0(x), \ f_1(x), \ \ldots, \ f_p(x).$$

In the theory of approximate transformation groups, one considers ordered sets of smooth vector-functions depending on $x$'s and a group parameter $a$:

$$f_0(x, a), \ f_1(x, a), \ \ldots, \ f_p(x, a)$$

with coordinates

$$f_0^i(x, a), \ f_1^i(x, a), \ \ldots, \ f_p^i(x, a), \quad i = 1, \ldots, n.$$

Let us define the one-parameter family $G$ of *approximate transformations*

$$\bar{x}^i \approx f_0^i(x,a) + \varepsilon f_1^i(x,a) + \cdots + \varepsilon^p f_p^i(x,a), \quad i = 1, \ldots, n, \quad (1.6.1)$$

of points $x = (x^1, \ldots, x^n) \in R^n$ into points $\bar{x} = (\bar{x}^1, \ldots, \bar{x}^n) \in R^n$ as the class of invertible transformations

$$\bar{x} = f(x,a,\varepsilon) \quad (1.6.2)$$

with vector-functions $f = (f^1, \ldots, f^n)$ such that

$$f^i(x,a,\varepsilon) \approx f_0^i(x,a) + \varepsilon f_1^i(x,a) + \cdots + \varepsilon^p f_p^i(x,a).$$

Here $a$ is a real parameter, and the following condition is imposed:

$$f(x,0,\varepsilon) \approx x.$$

Furthermore, it is assumed that the transformation (1.3.2) is defined for any value of $a$ from a small neighborhood of $a = 0$, and that, in this neighborhood, the equation $f(x,a,\varepsilon) \approx x$ yields $a = 0$.

**Definition 1.6.1** The set of transformations (1.6.1) is called a one-parameter approximate transformation group if

$$f(f(x,a,\varepsilon),b,\varepsilon) \approx f(x,a+b,\varepsilon)$$

for all transformations (1.6.2).

**Remark 1.6.1** Here, unlike the classical Lie group theory, $f$ does not necessarily denote the same function at each occurrence. It can be replaced by any function $g \approx f$ (see the next example).

**Example 1.6.1** Let us take $n = 1$ and consider the functions

$$f(x,a,\varepsilon) = x + a\left(1 + \varepsilon x + \frac{1}{2}\varepsilon a\right)$$

and

$$g(x,a,\varepsilon) = x + a(1 + \varepsilon x)\left(1 + \frac{1}{2}\varepsilon a\right).$$

They are equal in the first order of precision, namely:

$$g(x,a,\varepsilon) = f(x,a,\varepsilon) + \varepsilon^2 \varphi(x,a), \quad \varphi(x,a) = \frac{1}{2}a^2 x,$$

and satisfy the approximate group property. Indeed,

$$f(g(x,a,\varepsilon),b,\varepsilon) = f(x,a+b,\varepsilon) + \varepsilon^2 \phi(x,a,b,\varepsilon),$$

where

$$\phi(x,a,b,\varepsilon) = \frac{1}{2}a(ax + ab + 2bx + \varepsilon abx).$$

The *generator of an approximate transformation group* $G$ given by (1.6.2) is the class of first-order linear differential operators

$$X = \xi^i(x, \varepsilon) \frac{\partial}{\partial x^i} \tag{1.6.3}$$

such that

$$\xi^i(x, \varepsilon) \approx \xi_0^i(x) + \varepsilon \xi_1^i(x) + \cdots + \varepsilon^p \xi_p^i(x),$$

where the vector fields $\xi_0, \xi_1, \ldots, \xi_p$ are given by

$$\xi_\nu^i(x) = \left. \frac{\partial f_\nu^i(x, a)}{\partial a} \right|_{a=0}, \quad \nu = 0, \ldots, p; \ i = 1, \ldots, n.$$

In what follows, an approximate group generator

$$X \approx \left( \xi_0^i(x) + \varepsilon \xi_1^i(x) + \cdots + \varepsilon^p \xi_p^i(x) \right) \frac{\partial}{\partial x^i}$$

is written simply

$$X = \left( \xi_0^i(x) + \varepsilon \xi_1^i(x) + \cdots + \varepsilon^p \xi_p^i(x) \right) \frac{\partial}{\partial x^i}. \tag{1.6.4}$$

In theoretical discussions, approximate equalities are considered with an error $o(\varepsilon^p)$ of an arbitrary order $p \geq 1$. However, in the most of applications the theory is simplified by letting $p = 1$.

### 1.6.2  Approximate Lie Equations

Consider one-parameter approximate transformation groups in the first order of precision. Let

$$X = X_0 + \varepsilon X_1 \tag{1.6.5}$$

be a given approximate operator, where

$$X_0 = \xi_0^i(x) \frac{\partial}{\partial x^i}, \quad X_1 = \xi_1^i(x) \frac{\partial}{\partial x^i}.$$

The corresponding approximate transformation group of points $x$ into points $\bar{x} = \bar{x}_0 + \varepsilon \bar{x}_1$ with the coordinates

$$\bar{x}^i = \bar{x}_0^i + \varepsilon \bar{x}_1^i \tag{1.6.6}$$

is determined by the following equations:

$$\frac{d\bar{x}_0^i}{da} = \xi_0^i(\bar{x}_0), \quad \bar{x}_0^i \big|_{a=0} = x^i, \quad i = 1, \ldots, n, \tag{1.6.7}$$

$$\frac{d\bar{x}_1^i}{da} = \sum_{k=1}^n \left. \frac{\partial \xi_0^i(x)}{\partial x^k} \right|_{x=\bar{x}_0} \bar{x}_1^k + \xi_1^i(\bar{x}_0), \quad \bar{x}_1^i \big|_{a=0} = 0. \tag{1.6.8}$$

The equations (1.6.7)–(1.6.8) are called the *approximate Lie equations*.

**Example 1.6.2** Let $n = 1$ and let

$$X = (1 + \varepsilon x)\frac{\partial}{\partial x}.$$

Here $\xi_0(x) = 1$, $\xi_1(x) = x$, and equations (1.6.7)–(1.6.8) are written:

$$\frac{d\bar{x}_0}{da} = 1, \quad \bar{x}_0\big|_{a=0} = x,$$
$$\frac{d\bar{x}_1}{da} = \bar{x}_0, \quad \bar{x}_1\big|_{a=0} = 0.$$

Its solution has the form

$$\bar{x}_0 = x + a, \quad \bar{x}_1 = ax + \frac{a^2}{2}.$$

Hence, the approximate transformation group is given by

$$\bar{x} \approx x + a + \varepsilon\left(ax + \frac{a^2}{2}\right).$$

**Example 1.6.3** Let $n = 2$ and let

$$X = (1 + \varepsilon x^2)\frac{\partial}{\partial x} + \varepsilon xy\frac{\partial}{\partial y}.$$

Here $\xi_0(x, y) = (1, 0)$, $\xi_1(x, y) = (x^2, xy)$, and (1.6.7)–(1.6.8) are written:

$$\frac{d\bar{x}_0}{da} = 1, \quad \frac{d\bar{y}_0}{da} = 0, \quad \bar{x}_0\big|_{a=0} = x, \quad \bar{y}_0\big|_{a=0} = y,$$
$$\frac{d\bar{x}_1}{da} = (\bar{x}_0)^2, \quad \frac{d\bar{y}_1}{da} = \bar{x}_0\bar{y}_0, \quad \bar{x}_1\big|_{a=0} = 0, \quad \bar{y}_1\big|_{a=0} = 0.$$

The integration gives the following approximate transformation group:

$$\bar{x} \approx x + a + \varepsilon\left(ax^2 + a^2x + \frac{a^3}{3}\right), \quad \bar{y} \approx y + \varepsilon\left(axy + \frac{a^2}{2}y\right).$$

### 1.6.3 Approximate Symmetries

Let $G$ be a one-parameter approximate transformation group given by

$$\bar{z}^i \approx f(z, a, \varepsilon) \equiv f_0^i(z, a) + \varepsilon f_1^i(z, a), \quad i = 1, \ldots, N. \qquad (1.6.9)$$

An approximate equation

$$F(z, \varepsilon) \equiv F_0(z) + \varepsilon F_1(z) \approx 0 \qquad (1.6.10)$$

is said to be *approximately invariant* with respect to $G$ if

$$F(\bar{z}, \varepsilon) \approx (F(f(z, a, \varepsilon), \varepsilon) = o(\varepsilon)$$

whenever $z = (z^1, \ldots, z^N)$ satisfies (1.6.10).

If $z = (x, u, u_{(1)}, \ldots, u_{(k)})$, then (1.6.10) becomes an approximate differential equation of order $k$, and $G$ is an *approximate symmetry group* of this differential equation.

For example, the second-order equation

$$y'' - x - \varepsilon y^2 = 0 \tag{1.6.11}$$

has no exact point symmetries if $\varepsilon \neq 0$ is regarded as a constant coefficient, and hence cannot be integrated by the Lie method. Moreover, this equation cannot be integrated by quadrature. However, it possesses approximate symmetries if $\varepsilon$ is treated as a small parameter, e.g.

$$\begin{aligned}
X_1 &= \frac{\partial}{\partial y} + \frac{\varepsilon}{3}\left[2x^3\frac{\partial}{\partial x} + \left(3yx^2 + \frac{11}{20}x^5\right)\frac{\partial}{\partial y}\right], \\
X_2 &= x\frac{\partial}{\partial y} + \frac{\varepsilon}{6}\left[x^4\frac{\partial}{\partial x} + \left(2yx^3 + \frac{7}{30}x^6\right)\frac{\partial}{\partial y}\right].
\end{aligned} \tag{1.6.12}$$

The operators (1.6.12) span a two-dimensional approximate Lie algebra and can be used for consecutive integration of (1.6.11) (see [8], Sect. 12.4).

For a detailed discussion of *approximate symmetries* of differential equations with a small parameter as well as numerous examples we refer the reader to [7], Chaps. 2 and 9, and to the references therein.

# References

1. Bäcklund, A.V.: Ueber Flächentransformationen. Math. Ann. **IX**, 297–320 (1876)
2. Baikov, V.A., Gazizov, R.K., Ibragimov, N.H.: Approximate symmetries. Math. Sb. **136(178)**(3), 435–450 (1988). English transl.: Math. USSR Sb. **64**(2), (1989)
3. Ibragimov, N.H.: Sur l'équivalence des équations d'évolution, qui admettent une algébre de Lie–Bäcklund infinie. C. R. Acad. Sci. Paris Sér. I **293**, 657–660 (1981)
4. Ibragimov, N.H.: Transformation Groups in Mathematical Physics. Nauka, Moscow (1983). English transl.: Transformation Groups Applied to Mathematical Physics. Riedel, Dordrecht (1985)
5. Ibragimov, N.H.: Group analysis of ordinary differential equations and the invariance principle in mathematical physics (for the 150th anniversary of Sophus Lie). Usp. Mat. Nauk **47**(4), 83–144 (1992) English transl.: Russ. Math. Surv. **47**(2), 89–156 (1992)
6. Ibragimov, N.H. (ed.): CRC Handbook of Lie Group Analysis of Differential Equations, vol. 2: Applications in Engineering and Physical Sciences. CRC Press, Boca Raton (1995)
7. Ibragimov, N.H. (ed.): CRC Handbook of Lie Group Analysis of Differential Equations, vol. 3: New Trends in Theoretical Developments and Computational Methods. CRC Press, Boca Raton (1996)
8. Ibragimov, N.H.: Elementary Lie Group Analysis and Ordinary Differential Equations. Wiley, Chichester (1999)
9. Ibragimov, N.H. (ed.): Lie Group Analysis: Classical Heritage. ALGA Publications, Karlskrona (2004)
10. Ibragimov, N.H.: A Practical Course in Differential Equations and Mathematical Modelling, 3rd edn. ALGA Publications, Karlskrona (2006)
11. Ibragimov, N.H.: Optimal system of invariant solutions for the Burgers equation. In: Lecture at the MOGRAN-12 Conference, Porto, July 28–31, 2008

12. Lie, S.: Über die Integration durch bestimmte Integrale von einer Klasse linearer partieller Differentialgleichungen. Arch. Math. **6**(3), 328–368 (1881). English transl. in: Ibragimov, N.H. (ed.) CRC Handbook of Lie Group Analysis of Differential Equations, vol. 2: Applications in Engineering and Physical Sciences. CRC Press, Boca Raton (1995). Reprinted also in: Ibragimov, N.H. (ed.) Lie Group Analysis: Classical Heritage. ALGA Publications, Karlskrona (2004)
13. Lie, S.: Vorlesungen über Differentialgleichungen mit bekannten infinitesimalen Transformationen. Bearbeited und herausgegeben von Dr. G. Scheffers. Teubner, Leipzig (1891)
14. Olver, F.W.J.: Asymptotics and Special Functions. Academic Press, San Diego (1974)
15. Olver, P.J.: Applications of Lie Groups to Differential Equations, Springer, New York (1986). 2nd edn. (1993)
16. Ovsyannikov, L.V.: Group Properties of Differential Equations. Siberian Branch, USSR Academy of Sciences, Novosibirsk (1962) (in Russian)
17. Polyanin, A.D., Zaitsev, V.F.: Handbook of Exact Solutions for Ordinary Differential Equations. CRC Press, Boca Raton (1995)

# Chapter 2
# Introduction to Group Analysis and Invariant Solutions of Integro-Differential Equations

> *The method is a technique which I have applied twice.*
> *Maxim of a traditional professor in mathematics.*
> G. Polya

In this chapter we give an introduction into applications of group analysis to equations with nonlocal operators, in particular, to integro-differential equations. The first section of this chapter contains a retrospective survey of different methods for constructing symmetries and finding invariant solutions of such equations. The presentation of the methods is carried out using simple model equations. In the next section, the classical scheme of the construction of determining equations of an admitted Lie group is generalized for equations with nonlocal operators. In the concluding sections of this chapter, the developed regular method of obtaining admitted Lie groups is illustrated by applications to some known integro-differential equations.

## 2.1 Integro-Differential Equations in Mathematics and in Applications

Equations with nonlocal operators include integro-differential equations (IDE), delay differential equations, stochastic differential equations and some other types of less-known equations. They have been intensively studied for a long time already, in mathematics and in numerous scientific and engineering applications.

The most known integro-differential equations are kinetic equations (KE) which form the basis in the kinetic theories of rarefied gases, plasma, radiation transfer, coagulation. The Boltzmann kinetic equation [10] in rarefied gas dynamics, the Vlasov and Landau equations in plasma physics [2], and the Smolukhovsky equation in coagulation theory [71] are widely used and have become classical. Numerous generalizations of these equations are also used in other applications. Brief outlines of delay and stochastic differential equations are presented in Chaps. 5 and 6.

Y.N. Grigoriev et al., *Symmetries of Integro-Differential Equations,*
Lecture Notes in Physics 806,
DOI 10.1007/978-90-481-3797-8_2, © Springer Science+Business Media B.V. 2010

The kinetic equations describe the time evolution of a distribution function (DF) of some interacting particles such as gas molecules, ions, electrons, aerosols, etc. DF has the meaning of a nonnormalized probability density function defined on the space of dynamical variables of particles. A large number of independent variables and the presence of complicated integral operators are typical features of KEs. KEs for dynamical systems with strong pair particle interaction include special operators which are called collision integrals. In general, they are integral operators with quadratic nonlinearity and multiple kernels as in the Boltzmann and Smolukhovsky equations. For systems where collective (averaged) particle interactions are of principal importance, the nonlocal operators have the form of functionals of DF, as for example, in the Vlasov equation for collisionless plasma or in the Bhatnagar–Gross–Krook equation in rarefied gas dynamics [12]. These peculiarities create large difficulties for investigation of integro-differential equations by both analytical and numerical methods. Starting with the classical paper [48], partial simplification of these difficulties was done by reducing the integro-differential equations to infinite systems of first order differential equations for power moments of DF. Such systems are derived by integration of the original integro-differential equation with power weights with respect to some dynamical variables. Using certain asymptotical procedures [25] one can transform infinite systems for moments into hydrodynamic type finite partial differential equation systems such as the Navier–Stokes system for the Boltzmann equation or the system of ideal magnetic hydrodynamics for the Vlasov–Maxwell system. The mathematical theory of these systems has been independently developed from the studies of the corresponding integro-differential equations.

## 2.2 Survey of Various Approaches or Finding Invariant Solutions

In pure mathematical theories and especially in applied disciplines a special attention is given to the study of invariant solutions of integro-differential equations which are directly associated with fundamental symmetry properties of these equations. In Chap. 1 an application of the classical Lie group theory for finding invariant solutions of differential equations was presented. Group analysis in this case is an universal tool for calculating complete sets of searched symmetries. However a direct transference of the known scheme of the group analysis method on integro-differential equations is impossible. As shown since the first work in this way [28] (see also [29]) the main obstacle consists in a presence of nonlocal integral operators. Several approaches to this problem were worked out during a long history of studying invariant (self-similar) solutions of IDEs. The main of these approaches can be classified as follows:

(1) Use of a presentation of a solution or an admitted Lie group of transformations on the basis of a priori simplified assumptions;
(2) Investigation of infinite systems of differential equations for power moments;

(3) Transformation of an original integro-differential equation into a differential equation;
(4) Direct derivation of a Lie group of transformations through corresponding determining equations and construction of a representation of invariant solutions of IDE.

Methods of the first and fourth groups one can characterize as direct methods because they deal directly with an original IDE. At the same time the methods of the second and third groups are indirect. They are based on the replacement of a considered integro-differential equation by an infinite system of differential equations or by a single differential equation. This allows one to analyze derived equations using the standard methods of the classical Lie group theory outlined in Chap. 1.

In the present section a brief survey of all these approaches is given. Each method is illustrated with a simple (model) integro-differential equation with minimal number of variables. It allows us to explain an essence of the method without too cumbersome calculations. The most noticeable results obtained in corresponding frameworks are annotated with references.

## 2.2.1 Methods Using a Presentation of a Solution or an Admitted Lie Group

Methods of this type have an heuristic character. Possibilities of their universalization are restricted. Just to them one can relate epigraph of the chapter. They have no direct relations with group theoretical analysis. However, these methods intuitively use some symmetry properties of equations. This allows one to choose a form of a solution or an admitted transformation. It is worth to note that most known invariant solutions of IDEs for today were obtained applying these methods.

**Local-Equilibrium or Stationary Solutions** Historically the first approach of finding invariant solutions of integro-differential (kinetic) equations was based on splitting original equation in two simpler equations [10, 48]. One of these equations allows one to define a structure of a seeking solution. Consistence with another equation provides an explicit form of the solution. Using this method (local) equilibrium and stationary solutions of some kinetic equations were obtained. Here an application of this approach to basic types of integro-differential kinetic equations is considered.

The Kac equation [38] is the simplest model of the full Boltzmann kinetic equation. This equation is

$$\frac{\partial f}{\partial t} + v\frac{\partial f}{\partial x} + F\frac{\partial f}{\partial v} = J(f, f), \tag{2.2.1}$$

where

$$J(f, f) = \int\limits_{-\infty}^{\infty} dw \int\limits_{-\pi}^{\pi} d\theta g(\theta)[f(v')f(w') - f(v)f(w)]. \tag{2.2.2}$$

Here $f(t, v, x)$ is the distribution function (DF), $t \in \mathbb{R}_+^1$, $v, x \in \mathbb{R}^1$, $J(f, f)$ is the collision operator (integral), $F$ is an external force, $g(\theta) = g(-\theta)$ is a kernel associated with details of particle interaction subject to the normalization condition

$$\int\limits_{-\pi}^{\pi} g(\theta) \, d\theta = 1.$$

For the sake of brevity only the velocity arguments of DF are saved in the integrand of (2.2.2). In this case the function $g(\theta)$ corresponds to the Maxwell molecular model [25]. The collision transformation $(v, w) \to (v', w')$ is given by the group of rotations in $R^2 = R^1 \times R^1$ (see (1.1.2)) with the matrix representation $A$

$$(v', w') = (v, w)A, \quad A = \begin{pmatrix} \cos \theta & -\sin \theta \\ \sin \theta & \cos \theta \end{pmatrix}.$$

Separating (2.2.1) in two parts, the form of local equilibrium solutions (so-called Maxwellians) is obtained from the equation $J(f, f) = 0$. This equation is satisfied for any function $g(\theta)$ if and only if

$$f(v')f(w') - f(v)f(w) = 0, \tag{2.2.3}$$

or, that is the same,

$$\ln f(v') + \ln f(w') = \ln f(v) + \ln f(w).$$

This means that $\ln f(v)$ is a summation invariant of the group of rotations in $R^2$. Using the infinitesimal generator (1.1.7) $X = w \partial_v - v \partial_w$ of the group, one obtains from $XI = 0$ that in this case the unique summation invariant is $v^2 + w^2 = v'^2 + w'^2$. This gives us that the local Maxwellian solutions of (2.2.1) have the form

$$f_M(t, v, x) = a(t, x) \exp[-b(t, x)v^2]. \tag{2.2.4}$$

It is worth to emphasize a crucial step which consists here in solving functional equation (2.2.3). In turn, the solution is defined by summation invariants of the group of transformations corresponding to a collision interaction. For example, in the case of monatomic gas we deal with the group of rotations in $R^6 = R^3 \times R^3$ which has four such invariants [25].

The function (2.2.4) has also to satisfy the equation

$$\frac{\partial f_M}{\partial t} + v \frac{\partial f_M}{\partial x} + F \frac{\partial f_M}{\partial v} = 0.$$

For example, if the force $F = -\varphi'$ is conservative with the potential $\varphi(x)$, then $b = \mathrm{const}$, $a = C \exp(-2b\varphi)$ and the well-known Maxwell–Boltzmann distribution $f_M(v, x) = C \exp[-b(v^2 + 2\varphi)]$ in potential field is obtained.[1]

The local Maxwellian solutions of the full Boltzmann equation were completely studied using the outlined method by outstanding scientists: J.C. Maxwell [48],

---

[1] The complete study of local Maxwellian solutions of (2.2.1) done in [21].

L. Boltzmann [10], T. Carleman [14], H. Grad [27]. The local-equilibrium solutions for kinetic equations with similar collision integrals such as the linear Boltzmann equation in the neutron transfer theory [25], the Landau kinetic equation in the plasma physics [2], the Wang Chang–Uhlenbeck equation in the kinetic theory of polyatomic gases [25] and others were constructed using similar approach.

There exists a wide class of integro-differential equations which include integral operators in the form of functionals depending on their solutions. In particular, kinetic equations with a self-consistent field (so-called Vlasov-type equations) belong to this class. These equations are used in plasma physics, gravitational astrophysics, theory of nonlinear waves and others. In this case such equations have the form of a first order partial differential equation with associative equations for functionals. According to the theory of differential equations their general solutions are arbitrary differentiable functions of first integrals. This property allows one to find invariant solutions of some simple problems.

To illustrate this approach let us consider the one-dimensional problem of equilibrium of a plane gravitating homogeneous layer [59]. The problem is described by the Vlasov–Poisson system:

$$v\frac{\partial f}{\partial x} + F\frac{\partial f}{\partial v} = 0, \tag{2.2.5}$$

$$\frac{d^2\varphi}{dx^2} = C. \tag{2.2.6}$$

Here $f(v, x)$ is the distribution function of gravitating particles, $v \in R^1$ is the particle velocity, $x \in [-1, 1]$ is the space coordinate, $F = -\varphi'$ is the gravity force, $\varphi(x)$ is the gravitational potential. The density of particles $\rho(x)$ is the zeroth-order moment of the DF:

$$\rho(x) = \int f(v, x)\,dv. \tag{2.2.7}$$

Since the density is constant along a layer, it can be written as $\rho(x) = \rho_0 H(1 - x^2)$, where $H$ is the unit Heaviside step-function. The right hand side of (2.2.6) is constant $C = \rho_0$. Then $F(x) = -x$ and the general solution of (2.2.5) is $f = f_0(E)$, where the first integral $E = v^2/2 + x^2/2$ is the energy invariant of the particle motion. It is also necessary to satisfy the self-consistency condition (2.2.7). In fact, one has to solve the integral equation of the first kind

$$\int f_0(E)\,dv = \rho_0 H(1 - x^2).$$

The last equation can be transformed into the Abel equation by the substitution $y = 1 - 2E$:

$$\int_0^z \frac{f_0(y)dy}{\sqrt{z - y}} = \rho_0 H(z), \qquad z = 1 - x^2.$$

The Abel equation is invertible for an arbitrary right hand side [72]:

$$f_0(y) = \frac{1}{\pi} \frac{d}{dy} \int_0^y \frac{\rho_0 H(z) dz}{\sqrt{y-z}}.$$

Finally, one obtains

$$f_0(E) = \frac{\rho_0 H(1-2E)}{\pi \sqrt{1-2E}}.$$

Invariant solutions were similarly obtained for gravitating problems with cylindrical and spherical symmetries (see references in [59]). It is obvious that this method can also be used in other applications of the Vlasov-type equations with two independent variables. In particular, the one-dimensional dynamics of collisionless plasma with a neutralizing background and a potential field is described by the following system

$$\frac{\partial f}{\partial t} + v \frac{\partial f}{\partial x} + F \frac{\partial f}{\partial v} = 0, \tag{2.2.8}$$

$$\frac{\partial F}{\partial x} = 1 - \int_{-\infty}^{\infty} dv \, f, \qquad \frac{\partial F}{\partial t} = \int_{-\infty}^{\infty} dv \, fv. \tag{2.2.9}$$

From [1] it follows that there exists some transformation, which maps (2.2.8), (2.2.9) to the above stationary Vlasov–Poisson system. Then, one can derive nonstationary solutions of the Vlasov–Maxwell system (2.2.8), (2.2.9) starting from the stationary solutions.

## A Priori Choice of Invariant Transformations

1. Nikolskii's transformations.

First time this approach was systematically applied to the Boltzmann integro-differential equation by A.A. Nikolskii in the series of papers [51–53]. Transformations obtained by this approach provide nonstationary space-dependent solutions from space-homogeneous.

Let us illustrate the Nikolskii approach using the Kac equation (2.2.1). In the space-homogeneous case and in absence of the external force $F$ it becomes

$$\frac{\partial f(t, v)}{\partial t} = J(f, f). \tag{2.2.10}$$

Assume that $f_h(t, v)$ is a solution of (2.2.10). The Nikolskii transformation is

$$f_s(t, x, v) = f_h(\bar{t}, \bar{v}), \tag{2.2.11}$$

where

$$\bar{t} = \tau(t), \qquad \bar{v} = (1 + t/t_0)\left(v - \frac{x}{t + t_0}\right). \tag{2.2.12}$$

Here $\tau(t)$ is a temporarily unknown function. One can consider the quantity $c = v - \frac{x}{t+t_0}$ as the heat (eigen) microscopic velocity of a particle, and the quantity

$$U = \frac{x}{t + t_0}$$

as the macroscopic velocity of a continuum (model gas) in the space position $x$. Flows with this velocity distribution in the framework of the one-dimensional ideal gas dynamics were studied by L.I. Sedov [61]. For $t, t_0 > 0$ it is an expansion flow of a gas; if $t_0 < 0$ it is a compression flow. Therefore the solution (2.2.11), (2.2.12) is called "expansion–compression" motions of a model gas. This means that the distribution function $f_s$ of eigen velocities is the same at each space point at any given instant.

Substitution of (2.2.11) into the left hand side of (2.2.1) with $F(x) = 0$ gives

$$\frac{\partial f_s}{\partial t}(t, x, v) + v \frac{\partial f_s}{\partial x}(t, x, v) = \frac{d\tau}{dt}(t) \frac{\partial f_h}{\partial t}(\bar{t}, \bar{v}), \qquad (2.2.13)$$

where $(\bar{t}, \bar{v})$ are defined by (2.2.12). Taking into account that $f_h(t, v)$ is a solution of (2.2.10), one can write

$$\frac{d\tau}{dt}(t) \frac{\partial f_h}{\partial t}(\bar{t}, \bar{v})$$

$$= \frac{d\tau}{dt}(t) \int_{-\infty}^{\infty} d\bar{w} \int_{-\pi}^{\pi} d\theta g(\theta)[f_h(\bar{v}')f_h(\bar{w}') - f_h(\bar{v})f_h(\bar{w})], \qquad (2.2.14)$$

where $\bar{v}' = \bar{v}\cos\theta + \bar{w}\sin\theta$, $\bar{w}' = -\bar{v}\sin\theta + \bar{w}\cos\theta$.

By virtue of linearity of the collision transformation for dilations of the velocity space we have

$$(\lambda v', \lambda w') = (\lambda v, \lambda w)A.$$

Hence, the collision integral under such dilations is transformed as follows

$$J(f, f)(\lambda v) = \lambda J(f, f)(v). \qquad (2.2.15)$$

Let us additionally assume that the studied class of distribution functions $f_h(t, v)$ leaves the collision integral invariant with respect to the translations of the velocity space

$$\bar{f}_h(t, v) = f_h(t, v - a).$$

This property corresponds [16] to the physical meaning of the distribution function as the particle number density in the velocity space. In this functional class the collision integral $J(f, f)$ has the property

$$J(\bar{f}_h, \bar{f}_h)(v) = J(f_h, f_h)(v - a). \qquad (2.2.16)$$

Sequentially exploiting the properties of the collision integral (2.2.15) and then (2.2.16), the equation (2.2.13) becomes

$$\frac{\partial f_s}{\partial t} + v \frac{\partial f_s}{\partial x} = (1 + t/t_0)\frac{d\tau}{dt} J(f_s, f_s). \qquad (2.2.17)$$

Hence, the function $f_s(t, x, v)$ determined by (2.2.11), (2.2.12) is a solution of the equation

$$\frac{\partial f}{\partial t} + v\frac{\partial f}{\partial x} = J(f, f)$$

if and only if the unknown function $\tau(t)$ satisfies differential equation

$$\frac{d\tau}{dt}(1 + t/t_0) = 1.$$

Choosing $\tau(0) = 0$, one obtains that $\tau(t) = t_0 \ln(1 + t/t_0)$ for any positive $t$. If the factor in front of $J(f_s, f_s)$ in (2.2.17) is chosen as an arbitrary constant, then (2.2.12) is an equivalence transformation [56].

It is known [38] that for $t \to \infty$ a solution of the space homogeneous equation (2.2.10) with arbitrary initial data converges to the absolute Maxwellian distribution $f_M$.

One can note that in an expansion flow for $t, t_0 > 0$ the equilibrium distribution is reached

$$\lim_{t \to \infty} f_s(t, x, v) = f_M(v).$$

Whereas in an compression flow (where $t_0 < 0$) one has for $t \to -0$ that

$$f_s(0, x, v) = f_h\left(\tau(0), v - \frac{x}{t_0}\right) \neq f_M(v),$$

and the equilibrium distribution is not achieved (see [52]).

In many IDEs the differential operator has a similar form. If the collision integral possesses similar invariant properties, then Nikolskii's transformation can also be applied. Here it can also be mentioned the linear Boltzmann equation [25], the Landau equation [2] and some others. Unfortunately, as a rule, solutions of space homogeneous equations excepting stationary equilibrium solutions are unknown.

## 2. The Bobylev approach.

All methods for constructing invariant solutions of IDEs presented in this subsection have ad-hoc character. This means that they are not universal and, hence, have a confined field of applications. As a rule, such methods are based on intuitive windfalls rather than on systematic approach. The most outstanding results in the frameworks of this direction were derived by Bobylev [5–7][2] for the Boltzmann kinetic equation for Maxwell molecules.

Here the windfall was the Fourier transform of the Boltzmann equation (BE) with respect to the velocity variables. The transformation drastically simplified an investigation of mathematical properties of BE. This has allowed one not only to obtain a new nontrivial symmetry of BE but also to complete a relaxation theory of a Maxwellian gas.

---

[2]Some generalizations of the Bobylev approach were also done in [24].

Let us demonstrate the Bobylev approach on the space homogeneous Kac model as was done in [32]. The Cauchy problem for the distribution function $f(t, v)$ has the form

$$\frac{\partial f}{\partial t} = \int\limits_{-\infty}^{\infty} dw \int\limits_{-\pi}^{\pi} d\theta g(\theta)[f(v')f(w') - f(v)f(w)], \qquad (2.2.18)$$

$$f(0, v) = f_0(v). \qquad (2.2.19)$$

The equilibrium solution of (2.2.18) when $t \to \infty$ is the absolute Maxwellian distribution

$$f_M(v) = \frac{1}{\sqrt{2\pi}} \exp(-v^2/2). \qquad (2.2.20)$$

The problem (2.2.18), (2.2.19) possesses the mass and energy conservation laws of the forms

$$\int\limits_{-\infty}^{\infty} f(t, v) \, dv = \int\limits_{-\infty}^{\infty} f_0(v) \, dv = 1,$$

$$\int\limits_{-\infty}^{\infty} v^2 f(t, v) \, dv = \int\limits_{-\infty}^{\infty} v^2 f_0(v) \, dv = 1. \qquad (2.2.21)$$

For an arbitrary integrable function $\psi(v)$ and the collision integral (2.2.2) the integral identity takes place

$$I(\psi) = \int\limits_{-\infty}^{\infty} dv \psi(v) J(f, f)$$

$$= \int\limits_{-\infty}^{\infty} \int\limits_{-\infty}^{\infty} \int\limits_{-\pi}^{\pi} g(\theta)(\psi(v') - \psi(v)) f(v) f(w) \, dv dw d\theta. \qquad (2.2.22)$$

The direct and inverse Fourier transforms are defined as follows

$$\varphi(k) = \int\limits_{-\infty}^{\infty} f(v)e^{-ikv} \, dv. \qquad (2.2.23)$$

$$f(v) = (2\pi)^{-1} \int\limits_{-\infty}^{\infty} \varphi(k)e^{ikv} \, dk. \qquad (2.2.24)$$

Applying the direct transform (2.2.23) to (2.2.18) and taking into account identity (2.2.22), one can derive the Fourier representation of the Cauchy problem (2.2.18), (2.2.19):

$$\frac{\partial \varphi(t, k)}{\partial t} = \hat{J}(\varphi, \varphi), \quad \varphi(0, k) = \Phi(k), \qquad (2.2.25)$$

where

$$\hat{J}(\varphi, \varphi) = \int_{-\pi}^{\pi} d\theta g(\theta)[\varphi(k \cos\theta)\varphi(k \sin\theta) - \varphi(k)\varphi(0)],$$

and

$$\Phi(k) = \int_{-\infty}^{\infty} f_0(v)e^{-ikv}\, dv. \tag{2.2.26}$$

Note that an essential simplification of the collision term occurred:[3] the collision term contains a single integral over the collision parameter $\theta$.

The Fourier transform of the equilibrium solution (2.2.20) is

$$\varphi_M(k) = \exp\left(-\frac{k^2}{2}\right). \tag{2.2.27}$$

The conservation laws (2.2.21) in terms of Fourier transforms become

$$\varphi(t, 0) = \Phi(0) = 1, \quad \left.\frac{\partial^2 \varphi(t, k)}{\partial k^2}\right|_{k=0} = \left.\frac{\partial^2 \Phi(k)}{\partial k^2}\right|_{k=0} = -1. \tag{2.2.28}$$

One can easily verify that (2.2.25) admits some simple groups of transformations. In fact, there is a group of translations of the time $\bar{t} = t + a$ . The corresponding infinitesimal generator of this group is $X_1 = \partial_t$.

It is necessary to point out that each transformation in the $k$-space has a corresponding representation in the original $v$-space. In such a way there is a dilation group in the $k$-space

$$\bar{k} = e^a k, \qquad X_2 = k\partial_k. \tag{2.2.29}$$

This transformation leads to the change of variables in the $v$-space:

$$\bar{f}(t, v) = e^{-a} f(t, e^{-a}v).$$

This property corresponds to the transformation defined by the infinitesimal generator:

$$Y_2 = v\partial_v + f\partial_f.$$

The Bobylev symmetry of (2.2.25) is defined by the formula

$$\bar{\varphi}(t, k) = \exp\left(-\frac{ak^2}{2}\right)\varphi(t, k). \tag{2.2.30}$$

This symmetry corresponds to the infinitesimal generator $X_3 = -\frac{k^2}{2}\varphi\partial_\varphi$.

---

[3]More impressive simplification the Fourier transform gives for the full Boltzmann equation with Maxwell molecules: the five-fold collision integral is reduced to a two-fold integral [7]. Unfortunately for other power-like molecular potentials Fourier transform does not give simplifications [33].

The invariance of (2.2.25) with respect to the change (2.2.30) is easily ascertained. Because the existence of an inverse Fourier transform requires that $a \geq 0$, the transformation (2.2.30) determines a semigroup. Using (2.2.24) and the convolution theorem, one can obtain the corresponding semigroup in the $v$-space:[4]

$$\bar{f}(t, v) = \frac{1}{\sqrt{2\pi a}} \int_{-\infty}^{\infty} f(t, w) dw \exp\left[-\frac{(v-w)^2}{2a}\right]. \qquad (2.2.31)$$

Here corresponding an infinitesimal generator is the one-dimensional Laplace operator

$$Y_5 = \frac{1}{2}\partial_{vv}.$$

The invariant solution of the problem (2.2.25) which is consistent from the physical point of view has to satisfy the initial conditions (2.2.26), the conservation laws (2.2.28) and has to converge to $\varphi_M(k)$ (2.2.20) for $t \to \infty$. Taking into account these demands, the invariant solution similar to the well-known BKW-mode [5] is constructed in the following way.[5]

To reduce the number of independent variables and to use simultaneously the new symmetry (2.2.30) one can seek for a solution in the form

$$\varphi(k, t) = \exp\left(-\frac{ak^2}{2}\right)\Psi(x), \qquad x = \tau(t)k, \qquad (2.2.32)$$

where $\tau(t)$ is determined later. Substituting the presentation (2.2.32) into (2.2.25) and taking into account its invariance under the transformation (2.2.30), one obtains

$$\frac{d\tau}{dt}\frac{1}{\tau}x\frac{d\Psi}{dx} = \hat{J}(\Psi, \Psi).$$

To separate variables here it is necessary to set

$$\frac{d\tau}{dt}\frac{1}{\tau} = c.$$

The last equation determines the function $\tau(t) = \theta_0 \exp(ct)$, where $c$ and $\theta_0$ are arbitrary constants. To satisfy the initial conditions one has to require

$$\varphi(k, 0) = \exp\left(-\frac{ak^2}{2}\right)\Psi(\theta_0 k) = \Phi(k).$$

Hence, the representation of the invariant solution (2.2.32) becomes

$$\varphi(k, t) = \exp\left[\frac{1}{2}a(x^2 - k^2)\right]\Phi(x). \qquad (2.2.33)$$

---

[4]The invariance of the Boltzmann equation with isotropic Maxwell molecular model with respect to semigroup (2.2.31) was discovered in [50] but it was not used by the author for constructing invariant solutions and for a long time this result was lost.

[5]The authors of BKW-mode [42] used a much more long and intricate approach.

Since $\Phi(0) = 1$, for asymptotic convergence of (2.2.33) to the equilibrium solution (2.2.27), it is sufficient to accept that $a = 1$ and $c < 0$. Simultaneously this solution satisfies the mass conservation law. The energy conservation law will be automatically satisfied after constructing the solution in the explicit form.

One can check that the invariant solution (2.2.33) is determined by the infinitesimal generator $X = -c^{-1}X_1 + X_2 - X_5$. In fact, solving the first-order partial differential equation

$$X(I) \equiv -c^{-1}\frac{\partial I}{\partial t} + k\frac{\partial I}{\partial k} - k^2\varphi\frac{\partial I}{\partial \varphi} = 0,$$

one derives two independent integrals $I_1 = k\theta_0 \exp(ct) = k\tau(t)$ and $I_2 = \varphi \times \exp(k^2/2)$ which are *independent invariants* (see Chap. 1). Since for constructing the invariant solution one requires that

$$I_2 = h(I_1),$$

one has the representation of the invariant solution $\varphi = \exp(-k^2/2)h(x)$. Finally to satisfy the imposed demands it is sufficient to set $h(x) = \exp(x^2/2)\Phi(x)$.

Substitution of the presentation (2.2.33) into (2.2.25) gives the factor-equation

$$cx\left(\frac{d\Phi}{dx} + x\Phi\right) = \hat{J}(\Phi, \Phi). \qquad (2.2.34)$$

To find the BKW-mode one uses the Taylor expansion

$$\Phi(x) = 1 + \sum_{n=1}^{\infty}\frac{c_n}{n!}x^n, \qquad (2.2.35)$$

where the choice $c_0 = 1$ explicitly accomplishes the mass conservation law.

After substitution (2.2.35) into (2.2.34) one obtains a specific nonlinear spectral problem for the coefficients $c_n$. Even coefficients $c_{2k}$ ($n = 2k$) are separately determined from closed subsystem. In particular, $c_2 = -1$ and the energy conservation law is satisfied. Some resonance property of even eigen values allows to cut the series (2.2.35) and find a solution in the form

$$\Phi(x) = 1 - x^2, \quad x = k\tau(t) \equiv k\theta_0 \exp(ct), \quad c = -\frac{1}{8}\int_{-\pi}^{\pi} d\theta g(\theta)\sin^2 2\theta.$$

Applying the inverse Fourier transform to (2.2.33), one can derive the explicit expression of the BKW-mode[6] in the $v$-space:

$$f(t, v) = \frac{1}{\sqrt{2\pi(1 - \lambda(t))}}\left[1 + \frac{\lambda(t)}{2(1 - \lambda(t))}\left(\frac{v^2}{1 - \lambda(t)} - 1\right)\right]\exp\left[-\frac{v^2}{2(1 - \lambda(t))}\right],$$

where $\lambda(t) = \tau^2(t)$ and $0 < \theta_0^2 < 2/3$.

---

[6]It is worth to note that the Fourier transform of the Boltzmann equation and the explicit solution rediscovered in [6, 7, 42] were first derived in unknown MS thesis of R. Krupp (see Ref. [15]).

## 3. Scaling conjecture.

In the work of the authors [28] some generalization of known symmetry properties of the Boltzmann equation and its models was proposed. In application to the Kac model in absence of an external force $F$

$$\frac{\partial f}{\partial t} + v \frac{\partial f}{\partial x} = J(f, f), \tag{2.2.36}$$

the admitted Lie group $G$ of transformations $T_a$ was sought in the form

$$\bar{f} = \psi(\bar{t}, \bar{x}, a) f, \quad t = q(\bar{t}, \bar{x}, a), \quad x = h(\bar{t}, \bar{x}, a),$$
$$v = r(\bar{t}, \bar{x}, a) \bar{v}. \tag{2.2.37}$$

Here $\{f, t, x, v\}$ and $\{\bar{f}, \bar{t}, \bar{x}, \bar{v}\}$ are original and transformed variables, respectively, $\psi, h, \theta, r, p$ are unknown functions which define the sought group $G$ with the group parameter $a$. These functions have necessarily to satisfy the main group superposition property in the form

$$T_b T_a = T_{a+b}, \tag{2.2.38}$$

and the identity property for the group parameter $a = 0$:

$$\psi(\bar{t}, \bar{x}, 0) = 1, \quad q(\bar{t}, \bar{x}, 0) = \bar{t}, \quad h(\bar{t}, \bar{x}, 0) = \bar{x},$$
$$r(\bar{t}, \bar{x}, 0) \bar{v} = \bar{v}. \tag{2.2.39}$$

The Lie group of transformations $G$ is said to be admitted by (2.2.36) or (2.2.36) admits the group $G$ if transformations (2.2.37) convert every solution of (2.2.36) into a solution of the same equation. This means that if a function $f(t, x, v)$ is a solution of (2.2.36), then the function

$$\bar{f}(\bar{t}, \bar{x}, \bar{v}, a) = \psi(\bar{x}, \bar{t}, a) f(q(\bar{x}, \bar{t}, a), h(\bar{x}, \bar{t}, a), r(\bar{t}, \bar{x}, a) \bar{v}) \tag{2.2.40}$$

satisfies the equation

$$\frac{\partial \bar{f}}{\partial \bar{t}} + \bar{v} \frac{\partial \bar{f}}{\partial \bar{x}} = J(\bar{f}, \bar{f}). \tag{2.2.41}$$

By virtue of the properties of the collision integral (2.2.16) and (2.2.37), one can show that

$$J(\bar{f}, \bar{f}) = g(\bar{t}, \bar{x}, a) J(f, f) \tag{2.2.42}$$

with some function $g(\bar{t}, \bar{x}, a)$.

Calculating the derivatives of the function $\bar{f}(\bar{t}, \bar{x}, \bar{v}, a)$ (2.2.40) and the collision integral $J(\bar{f}, \bar{f})$, one gets

$$\frac{\partial \bar{f}}{\partial \bar{t}} = \frac{\partial \psi}{\partial \bar{t}} f + \psi \left( \frac{\partial f}{\partial t} \frac{\partial q}{\partial \bar{t}} + \frac{\partial f}{\partial x} \frac{\partial h}{\partial \bar{t}} + \frac{\partial f}{\partial v} \frac{\partial r}{\partial \bar{t}} v \right),$$

$$\frac{\partial \bar{f}}{\partial \bar{x}} = \frac{\partial \psi}{\partial \bar{x}} f + \psi \left( \frac{\partial f}{\partial t} \frac{\partial q}{\partial \bar{x}} + \frac{\partial f}{\partial x} \frac{\partial h}{\partial \bar{x}} + \frac{\partial f}{\partial v} \frac{\partial r}{\partial \bar{x}} v \right),$$

$$J(\bar{f}, \bar{f})(\bar{t}, \bar{x}, \bar{v}, a) = \frac{\psi^2(\bar{t}, \bar{x}, \bar{v}, a)}{r(\bar{t}, \bar{x}, \bar{v}, a)} J(f, f)(t, x, v),$$

where $(t, x, v)$ are defined by (2.2.37). Since the function $f(t, x, v)$ is a solution of the Kac equation (2.2.36), the collision integral $J(f, f)$ can be exchanged with the left hand side of this equation. This gives that

$$J(\bar{f}, \bar{f}) = \frac{\psi^2}{r}\left(\frac{\partial f}{\partial t} + r\bar{v}\frac{\partial f}{\partial x}\right).$$

Taking into account that $f(t, x, v)$ is an arbitrary solution of (2.2.36) one can split the derived equation with respect to $f$ and its derivatives:

$$f : \frac{\partial \psi}{\partial \bar{t}} + \bar{v}\frac{\partial \psi}{\partial \bar{x}} = 0,$$

$$\frac{\partial f}{\partial t} : \frac{\partial q}{\partial \bar{t}} + \bar{v}\frac{\partial q}{\partial \bar{x}} - \frac{\psi}{r} = 0,$$

$$\frac{\partial f}{\partial x} : \frac{\partial h}{\partial \bar{t}} + \bar{v}\frac{\partial h}{\partial \bar{x}} - r\bar{v}\frac{\psi}{r} = 0,$$

$$\frac{\partial f}{\partial v} : \frac{\partial r}{\partial \bar{t}}\bar{v} + \frac{\partial p}{\partial \bar{t}} + \bar{v}\frac{\partial r}{\partial \bar{x}}\bar{v} = 0.$$

Additional splitting of these equations with respect to the variable $\bar{v}$ gives the equations

$$\frac{\partial \psi}{\partial \bar{t}} = 0, \quad \frac{\partial \psi}{\partial \bar{x}} = 0, \tag{2.2.43}$$

$$\frac{\partial q}{\partial \bar{t}} - \frac{\psi}{r} = 0, \quad \frac{\partial q}{\partial \bar{x}} = 0, \tag{2.2.44}$$

$$\frac{\partial h}{\partial \bar{t}} = 0, \quad \frac{\partial h}{\partial \bar{x}} - \psi = 0, \tag{2.2.45}$$

$$\frac{\partial r}{\partial \bar{t}} = 0, \quad \frac{\partial r}{\partial \bar{x}} = 0. \tag{2.2.46}$$

From (2.2.43) one has that $\psi = \psi(a)$. The general solution of (2.2.45) is

$$h(\bar{t}, \bar{x}, a) = \bar{x}\psi(a) + c_1(a)$$

with an arbitrary function $c_1(a)$. Equations (2.2.46) define that

$$r = r(a).$$

The general solution of (2.2.44) is

$$q(\bar{t}, \bar{x}, a) = \bar{t}\frac{\psi(a)}{r(a)} + c_2(a),$$

where $c_2(a)$ is an arbitrary function.

Thus, using the properties of the collision integral (2.2.15), one derives that the form of admitted transformations (2.2.37) is

$$\bar{f} = \psi(a)f, \quad \bar{t} = \bar{t}\frac{\psi(a)}{r(a)} + c_2(a),$$
$$\bar{x} = \bar{x}\psi(a) + c_1(a), \quad \bar{v} = r(a)\bar{v}. \tag{2.2.47}$$

The identity conditions (2.2.39) of transformations (2.2.47) at $a = 0$ impose the additional relations

$$\psi(0) = 1, \quad c_1(0) = 0, \quad r(0) = 1, \quad c_2(0) = 0. \tag{2.2.48}$$

The requirement to satisfy the main Lie group property (2.2.38) for the variable $f$ and $v$ leads to the conditions

$$\psi(a)\psi(b) = \psi(a+b), \quad r(a)r(b) = r(a+b). \tag{2.2.49}$$

Using (2.2.48), the general solutions of these equations are

$$\psi(a) = \exp(\hat{c}_1 a), \quad r(a) = \exp(\hat{c}_2 a),$$

where $\hat{c}_1$ and $\hat{c}_2$ are arbitrary constants. Hence, transformations (2.2.47) become

$$\bar{f} = \exp(\hat{c}_1 a) f(x, v, t), \quad \bar{x} = (x - c_1(a)) \exp(-\hat{c}_1 a),$$
$$\bar{v} = v \exp(-\hat{c}_2 a), \quad \bar{t} = (t - c_2(a)) \exp[-(\hat{c}_1 - \hat{c}_2)a)]. \tag{2.2.50}$$

Since there is one-to-one correspondence between an infinitesimal generator and a Lie group, the undefined functions $c_1(a)$ and $c_2(a)$ in (2.2.50) can be found from the system of Lie equations.

Recall that the coefficients of the admitted generator of the Lie group $G$

$$X = \xi^t \frac{\partial}{\partial t} + \xi^x \frac{\partial}{\partial x} + \xi^v \frac{\partial}{\partial v} + \zeta^f \frac{\partial}{\partial f}$$

are defined by the formulae

$$\xi^t = \frac{d\bar{t}}{da}\bigg|_{a=0} = -c_2'(0) - (t - c_2(0))(\hat{c}_1 - \hat{c}_2),$$

$$\xi^x = \frac{d\bar{x}}{da}\bigg|_{a=0} = -c_1'(0) - \hat{c}_1(x - c_1(0)),$$

$$\xi^v = \frac{d\bar{v}}{da}\bigg|_{a=0} = -\hat{c}_2 v,$$

$$\zeta^f = \frac{d\bar{f}}{da}\bigg|_{a=0} = \hat{c}_1 f.$$

By virtue of (2.2.48), one obtains that

$$\xi^t = -c_2'(0) - t(\hat{c}_1 - \hat{c}_2),$$
$$\xi^x = -c_1'(0) - \hat{c}_1 x,$$
$$\xi^v = -v\hat{c}_2,$$
$$\zeta^f = \hat{c}_1 f.$$

Thus, one has the basis of admitted generators

$$\begin{aligned}
\hat{c}_1 : \quad & X_4 = f\partial_f - t\partial_t - x\partial_x, \\
\hat{c}_2 : \quad & X_3 = v\partial_v - t\partial_t, \\
c_1'(0) : \quad & X_1 = \partial_x, \\
c_2'(0) : \quad & X_2 = \partial_t.
\end{aligned} \tag{2.2.51}$$

Now after finding the invariants of the group $X_i J = 0$ $(i = 1, \ldots, 5)$ by the usual way, one can obtain representations of invariant solutions.

It is seen that the integral transformation (2.2.31) is absent in transformations (2.2.50). However, as will be shown in Chap. 3 such simple *scaling conjecture* allows us [28] to define 11-parameter Lie algebra admitted by the full Boltzmann equation and all known extensions for some special cases of molecular potentials (see also [31]).

4. Teshukov's wave-type solutions.

It is worth to mention here one more approach which was developed by V.M. Teshukov. In [67] an extension of the theory of characteristics for systems of integro-differential equations was proposed. Using the generalized characteristics and Riemann invariants, simple waves of a system of integro-differential equations were determined.

The system of integro-differential equations describing evolution of rotational free-boundary flows of an ideal incompressible fluid in a shallow-water approximation is the following

$$hu_t + uu_x + vu_y + gh = 0, \quad u_x + v_y = 0,$$

$$h_t + \left( \int_0^h u\, dy \right)_x = 0. \tag{2.2.52}$$

Here $(u, v)$ is the fluid-velocity vector, $h$ is the layer depth, $g$ is the gravitational acceleration, $x$ and $y$ are the Cartesian plane coordinates, and $t$ is time. The impenetration condition $v(x, 0, t) = 0$ is satisfied at the layer bottom. Equations (2.2.52) are considered in the Eulerian–Lagrangian coordinates $x', \lambda, t'$, where

$$x = x', \quad t = t', \quad y = \Phi(x', \lambda, t'),$$

and $\Phi = \Phi(x', \lambda, t')$ is the solution of the Cauchy problem

$$\Phi_t + u(x, \Phi, t)\Phi_x = v(x, \Phi, t), \quad \Phi(x, \lambda, 0) = \Phi_0(x, \lambda).$$

In the new coordinates (2.2.52) become

$$u_t(x, \lambda, t) + u(x, \lambda, t)u_x(x, \lambda, t) + g\int_0^1 H_x(x, v, t)\, dv = 0,$$

$$H_t(x, \lambda, t) + (u(x, \lambda, t)H(x, \lambda, t))_x = 0,$$

where the prime is omitted and $H(x, \lambda, t) = \Phi_\lambda(x, \lambda, t) > 0$.

Solutions of the simple wave type are sought in the form

$$u = U(\alpha(x, t), \lambda), \quad H = P(\alpha(x, t), \lambda),$$

where $\alpha(x, t)$ is a function of two variables. The functions $U(\alpha, \lambda)$, $P(\alpha, \lambda)$ have to satisfy the equations

$$(u(\alpha, \lambda) - k)u_\alpha(\alpha, \lambda) + g \int\limits_0^1 P_\alpha(\alpha, \mu)\, d\mu = 0,$$

$$(u(\alpha, \lambda) - k)P_\alpha(\alpha, \lambda) + P(\alpha, \lambda)u_\alpha(\alpha, \lambda) = 0,$$

where $k = -\alpha_t/\alpha_x$. The existence of simple waves, their properties and extensions for other systems of integro-differential equations were studied in [17, 68–70].

## 2.2.2 Methods of Moments

The method of moments for finding symmetries of integro-differential equations is based on the idea to use an infinite system of partial differential equations which is equivalent to the original integro-differential system of equations. The general idea of consideration such a system goes back to the pioneering paper [48] where the Boltzmann equation was studied by using the power moments defined on a solution of the Boltzmann equation.

The moment method for obtaining symmetries consists of the following steps. A finite subsystem of $N$ moment equations is chosen. Applying the classical group analysis method developed for partial differential equations to the chosen subsystem, one finds the admitted Lie group (algebra) of this subsystem. Expanding the subsystem and letting $N \to \infty$, the intersection of all calculated Lie groups is carried out. The final step consists of returning the obtained symmetries for the moment representation to the symmetries of the original integro-differential equations.

The first application of this method was done in [64] for the system of the Vlasov–Maxwell collisionless plasma equations.

It is worth to notice that among the indirect methods of studying symmetries of IDEs, the method of moments is the most universal ones, despite of the substantial restrictions of its applications.

Let us demonstrate this approach by the simple model Kac equation (2.2.1). The power moments for this model are defined as:

$$M_n = \int v^n f\, dv, \quad v \in R^1 \qquad (n = 0, 1, \ldots).$$

Multiplying (2.2.1) with $v^n$ and integrating it with respect to $v$, one obtains on the left hand side the expression

$$\frac{\partial M_n}{\partial t} + \frac{\partial M_{n+1}}{\partial x}.$$

This expression represents two terms which are typical for the moment system of a kinetic equation. For integration of the right hand side one can use the following integral identity for the collision integral (2.2.2):

$$I(v^n) = \int_{-\infty}^{\infty} dv v^n J(f, f)$$

$$= \frac{1}{2} \int_{-\infty}^{\infty} \int_{-\infty}^{\infty} \int_{-\pi}^{\pi} g(\theta)[v'^n + w'^n - v^n - w^n] f(v) f(w) \, dv \, dw \, d\theta,$$

$$(2.2.53)$$

where $v' = v\cos\theta + w\sin\theta$ and $w' = w\cos\theta - v\sin\theta$. Integrating the moment system for the Kac equation (2.2.1) is obtained

$$\frac{\partial M_n}{\partial t} + \frac{\partial M_{n+1}}{\partial x} - \Lambda_n M_0 M_n = \sum_{m=1}^{n-1} H_{m,n-m} M_m M_{n-m} \quad (n = 0, 1, \ldots),$$

$$(2.2.54)$$

where

$$\Lambda_{2k} = \int_{-\pi}^{\pi} g(\theta)[\cos^{2k}\theta + \sin^{2k}\theta - 1 - \delta_{k0}] \, d\theta,$$

$$\Lambda_{2k+1} = \int_{-\pi}^{\pi} g(\theta)[\cos^{2k+1}\theta - 1] \, d\theta \qquad (k = 0, 1, \ldots),$$

$$H_{m,n-m} = \frac{1}{2} C_n^k \int_{-\pi}^{\pi} g(\theta)[\cos^m\theta \sin^{n-m}\theta + (-1)^m \sin^m\theta \cos^{n-m}\theta] \, d\theta.$$

It is seen that for any $N$ the last equation of the $N$-order system contains the moment $M_{N+1}$. Hence each truncated subsystem is unclosed. However this does not impede one to find a symmetry.

Applying the classical group analysis method to this system, and solving the determining equations, one obtains that the admitted generator is

$$X^{(3)} = k_1 X_1 + k_2 X_2 + k_3 Y_3^{(3)} + k_4 Y_4^{(3)} + p_1(t) \partial_{M_2} + (q_1(t, x) - x p_1'(t)) \partial_{M_3},$$

where

$$X_1 = \partial_t, \quad X_2 = \partial_x,$$

$$Y_3^{(3)} = x \partial_x + M_1 \partial_{M_1} + 2 M_2 \partial_{M_2} + 3 M_3 \partial_{M_3}, \qquad (2.2.55)$$

$$Y_4^{(3)} = t \partial_t - M_0 \partial_{M_0} - 2 M_1 \partial_{M_1} - 3 M_2 \partial_{M_2} - 4 M_3 \partial_{M_3}.$$

The part of system (2.2.54) including the fourth moment $M_4$ consists of the equations

$$\frac{\partial M_0}{\partial t} + \frac{\partial M_1}{\partial x} = 0,$$

$$\frac{\partial M_1}{\partial t} + \frac{\partial M_2}{\partial x} - \Lambda_1 M_0 M_1 = 0,$$

$$\frac{\partial M_2}{\partial t} + \frac{\partial M_3}{\partial x} = 0,$$

$$\frac{\partial M_3}{\partial t} + \frac{\partial M_4}{\partial x} - \Lambda_3 M_0 M_3 = (H_{1,2} + H_{2,1}) M_1 M_2.$$

(2.2.56)

Notice that: (a) system (2.2.56) contains (2.2.55) as a subsystem; (b) the set of derivatives for splitting the determining equations of system (2.2.56) contains the set of derivatives for splitting the determining equations of system (2.2.55). Because of these two properties, the generator admitted by system (2.2.56) can be obtained by expanding the operator $X_3^{(3)}$ on the space of the variables $t, x, M_0, M_1, M_2, M_3$ and $M_4$:

$$X^{(4)} = k_1 X_1 + k_2 X_2 + k_3 Y_3^{(3)} + k_4 Y_4^{(3)} + p_2(t, M_4) \partial_{M_2}$$
$$+ (q_2 - x p_{2t} \partial_t) \partial_{M_3} + \zeta \partial_{M_4},$$

where $p_2 = p_2(t, M_4)$, $q_2 = q_2(t, x, M_4)$ and $\zeta = \zeta(t, x, M_1, M_2, M_3, M_4)$. Applying this operator to system (2.2.56) one obtains that

$$p_2 = 0, \quad q_2 = 0$$

and

$$\zeta = (4k_3 - 5k_4) M_4 + q_3(t).$$

This means that the admitted generator of system (2.2.56) is

$$X^{(4)} = k_1 X_1 + k_2 X_2 + k_3 Y_3^{(4)} + k_4 Y_4^{(4)} + q_3(t) \partial_{M_4},$$

where

$$Y_3^{(4)} = Y_3^{(3)} + 4M_4 \partial_{M_4}$$
$$= x \partial_x + M_1 \partial_{M_1} + 2M_2 \partial_{M_2} + 3M_3 \partial_{M_3} + 4M_4 \partial_{M_4},$$
$$Y_4^{(4)} = Y_4^{(3)} - 5M_4 \partial_{M_4}$$
$$= t \partial_t - M_0 \partial_{M_0} - 2M_1 \partial_{M_1} - 3M_2 \partial_{M_2} - 4M_3 \partial_{M_3} - 5M_4 \partial_{M_4}.$$

During calculations the following condition was used

$$H_{1,2} + H_{2,1} \neq 0.$$

One can check that if $H_{1,2} + H_{2,1} = 0$, then the operator $X^{(4)}$ is also admitted by system (2.2.56).

Proceeding by this way, one obtains that the only generator which is admitted by all finite subsystems of (2.2.54) is

$$X = k_1 X_1 + k_2 X_2 + k_3 Y_3 + k_4 Y_4,$$

where

$$Y_3 = x\partial_x + \sum_{k=1}^{\infty} k M_k \partial_{M_k}, \quad Y_4 = t\partial_t - \sum_{k=0}^{\infty} (k+1) M_k \partial_{M_k}.$$

The operator $X$ is more convenient to rewrite in the form

$$X = k_1 X_1 + k_2 X_2 + k_3 X_3 + k_4 X_4,$$

where

$$X_3 = Y_4, \quad X_4 = Y_3 + Y_4 = x\partial_x + t\partial_t - \sum_{k=0}^{\infty} M_k \partial_{M_k}.$$

Let us define corresponding generators in the space of the original variables $(t, x, v, f)$.

Consider the generator

$$X_3 = t\partial_t - \sum_{k=0}^{\infty} (k+1) M_k \partial_{M_k}.$$

It is necessary to obtain the corresponding group of transformations in an explicit form. Solving the Lie equations one has

$$\bar{t} = te^a, \quad \bar{x} = x, \quad \bar{M}_k = M_k e^{-(k+1)a} \quad (k = 0, 1, 2, \ldots). \qquad (2.2.57)$$

It is logical to assume that the variables $v$ and $f$ are also scaled in the space of the variables $t, x, v, f$:

$$\bar{v} = ve^{\alpha a}, \quad \bar{f} = fe^{\beta a}.$$

Using this change, the transformed function and the transformed moments are determined by the formulae

$$\bar{f}(\bar{t}, \bar{x}, \bar{v}) = f(\bar{t}e^{-a}, \bar{x}, \bar{v}e^{-\alpha a})e^{\beta a},$$

$$\bar{M}_k = \int_{-\infty}^{\infty} \bar{v}^k \bar{f}(\bar{v}) \, d\bar{v} = e^{\beta a} \int_{-\infty}^{\infty} \bar{v}^k f(\bar{v}e^{-\alpha a}) \, d\bar{v} \qquad (2.2.58)$$

$$= e^{(\beta + (k+1)\alpha)a} \int_{-\infty}^{\infty} v^k f(v) \, dv = M_k e^{(\beta + (k+1)\alpha)a}.$$

Thus, comparing with (2.2.57), one gets

$$\beta + \alpha = -1, \quad \alpha = -1.$$

This gives the generator

$$X_3 = t\partial_t - v\partial_v.$$

Similar to the previous generator one obtains for the generator

$$X_4 = x\partial_x + t\partial_t - \sum_{k=0}^{\infty} M_k \partial_{M_k}$$

that

$$\bar{t} = te^a, \quad \bar{x} = xe^a, \quad \bar{M}_k = M_k e^{-a} \quad (k = 0, 1, 2, \ldots). \qquad (2.2.59)$$

Comparing this with (2.2.58), one finds

$$\beta + \alpha = -1, \quad \alpha = 0.$$

This gives the generator

$$X_4 = x\partial_x + t\partial_t - f\partial_f.$$

Therefore, the Kac equation (2.2.1) admits the Lie group with the generators:[7]

$$X_1 = \partial_t, \quad X_2 = \partial_x, \quad X_3 = t\partial_t - v\partial_v, \quad X_4 = x\partial_x + t\partial_t - f\partial_f.$$

Starting with [64] (see also [65]) the moment method was applied to Vlasov-type equations such as different modifications of the Benney equation [41], where a transition to a moment system is natural. In order to use the classical group analysis method it is necessary that each finite subsystem of a moment system contains a finite number of moments. Taking into account this property one can mention the papers [12, 13] where the moment method was used for the group analysis of the Bhatnagar–Gross–Krook (BGK) kinetic equation of rarefied gas dynamics. In the simplest model case this equation takes the form

$$\frac{\partial f}{\partial t} + v\frac{\partial f}{\partial x} = \nu(f_0 - f). \qquad (2.2.60)$$

Here as in (2.2.1), the distribution function is $f = f(t, v, x), t \in \mathbb{R}_+^1, v, x \in \mathbb{R}^1$. The local Maxwellian distribution

$$f_0 = n\left(\frac{1}{2\pi T}\right)^{1/2} \exp\left[-\frac{(v - V)^2}{2T}\right]$$

is defined through the moments of an unknown solution

$$n = \int dv\, f, \quad V = \frac{1}{n}\int dv\, v\, f, \quad T = \frac{1}{n}\int dv\, (v - V)^2 f.$$

Equations similar to the BGK-equation with the so-called relaxation collision integral are also considered in the kinetic theory of molecular gases (the Landau–Teller equation [45]), in the plasma physics, etc. For these equations, a finite subsystem for power moments contains a finite set of moments. However, in the general case of dissipative kinetic equations such as the Boltzmann equation, the Smolukhovsky equation and others this property is exceptional. For example, the Boltzmann equation only has this property for Maxwellian-type molecular interaction. As noted, this case of the Boltzmann equation can be modeled by the Kac equation. The application of the group analysis method to the moment system corresponding to the Kac equation has been demonstrated above. For arbitrary intermolecular potentials, each moment equation contains an infinite number of moments. For this reason, in the general case the difficulty of constructing an admitted

---

[7]This Lie group coincides with the group obtained by using the scaling conjecture (compare with (2.2.51)).

Lie group for such a system is equally difficult as the direct integration of the moment system as a whole.

Other difficulties related with finding an admitted Lie group of transformations using moment equations consist of some problems of inverse transition from a Lie group of transformations for the moment system to the corresponding Lie group of the original equation. In all known cases [12, 13, 41, 64] one deals with the Lie group of scaling transformations similar to the example for the Kac equation considered above. The scaling transformations are naturally carried out on the original variables $v$, $f$. However, for more complicated transformations such a transition may be not as easy.

It is clear that the form of a moment system and its Lie group depend on a moment representation. As an example for the Boltzmann equation with Maxwell molecules (also for the Kac model (2.2.1)) an alternative to the power moments can be presented by the Fourier coefficients of the expansion of the distribution function in Hermitian polynomials. In general there are no results on relations between these possible approaches.

Moreover, as a rule there are no rigorous proofs of equivalence between an original kinetic equation and the corresponding moment system. In some cases the Lie group obtained by the moment method coincides with the Lie group calculated by the regular method [29] applied to the original equations. For example, this happens for the 4-parameter Lie group derived for the moment system of the Vlasov equation [64] and for the Vlasov equation [30]. The 11-parameter Lie group of the Boltzmann equation with arbitrary power potential found in [12, 13] and the Lie group calculated directly from the equation [28] also coincide. At the same time as shown in [37], the finite Lie group calculated in [41] using the moment method for the Benney equation is not complete. Since the Benney equation possesses [44] an infinite set of conservation laws, one can expect that the finite dimension of the derived Lie algebra contradicts the infinite set of conservation laws. This inconsistency was considered in detail in [37] (see also Chap. 4).

These remarks show that in finding symmetries of IDEs, the relatively universal moment method cannot be a valuable alternative to the regular method which is constructed as a generalization of the classical Lie method for differential equations.

### 2.2.3 Methods Using a Transition to Equivalent Differential Equations

The idea of these approaches is quite obvious. However, its realization in each case has very individual features. Therefore the survey of these approaches is restricted here by several examples. In spite of this restriction any of the chosen examples illustrates a technique which is used at least in two papers.

**Vlasov-Type Equations as First-Order Partial Differential Equations** There exists the possibility of a direct application of the classical group analysis (see

Chap. 1) for finding invariant solutions of the Vlasov-type kinetic equations. The idea of this application is related with the following.

It is well-known [20] that the Lie group admitted by the first-order quasilinear partial differential equation

$$u_t + a_i(x, u)u_{x_i} = b(x, u) \qquad (2.2.61)$$

coincides with the Lie group admitted by the characteristic system of ordinary differential equations of the quasilinear equation (2.2.61)

$$\frac{du}{dt} = b(x, u), \qquad \frac{dx_i}{dt} = a_i(x, u) \quad (i = 1, 2, \ldots, n).$$

Here $x = (x_1, x_2, \ldots, x_n)$.

Having this in minds let us separately consider the Vlasov kinetic equation (2.2.8) which can be rewritten in the form

$$\frac{\partial f}{\partial t} + \dot{x}\frac{\partial f}{\partial x} + F\frac{\partial f}{\partial \dot{x}} = 0, \qquad (2.2.62)$$

where $f = f(t, x, \dot{x})$, $F = F(t, x)$, and $\dot{x} = v$. Here the self-consistency of the force $F$ given by the Maxwell system (2.2.9) is temporarily neglected. Following [1] one makes the transition from the characteristic system of (2.2.62)

$$\frac{dt}{1} = \frac{dx}{\dot{x}} = \frac{d\dot{x}}{F} \qquad (2.2.63)$$

to the equivalent second-order ordinary differential equation

$$\Phi \equiv \frac{d^2x}{dt^2} - F(t, x) = 0.$$

According to the remark given above, it is clear that this equation admits the same Lie group as (2.2.62) and (2.2.63). In notations of Chap. 1 the infinitesimal criterion for the generator

$$X = \xi(t, x)\partial_t + \eta(t, x)\partial_x,$$

to be admitted by the equation $\Phi = 0$ is

$$X_{(2)}\Phi|_{\Phi=0} \equiv (\xi\Phi_t + \eta\Phi_x + \zeta_1\Phi_{\dot{x}} + \zeta_2\Phi_{\ddot{x}})|_{\Phi=0} = 0. \qquad (2.2.64)$$

Here $X_{(2)}$ is the second prolongation of the infinitesimal generator $X$, and the coefficients $\zeta_1$ and $\zeta_2$ are defined by the prolongation formulae. Calculations give that the determining equation (2.2.64) becomes

$$(\eta_x - 2\xi_t)F - \xi F_t - \eta F_x + \eta_{tt} + (2\eta_{tx} - \xi_{tt} - 3\xi_x F)\dot{x}$$
$$+ (\eta_{xx} - 2\xi_{tx})\dot{x}^2 - \xi_{xx}\dot{x}^3 = 0.$$

Splitting this determining equation with respect to powers of $\dot{x}$ one finds

$$\xi_{xx} = 0, \quad \eta_{xx} - 2\xi_{tx} = 0,$$
$$2\eta_{tx} - \xi_{tt} - 3\xi_x F = 0, \quad (\eta_x - 2\xi_t)F - \xi F_t - \eta F_x + \eta_{tt} = 0.$$

The general solution of the first two equations is [1]

$$\xi = xh_1(t) + h_2(t), \quad \eta = 2x^2 h_1'(t) + x h_3(t) + h_4(t),$$

where $h_i(t)$ ($i = 1, 2, 3, 4$) are arbitrary functions. Using the standard technique of constructing invariant solutions, for particular choices of the functions $h_i(t)$ ($i = 1, 2, 3, 4$) the Vlasov equation (2.2.62) is reduced to the stationary Vlasov equation in the new variables $\bar{f}, \bar{x}, V$:

$$\varphi_1 \frac{\partial \bar{f}}{\partial \bar{x}} + \varphi_2 \frac{\partial \bar{f}}{\partial V} = 0,$$

where $\varphi_1(\bar{x}, V), \varphi_2(\bar{x}, V)$ are some known functions. The last equation can be integrated only in a few particular cases. Notice also that these obtained solutions have to be consistent with the Maxwell system (2.2.9). A brief survey of these results one can find in [1]. It is clear that the presented approach is effective just for similar one-dimensional problems in plasma physics, gravitational astrophysics, etc., where the Vlasov-type equation with three independent variables appeared.

**Use of the Laplace Transform**   Successful applications of the Laplace and other integral transforms for reducing integro-differential equations to differential ones are restricted by some degenerated cases. As a rule these equations either possess a high symmetry in the phase space or present exact solvable models [23].

As a first example let us consider the Fourier-image of the spatially homogeneous and isotropic Boltzmann equation derived in [4]

$$\varphi_t(x, t) + \varphi(x, t)\varphi(0, t) - \int_0^1 \varphi(xs, t)\varphi(x(1 - s), t)\, ds = 0. \quad (2.2.65)$$

One can notice that any solution of (2.2.65) possesses the property $\varphi(0, t) = \text{const}$. This property corresponds to the mass conservation law of the Boltzmann equation.

The change $xs = y$ reduces (2.2.65) to the equation with the convolution-type integral:

$$x\varphi_t(x, t) + x\varphi(x, t)\varphi(0, t) - \int_0^x \varphi(y, t)\varphi(x - y, t)\, dy = 0. \quad (2.2.66)$$

In analysis of (2.2.66), one can assume that

$$\varphi(0, t) = 1. \quad (2.2.67)$$

Then applying the Laplace transform

$$u(z, t) = \mathscr{L}\{\varphi(x, t)\} = \int_0^\infty e^{-zx} \varphi(x, t)\, dx, \quad (2.2.68)$$

to (2.2.66) one comes to the partial differential equation[8]

$$\frac{\partial^2 u}{\partial z \partial t} + \frac{\partial u}{\partial z} + u^2 = 0. \tag{2.2.69}$$

Since (2.2.69) is a partial differential equation, one can apply to this equation the classical group analysis method. In fact, assuming that the infinitesimal generator of the admitted Lie group is

$$X = \tau \partial_t + \xi \partial_z + \eta \partial_u,$$

the determining equation of this Lie group is

$$\left(X^{(2)}\Psi\right)\big|_{(2.2.69)} = \left(\eta^{u_{zt}} + \eta^{u_z} + 2u\eta\right)\big|_{(2.2.69)} = 0. \tag{2.2.70}$$

Here the coefficients $\eta^{u_z}$ and $\eta^{u_{zt}}$ are defined by the prolongation formulae

$$\eta^{u_z} = D_z \eta - u_t D_z \tau - u_z D_z \xi, \quad \eta^{u_{zt}} = D_t \eta^{u_z} - u_{zt} D_t \tau - u_{zz} D_t \xi.$$

The general solution of the determining equation (2.2.70) is

$$X = c_1 Y_1 + c_2 Y_2 + c_3 Y_3 + c_4 Y_4,$$

where

$$Y_1 = \partial_t, \quad Y_2 = \partial_z, \quad Y_3 = -z\partial_z + u\partial_u, \quad Y_4 = e^t(-\partial_t + u\partial_u).$$

Notice that the original equation (2.2.65) admits the Lie algebra with the basis [28][9]

$$X_1 = \partial_t, \quad X_2 = x\varphi\partial_\varphi, \quad X_3 = x\partial_x, \quad X_4 = \varphi\partial_\varphi - t\partial_t.$$

The well-known solution of (2.2.65) is the BKW-solution [4, 42]: $\varphi = 6e^y(1-y)$, where $y = xe^{-t}$. This solution is an invariant solution of (2.2.65) under the Lie group of transformation corresponding to the subalgebra[10] $\{X_1 + X_3\}$.

Let us study the symmetries of (2.2.69) which inherit the symmetries of (2.2.65) and vice versa.

It is trivial to check that the transformations related with the generator $X_1$ in the space of the variables $(x, t, \varphi)$ are inherited in the space of the variables $(z, t, u)$.

The transformations corresponding to the generator $X_2$ map functions as

$$\bar{\varphi}(\bar{x}, \bar{t}) = e^{a\bar{x}} \varphi(\bar{x}, \bar{t}).$$

Hence the Laplace transform (2.2.68) maps solutions of (2.2.65) as follows

$$\bar{u}(\bar{z}, \bar{t}) = \mathscr{L}\{\bar{\varphi}(\bar{x}, \bar{t})\} = \int_0^\infty e^{-\bar{z}\bar{x}} \bar{\varphi}(\bar{x}, \bar{t})\, d\bar{x} = \int_0^\infty e^{-(\bar{z}-a)\bar{x}} \varphi(\bar{x}, \bar{t})\, d\bar{x} = u(\bar{z} - a, \bar{t}).$$

---

[8]This equation coincides with the equation obtained in [66] for the moment generating function of power moments of the original distribution function.

[9]Complete calculations using the regular method are presented in the next section.

[10]This solution is usually considered as invariant solution with respect to transformations corresponding to the subalgebra $\{X_2 - X_3 + c^{-1}X_1\}$ which is similar to $\{X_1 + X_3\}$.

This means that the symmetry corresponding to the generator $X_2$ becomes the symmetry corresponding to the generator $Y_2$.

Implementation of a similar procedure for the generator $X_3$ gives

$$\bar{\varphi}(\bar{x}, \bar{t}) = \varphi(e^{-a}\bar{x}, \bar{t}),$$

and the Laplace transform (2.2.68) maps solutions of (2.2.65) as follows

$$\bar{u}(\bar{z}, \bar{t}) = \mathcal{L}\{\bar{\varphi}(\bar{x}, \bar{t})\} = \int_0^\infty e^{-\bar{z}\bar{x}} \bar{\varphi}(\bar{x}, \bar{t}) d\bar{x} = \int_0^\infty e^{-\bar{z}\bar{x}} \varphi(e^{-a}\bar{x}, \bar{t}) d\bar{x}$$

$$= e^a \int_0^\infty e^{-e^a \bar{z} e^{-a}\bar{x}} \varphi(e^{-a}\bar{x}, \bar{t}) de^{-a}\bar{x} = e^a u(e^a \bar{z}, \bar{t}).$$

This relates the symmetry corresponding to the generator $X_3$ and the symmetry corresponding to the generator $Y_3$.

The heritage property fails for the generator $X_4 = \varphi \partial_\varphi - t \partial_t$, where the transformations are

$$\bar{\varphi}(\bar{x}, \bar{t}) = e^a \varphi(\bar{x}, e^a \bar{t}),$$

and the Laplace transforms of the functions $\mathcal{L}(\bar{\varphi})$ and $\mathcal{L}(\varphi)$ are related by the formula

$$\bar{u}(\bar{z}, \bar{t}) = \mathcal{L}\{\bar{\varphi}(\bar{x}, \bar{t})\} = \int_0^\infty e^{-\bar{z}\bar{x}} \bar{\varphi}(\bar{x}, \bar{t}) d\bar{x}$$

$$= e^a \int_0^\infty e^{-\bar{z}\bar{x}} \varphi(\bar{x}, e^a \bar{t}) d\bar{x} = e^a u(\bar{z}, e^a \bar{t}).$$

Thus the symmetry related to the generator $X_4 = \varphi \partial_\varphi - t \partial_t$ in the space of the variables $(z, t, u)$ becomes the symmetry corresponding to the generator

$$Y_5 = -t \partial_t + u \partial_u.$$

The last generator is not admitted by (2.2.69). It is explained by the restriction pressed by the condition (2.2.67): if $\varphi(0, t) = 1$, then $\bar{\varphi}(0, \bar{t}) = e^a \varphi(0, e^a \bar{t}) = e^a \neq 1$.

Let us analyze symmetry of the generator $Y_4 = e^t(-\partial_t + u \partial_u)$ admitted by (2.2.69). The transformations corresponding to this generator are

$$\bar{t} = t - \ln(1 + ae^t), \quad \bar{u} = (1 + ae^t)u.$$

These transformations map a function $u(z, t)$ into the function

$$\bar{u}(\bar{z}, \bar{t}) = \frac{1}{1 - ae^{\bar{t}}} u(\bar{z}, \bar{t} - \ln(1 - ae^{\bar{t}})).$$

The corresponding relations of the originals are

$$\bar{\varphi}(\bar{x}, \bar{t}) = \frac{1}{1 - ae^{\bar{t}}} \varphi(\bar{x}, \bar{t} - \ln(1 - ae^{\bar{t}})). \qquad (2.2.71)$$

These transformations of the function $\varphi(x, t)$ define the generator

$$X_5 = e^t(-\partial_t + \varphi\partial_\varphi).$$

Considering (2.2.71) at $\bar{x} = 0$, one gets

$$\bar{\varphi}(0, \bar{t}) = \frac{1}{1 - ae^{\bar{t}}}\varphi\left(0, \bar{t} - \ln(1 - ae^{\bar{t}})\right).$$

Because of the mass conservation law $\varphi(0, t) = \text{const}$ the operator $X_5$ is not admitted by (2.2.65).

One notices that differences of the Lie group admitted by (2.2.65) and the Lie group admitted by (2.2.69) come from the assumption (2.2.67). In fact, the direct application of the Laplace transformation to (2.2.66) leads it to the equation

$$\frac{\partial^2 u}{\partial z \partial t} + k\frac{\partial u}{\partial z} + u^2 = 0, \tag{2.2.72}$$

where $k = \varphi(0, t)$. Recall that according to the mass conservation mass law $\varphi(0, t) = \text{const}$. Because the functions $u(z, t)$ and $\varphi(x, t)$ are related by the Laplace transform, one can conclude that $k = \mathcal{L}^{-1}\{u(z, t)\}(0, t)$. Hence (2.2.72) is also a nonlocal equation and one cannot apply the classical group analysis method to this equation. This also explains the appearance of the new transformations.

Another way of applying the Laplace transform to (2.2.65) was proposed in [8]. Using the assumption (2.2.67) and the substitution $y = e^{-\lambda t}x$, the equation (2.2.66) is reduced to the equation

$$-\lambda y^2\frac{d\varphi(y)}{dy} + y\varphi(y) - \int_0^y \varphi(w)\varphi(y - w)\,dw = 0, \tag{2.2.73}$$

where $\lambda$ is constant. The Laplace transform $u(z) = \mathcal{L}\{\varphi(y)\}$ leads (2.2.73) into the second-order ordinary differential equation

$$\lambda z u'' + (2\lambda + 1)u' + u^2 = 0. \tag{2.2.74}$$

Considering $\lambda = 1/6$, and exploiting the substitution $v(p) = p^{-2} - p^{-3}u(p^{-1})$, the equation (2.2.74) was reduced in [8] to the equation defining the Weierstrass elliptic function [72]:

$$v'' = 6v^2. \tag{2.2.75}$$

In the simplest case of choice of the invariants of the Weierstrass function $g_2 = g_3 = 0$ one has $v(p) = (p - p_0)^{-2}$, where $p_0 > 1$ is constant. Returning to the original variables, one gets the solution

$$\varphi(y) = (1 - y/p_0)e^{y/p_0}.$$

This is the Fourier image of the known BKW-solution of the Boltzmann equation [6]. However, the transition to the differential equation (2.2.74) does not allow one to describe explicitly the class of invariant BKW-solutions in whole (compare with corresponding example in the next section).

Let us proceed here with application of the classical group analysis method to (2.2.74). For arbitrary $\lambda$ this equation admits the generator

$$Z_0 = -z\partial_z + u\partial_u.$$

Additional admitted generators occur for $\lambda$ satisfying the equation

$$(6\lambda - 1)(3\lambda + 2)(2\lambda + 3)(\lambda - 6) = 0.$$

These generators are

$$\lambda = 1/6 : Z_1 = z^2\partial_z + (2 - 3uz)\partial_u,$$
$$\lambda = -2/3 : Z_2 = \sqrt{z}\partial_z,$$
$$\lambda = -3/2 : Z_3 = 3z^{2/3}\partial_z - uz^{-1/3}\partial_u,$$
$$\lambda = 6 : Z_4 = z^{-7/6}3z^2\partial_z - (1 + 2uz)\partial_u.$$

The presence of two admitted generators allows one to use Lie's integration algorithm:[11] using canonical coordinates this algorithm reduces finding solutions of a second-order ordinary differential equation to quadratures. In fact, the use of canonical variables gives the changes

$$\lambda = 1/6 : u = z^{-1} - z^{-3}v, \quad p = z^{-1},$$
$$\lambda = -2/3 : u = v, \quad p = \sqrt{z},$$
$$\lambda = -3/2 : u = z^{-1/3}v, \quad p = z^{1/3},$$
$$\lambda = 6 : u = z^{-1} - z^{-2/3}v, \quad p = z^{1/6}.$$

In all of these cases (2.2.74) is reduced to the only equation (2.2.75). Since (2.2.75) is homogeneous, one can apply the substitution $v' = h(v)$. This substitution leads to the equation

$$h'h = 6v^2.$$

Integrating this equation, one obtains

$$h^2 = 4v^3 + c_1,$$

where $c_1$ is an arbitrary constant. Thus

$$v' = \gamma\sqrt{4v^3 + c_1} \quad (\gamma = \pm 1),$$

and the function $v(p)$ is found from the equation

$$\int \frac{dv}{\sqrt{4v^3 + c_1}} = \gamma p + c_2.$$

In particular, for $c_1 = 0$ one has

$$v = (\gamma p + c_2)^{-2}.$$

---

[11] See Chap. 1 for details.

This determines a particular solution of (2.2.74) for the chosen $\lambda$:

$$\lambda = 1/6 : u = \frac{1}{z} - \frac{1}{z(\gamma + c_2 z)^2},$$

$$\lambda = -2/3 : u = \frac{1}{(\gamma \sqrt{z} + c_2)^2},$$

$$\lambda = -3/2 : u = \frac{1}{z^{1/3}(\gamma z^{1/3} + c_2)^2},$$

$$\lambda = 6 : u = \frac{1}{z} - \frac{1}{z^{2/3}(\gamma z^{1/6} + c_2)^2}.$$

The particular solutions of (2.2.65) are obtained by applying the inverse Laplace transform to the found functions. It is worth to note that solutions for $\lambda < 0$ has no physical meaning for the original equation (2.2.65). The case $\lambda = 1/6$ was studied in [8]. In the case where $\lambda = 6$ it is difficult to find inverse Laplace transform.

Other examples of applications of integral transforms to small dimensional models of the Boltzmann equation one can find in [23, 46].

The use of the Laplace transform in the studies of more real kinetic equations one can find in the coagulation theory [71]. In fact, the Smolukhovsky kinetic equation of homogeneous coagulation is of the form

$$\frac{\partial f(t, v)}{\partial t} = \frac{1}{2} \int_0^v d v_1 \beta(v - v_1, v_1) f(t, v - v_1) f(t, v_1)$$

$$- f(t, v) \int_0^\infty d v_1 \beta(v, v_1) f(t, v_1). \qquad (2.2.76)$$

The Cauchy problem for this equation is considered with the following initial data

$$f(0, v) = f_0(v).$$

Application of the Laplace transform $F(z) = \mathscr{L}\{f(v)\}$ to (2.2.76) with the coagulation kernel $\beta(v, v_1) = b(v + v_1)$ gives one the first-order partial differential equation

$$\frac{\partial F(t, z)}{\partial t} + b\left((F(t, z) - F(t, 0))\frac{\partial F(t, z)}{\partial z} + M F(t, z)\right) = 0,$$

where

$$M = \int_0^\infty d v v f(t, v) = \text{const}$$

is the total mass of coagulating particles. The obtained equation can be integrated in an explicit form. However, the inverse Laplace transform of the derived solution is only possible for a few initial functions $f_0(v)$. More substantial results of a direct group analysis of (2.2.76) are presented in Chap. 3.

**Use of a Moment Generating Function**   This approach has a very restricted set of applications and just used in a few works which are devoted to invariant solutions of the spatially homogeneous and isotropic Boltzmann equation with isotropic scattering model [43, 54, 66]. The original interest of the study in [43] was the system of normalized power moments for the formulated case of the Boltzmann equation. As shown in [9] this system can be easily derived by the substitution of the Taylor expansion

$$\varphi(x, t) = \sum_{n=0}^{\infty} \frac{(-x)^n}{n!} M_n(t)$$

into (2.2.65). Such obtained system takes the form

$$\frac{d M_n}{dt} + M_n = \frac{1}{n+1} \sum_{k=0}^{n} M_k M_{n-k} \qquad (n = 0, 1, 2, \ldots). \qquad (2.2.77)$$

The moment generating function is introduced as follows

$$G(\xi, t) = \sum_{n=0}^{\infty} \xi^n M_n(t).$$

Multiplying (2.2.77) by $\xi^n$ and summing over all $n$, one finds

$$\frac{\partial G}{\partial t} + G = \sum_{n=0}^{\infty} \frac{\xi^n}{n+1} \sum_{k=0}^{n} M_k M_{n-k}.$$

Noting that

$$G^2 = \sum_{n=0}^{\infty} \xi^n \sum_{k=0}^{n} M_k M_{n-k},$$

the last equation can be transformed to the next differential equation

$$\frac{\partial^2(\xi G)}{\partial t \partial \xi} + \frac{\partial(\xi G)}{\partial \xi} = G^2. \qquad (2.2.78)$$

The change of variables

$$\xi = (z+1)^{-1}, \qquad \xi G = u(z, t)$$

leads (2.2.78) into

$$\frac{\partial^2 u}{\partial z \partial t} + \frac{\partial u}{\partial z} + u^2 = 0. \qquad (2.2.79)$$

This equation coincides with (2.2.69), but the variables $z$, $u$ in (2.2.79) and in (2.2.69) have a different origin.

Using further transformations of (2.2.79) and very complicated calculations, the invariant BKW-solution was also derived in [43]. Notice that in this approach the inverse transition to the distribution function is related with large difficulties.

In [66] the equation (2.2.79) was studied by the classical group analysis method as done above for (2.2.69). The same admitted Lie algebra with the basis of the generators $\{Y_1, \ldots, Y_4\}$ was obtained there. It is natural that the discrepancy between this Lie algebra and the admitted Lie algebra of the original equation was also noted. Studying this discrepancy, the authors showed that the class of the BKW-solutions is the only one which satisfies the mass conservation law $M_0(t) = 1$ ($\varphi(0, t) = 1$). Recall that for (2.2.79) this law corresponds to the condition

$$u(z = \infty, t) = 0.$$

It was also proposed in [66] to make use of other obtained there classes of invariant solutions of (2.2.79) to the spatially homogeneous and isotropic Boltzmann equation with some source term. In this case (2.2.79) has a nonzero function $\psi(z, t)$ in the right hand side and the determining equations impose conditions on the function $\psi(z, t)$.

Some years later the described above approach was directly applied in [54] to the spatially homogeneous and isotropic Boltzmann equation with a source term. Instead of nonautonomous equation (2.2.79) the slightly different equation

$$\frac{\partial^2 u}{\partial z \partial t} + M_0(t) \frac{\partial u}{\partial z} + u^2 = \sigma$$

was considered. This allowed the author to weaken the conditions imposed on the source function comparing with [66].

**Some Other Technique** In the framework of this subsection it is also worth to mention two more approaches which could pretend to be universal. Since they are based on very specific mathematical techniques, they are not widespread.

The method developed in [18] consists in reducing the original integro-differential equation to a system of boundary differential equations. As an example of such a transition one can consider the simple one-dimensional Gammershtein integral equation

$$u(x) = \int_a^b K(x, s, u(s)) \, ds, \tag{2.2.80}$$

where the kernel $K(x, s, u)$ is a given function and $x \in [a, b]$. The equivalent system of boundary differential equations is introduced as follows

$$v_s(x, s) = K(x, s, u(s)), \qquad v(x, a) = 0,$$
$$u(x) = v(x, b).$$

The new dependent variable $v$ is nonlocal because it depends on all values of a solution $u(x)$ on the interval $[a, b]$. For this reason one calls the derived system as a *covering* of (2.2.80).

In the more interesting case of the Smolukhovsky equation (2.2.76) which is considered in [18] the corresponding covering takes the form

$$u_{v_1}(v, v_1, t) - u_v(v, v_1, t) = \beta(v, v_1) f(t, v) f(t, v_1),$$

$$u(v, v_1, t) = -u(v_1, v, t),$$

$$w_{v_1}(v, v_1, t) = \beta(v, v_1) f(t, v), \quad w(v, 0, t) = 0.$$

Using homomorphisms of the intervals of the independent variables variation, the constructed covering is formally rewritten as another differential system. For this system a very complex generalization of the classical group analysis in the geometrical interpretation was developed. Its explanation here would be very long and it is omitted. One can only remark that there are many coverings for the same integro-differential equation. Because of that one can obtain different results using this approach.

More technically simple method of reducing integro-differential equations to differential ones was suggested in [28, 29]. In this method one uses Weil's fractional integrals and derivatives. The $v$-order ($v > 0$) integral is defined as

$$W_x^{-v} f(x) = \frac{1}{\Gamma(v)} \int\limits_x^\infty dy (y - x)^{v-1} f(y),$$

where $\Gamma(x)$ is the Euler gamma-function. Correspondingly, $\alpha$-order Weil's derivative is

$$W_x^\alpha f(x) = E^n W_x^{-(n-\alpha)} f(x), \quad n - 1 < \alpha < n, \quad E^n = (-1)^n \frac{d^n}{dx^n}.$$

For example, one can consider the spatially homogeneous and isotropic Boltzmann equation with asymptotic collision integral [29]

$$x^\alpha f_t(x, t) + f(x, t) - \int\limits_0^1 f(sx, t) f((1 - s)x, t) \, ds = 0.$$

Reducing it to the equation with the convolution-type integral and using the Laplace transform as was done for (2.2.65), one obtains

$$\int\limits_0^\infty dx e^{-zx} (x^{1+\alpha} f(x)) - F(z, t) - F^2(z, t) = 0.$$

In terms of Weil's derivatives one can rewrite the last equation in the form

$$W_z^{1+\alpha} F_t - F(z, t) - F^2(z, t) = 0.$$

Since some properties of fractional Weil's derivatives are analogical to the properties of usual derivatives, this representation can ease the search for the admitted dilation group.[12] Because for arbitrary $\alpha$ the operator $W_x^\alpha$ is nonlocal, for other transformations one needs a corresponding generalization of the classical group

---

[12] See also [11].

analysis scheme. A variant of such generalization with another definition of fractional derivatives was announced in [26].

In conclusion one can summarize that all methods of reducing integro-differential equations to differential equations are confronted with the same difficulties. Among them: the lack of universality, the complexity of direct and inverse transformations, the possible violation of homomorphism of admitted groups and others.

## 2.3 A Regular Method for Calculating Symmetries of Equations with Nonlocal Operators

The survey presented in the previous section gives a sufficiently complete idea about methods for finding invariant solutions of integro-differential equations. However it is worth to note that none of these methods allows one to be sure that a derived Lie group is the widest Lie group admitted by considered equations. There exists the only way to derive such result: it is necessary to develop a method for constructing determining equations defining a Lie group admitted by the studied integro-differential equations. Then the completeness of an obtained Lie group will be a corollary fact of the uniqueness of the general solution of the determining equations.

In this section a regular direct method of a complete group analysis of equations with nonlocal operators will be presented. In applications of group analysis to these equations it is necessary to pass the same successive stages as for differential equations. The central conception of an admitted Lie group of equations with nonlocal terms will be defined as a Lie group satisfying determining equations. In contrast to partial differential equations the property of an admitted Lie group to map any solution into a solution of the same equations will be not required, although the method developed for constructing the determining equations uses this property. In practice the algorithm for obtaining determining equations becomes no more difficult than for partial differential equations. The main difficulty consists of solving the determining equations because they also contain some nonlocal operators. As for partial differential equations splitting the determining equations helps to obtain their general solution. The splitting method can be based, for example, on the existence of the solution of a Cauchy problem. The realization of the splitting method depends on properties of a Cauchy problem of studied nonlocal equations. In the next section we demonstrate two different approaches.

As a rule considered equations or systems along nonlocal operators also include operators or equations with partial derivatives. Hence, the definition of an admitted Lie group for equations with nonlocal terms has to be consistent with the definition of an admitted Lie group of partial differential equations.

Since the definition of an admitted Lie group given for partial differential equations cannot be applied to equations with nonlocal terms, before giving a definition the concept of an admitted Lie group requires further discussion. This discussion assists in establishing a definition of an admitted Lie group for equations with nonlocal terms.

## 2.3.1  Admitted Lie Group of Partial Differential Equations

One of the definitions of a Lie group admitted by a system of partial differential equations $(S)$ is based on a knowledge of the solutions:[13] a Lie group is admitted by the system $(S)$ if any solution of this system is mapped into a solution of the same system. Two other definitions are based on the geometrical approach: equations are considered as manifolds. One of these definitions deals with the manifold defined by the system $(S)$. Another definition works with the extended frame of the system $(S)$: system $(S)$ and all its prolongations.[14] Notice that the definitions based on the geometrical approach have the following inadequacy. There are equations which have no solutions, however they have an admitted (in this meaning) Lie group. Although the geometrical approach has the advantage that it is simple in applications.

Here it should be also mentioned that different approaches have been developed for finite-difference equations. Review of these approaches can be found in [22] and in references therein.

The classical geometrical definition of an admitted Lie group deals with invariant manifolds: the group is admitted by the system of equations

$$(S) \qquad S(x, u, p) = 0 \qquad (2.3.1)$$

if the manifold defined by these equations is invariant with respect to this group. All functions are assumed enough times continuously differentiable, for example, of the class $C^\infty$. The manifold

$$(S) = \{(x, u, p) \mid S(x, u, p) = 0\},$$

defined by (2.3.1), is considered in the space $J^l$ of the variables

$$x = (x_1, x_2, \ldots, x_n), \quad u = (u^1, u^2, \ldots, u^m), \quad p = (p_\alpha^j) \ (j = 1, 2, \ldots, m; \ |\alpha| \le l).$$

Here and below the following notations are used:

$$p_\alpha^j = D^\alpha u^j, \quad D^\alpha = D_1^{\alpha_1} D_2^{\alpha_2} \ldots D_n^{\alpha_n},$$

$$\alpha = (\alpha_1, \alpha_2, \ldots, \alpha_n), \quad |\alpha| = \alpha_1 + \alpha_2 + \cdots + \alpha_n,$$

$$\alpha, i = (\alpha_1, \alpha_2, \ldots, \alpha_{i-1}, \alpha_i + 1, \alpha_{i+1}, \ldots, \alpha_n),$$

where $D_j$ is the operator of the total differentiation with respect to $x_j$ $(j = 1, 2, \ldots, n)$.

Any local Lie group of point transformations

$$\bar{x}_i = f^i(x, u; a), \quad \bar{u}^j = \varphi^j(x, u; a), \qquad (2.3.2)$$

---

[13]Definitions of an admitted Lie group of partial differential equations are discussed in [47], Chap. 6, Sect. 1, [55], Sect. 2.6, [35], Sect. 1.3, [36] (see also Chap. 1), Sect. 9.2, [62], [49], Sect. 6.1 and references therein.

[14]According to the Cartan–Kähler theorem, after a finite number of prolongations the system $(S)$ becomes either involutive or incompatible. Therefore, from the theory of compatibility point of view, there is no necessity for infinite prolongations of the system $(S)$.

is defined by the transformations of the independent and dependent variables[15] with the generator

$$X = \xi^i(x, u)\partial_{x_i} + \eta^j(x, u)\partial_{u^j},$$

where

$$\xi^i(x, u) = \frac{df^i}{da}(x, u; 0), \quad \eta^j(x, u) = \frac{d\varphi^j}{da}(x, u; 0).$$

Here $a$ is the group parameter.

Lie groups admitted in the sense of the geometrical approach have the property to transform any solution of the system of equations $(S)$ into a solution of the same system. This property can be taken as a definition of the admitted Lie group of partial differential equations $(S)$.

**Definition 2.3.1** A Lie group (2.3.2) is admitted by system $(S)$ if it maps any solution of $(S)$ into a solution of the same system.

This definition supposes that the system $(S)$ has at least one solution.

Recall that the determining equations for the admitted group are obtained as follows. Let a function $u = u_o(x)$ be given. Substituting it into the first part of transformation (2.3.2) and using the inverse function theorem one finds

$$x = g^*(\bar{x}, a). \tag{2.3.3}$$

The transformed function $u_a(\bar{x})$ is given by the formula

$$u_a(\bar{x}) = f^u(g^*(\bar{x}, a), u_o(g^*(\bar{x}, a)); a).$$

The transformed derivatives are $\bar{p}_\alpha^j(\bar{x}, a) = \varphi_\alpha^j(x, u_o(x), p(x); a)$, where $p(x)$ are derivatives of the function $u_o(x)$, $x$ is defined by (2.3.3), and the functions $\varphi_\alpha^j(x, u, p; a)$ are defined by the prolongation formulae. The prolongation formulae are obtained by requiring the tangent conditions

$$du^j - p_k^j dx_k = 0, \quad dp_\alpha^j - p_{\alpha,k}^j dx_k = 0, \tag{2.3.4}$$

to be invariant. For example, for the first order derivatives

$$d\bar{u}^j - \bar{p}_k^j d\bar{x}_k = \left((\varphi_{x_k}^j + \varphi_{u^i}^j p_k^i) - \bar{p}_s^j(f_{x_k}^s + f_{u^i}^s p_k^i)\right)dx_k = 0$$

or

$$\Phi - PF = 0,$$

where $\Phi$, $F$ and $P$ are matrices with the entries

$$\Phi_k^j = \varphi_{x_k}^j + \varphi_{u^i}^j p_k^i, \quad F_k^s = f_{x_k}^s + f_{u^i}^s p_k^i, \quad P_s^j = p_s^j$$

$$(s, k = 1, 2, \ldots, n; \ j = 1, 2, \ldots, m).$$

---

[15]For the sake of simplicity only a Lie group of point transformations is discussed. For tangent transformations the study is similar.

Since the matrix $F$ is invertible in a neighborhood of $a = 0$, one has

$$P = \Phi F^{-1}.$$

For higher order derivatives the prolongation formulae are obtained recurrently.

Let the function $u_o(x)$ be a solution of a system $(S)$. Because of the given definition any transformation of the admitted Lie group transforms any solution to a solution of the same system, the function $u_a(\bar{x})$ is also a solution of the system $(S)$:

$$\bar{S}(\bar{x}, a) = S(\bar{x}, u_a(\bar{x}), \bar{p}_\alpha^j(\bar{x}, a)) = 0.$$

In the last equations, instead of the independent variables $\bar{x}$, $a$ one can consider the independent variables $x$, $a$:

$$\bar{\bar{S}}(x, a) = \bar{S}(f^x(x, u_o(x); a), a).$$

Differentiating the functions $\bar{\bar{S}}(x, a)$ or $\bar{S}(\bar{x}, a)$ with respect to the group parameter $a$ and setting $a = 0$, one obtains the determining equations

$$\left( \frac{\partial}{\partial a} \bar{\bar{S}}(x, a) \right)(x, 0) = (XS)(x, u_o(x), p(x)) = 0 \qquad (2.3.5)$$

or

$$\left( \frac{\partial}{\partial a} \bar{S}(\bar{x}, a) \right)(\bar{x}, 0) = (\tilde{X}S)(x, u_o(x), p(x)) = 0. \qquad (2.3.6)$$

The operator $\tilde{X}$ is the canonical Lie–Bäcklund operator [34]

$$\tilde{X} = \bar{\eta}^j \partial_{u^j} + D^\alpha \bar{\eta}^j \partial_{p_\alpha^j}$$

equivalent to the generator $X$. Here

$$\bar{\eta}^j = \eta^j(x, u) - \xi^\beta(x, u) p_\beta^j.$$

Since the function $u_o(x)$ is a solution of the system $(S)$, the solutions of the determining equations (2.3.5) and (2.3.6) coincide.

For solving the determining equations one needs to know arbitrary elements. In the geometrical definitions the arbitrary elements are coordinates of the manifolds. In the case of the determining equations (2.3.5) or (2.3.6) for establishing the arbitrary elements one can use, for example, a knowledge of the existence of a solution of the Cauchy problem.

From one point of view the last definition (related to a solution) is more difficult for applications than the geometrical definitions. Although, from another point of view, this definition allows the construction of the determining equations for more general objects than differential equations: integro-differential equations, functional differential equations or even for more general type of equations.

### 2.3.2  The Approach for Equations with Nonlocal Operators

Let us consider an abstract system of integro-differential equations:

$$\Phi(x, u) = 0. \qquad (2.3.7)$$

Here as above $u$ is the vector of the dependent variables, $x$ is the vector of the independent variables. Assume that a one-parameter Lie group $G^1(X)$ of transformations

$$\bar{x} = f^x(x, u; a), \quad \bar{u} = f^u(x, u; a) \qquad (2.3.8)$$

with the generator

$$X = \eta^j(x, u)\partial_{u_j} + \xi^i(x, u)\partial_{x_i},$$

transforms a solution $u_0(x)$ of (2.3.7) into the solution $u_a(x)$ of the same equations. The transformed function $u_a(x)$ is

$$u_a(\bar{x}) = f^u(x, u(x); a),$$

where $x = \psi^x(\bar{x}; a)$ is substituted into this expression. The function $\psi^x(\bar{x}; a)$ is found from the relation $\bar{x} = f^x(x, u(x); a)$ using the inverse function theorem. Differentiating the equations $\Phi(x, u_a(x))$ with respect to the group parameter $a$ and considering the result for the value $a = 0$, one obtains the equations

$$\left(\frac{\partial}{\partial a}\Phi(x, u_a(x))\right)_{|a=0} = 0. \qquad (2.3.9)$$

For integro-differential equations one needs to have an existence of the inverse function defined on some interval. Because of the localness of the inverse function theorem this is one of the obstacles for applying to integro-differential equations the definition of an admitted Lie group based on a solution. However, notice that (2.3.9) coincide with the equations

$$(\bar{X}\Phi)(x, u_0(x)) = 0 \qquad (2.3.10)$$

obtained by the action of the canonical Lie–Bäcklund operator $\bar{X}$, which is equivalent to the generator $X$:

$$\bar{X} = \bar{\eta}^j \partial_{u^j},$$

where $\bar{\eta}^j = \eta^j(x, u) - \xi^i(x, u)p_i^j$. The actions of the derivatives $\partial_{u^j}$ and $\partial_{p_\alpha^j}$ are considered in terms of the Frechet derivatives. Equations (2.3.10) can be constructed without requiring the property that the Lie group should transform a solution into a solution. This allows the following definition of an admitted Lie group.

**Definition 2.3.2** A one-parameter Lie group $G^1$ of transformations (2.3.8) is a symmetry group admitted by (2.3.7) if $G^1$ satisfies (2.3.10) for any solution $u_0(x)$ of (2.3.7). Equations (2.3.10) are called the determining equations.

**Remark 2.3.1** For a system of differential equations (without integral terms) the determining equations (2.3.10) coincide with the determining equations (2.3.6).

The way of obtaining determining equations for integro-differential equations is similar (and not more difficult) to the way used for differential equations. Notice also that the determining equations of integro-differential equations are integro-differential.

The advantage of the given definition of an admitted Lie group is that it provides a constructive method for obtaining the admitted group. Another advantage of this definition is the possibility to apply it for seeking Lie–Bäcklund transformations,[16] conditional symmetries and other types of symmetries for integro-differential equations.

The main difficulty in obtaining an admitted Lie group consists of solving the determining equations. There are some methods for simplifying determining equations. As for partial differential equations the main method for simplification is their splitting. It should be noted that, contrary to differential equations, the splitting of integro-differential equations depends on the studied equations. Since the determining equations (2.3.10) have to be satisfied for any solution of the original equations (2.3.7), the arbitrariness of the solution $u_0(x)$ plays a key role in the process of solving the determining equations. The important circumstance in this process is the knowledge of the properties of solutions of the original equations. For example, one of these properties is the theorem of the existence of a solution of the Cauchy problem.

Along splitting determining equations there are some other ways to simplify them. For example, for the Vlasov-type or Benney kinetic equations a specific approach was proposed in [40]. The principal feature of this approach consists of treating equally the local and nonlocal variables in determining equations. It allows one to separate these equations in "local" and "nonlocal" parts. For solving local part of the determining equations the classical group analysis method is applied. As the result one gets a group generator which defines so-called *intermediate* symmetry. In the final step using the information adopted from intermediate symmetry the non-local determining equations are solved by special authors' procedure of variational differentiation (see Chap. 4 for details).

**Remark 2.3.2** A geometrical approach for constructing an admitted Lie group for integro-differential equations is applied in [18, 19].

## 2.4 Illustrative Examples

This section deals with two examples which illustrate the method developed in the previous section. In the first example the method is applied to the Fourier-image of the spatially homogeneous isotropic kinetic Boltzmann equation. This is an integro-differential equation which contains some nonlinear integral operator with respect to a so-called inner variable. The complete solution of the determining equation is

---

[16]There are some trivial examples of such applications for integro-differential equations.

given [28] by constructing necessary conditions for the coefficients of the admitted generator. These conditions are obtained by using a particular class of solutions of the original integro-differential equation. It is worth to note that the particular class of solutions allowed us to find the general solution of the determining equation.

Another example considered in this section is an application of the developed method to the equations describing one-dimensional motion of a viscoelastic continuum. The corresponding system of equations includes a linear Volterra integral equation of the second type. The method of solving the determining equations in this case differs from the previous example. The arbitrariness of the initial data in the Cauchy problem allows one to split the determining equations. Solving the split equations which are partial differential equations, one finds the general solution of the determining equations.

## 2.4.1 The Fourier-Image of the Spatially Homogeneous Isotropic Boltzmann Equation

In the case of the spatially homogeneous and isotropic Boltzmann equation corresponding distribution function $f(v, t)$ depends only on modulus of a molecular velocity $v$ and time $t$. The Fourier-image of the spatially homogeneous and isotropic Boltzmann equation was derived in [4]. The considered equation is (2.2.65):

$$\Phi \equiv \varphi_t(x, t) + \varphi(x, t)\varphi(0, t) - \int_0^1 \varphi(xs, t)\varphi(x(1 - s), t)\,ds = 0. \quad (2.4.1)$$

Here $\varphi(x, t) = \tilde{\varphi}(k^2/2, t)$, and the Fourier transform $\tilde{\varphi}(k, t)$ of the distribution function $f(v, t)$ is defined as

$$\tilde{\varphi}(k, t) = \frac{4\pi}{k} \int_0^\infty v \sin(kv) f(v, t)\,dv.$$

Further the existence of a solution of the Cauchy problem of (2.4.1) with the initial data

$$\varphi(x, t_0) = \varphi_0(x) \quad (2.4.2)$$

is used.[17]

By virtue of the initial conditions (2.4.2) and the equation (2.4.1), one can find the derivatives of the function $\varphi(x, t)$ at time $t = t_0$:

---

[17] See, for example, [9].

$$\varphi_t(x, t_0) = -\varphi_0(0)\varphi_0(x) + \int_0^1 \varphi_0(sx)\varphi_0((1-s)x)\,ds,$$

$$\varphi_{xt}(x, t_0) = -\varphi_0(0)\varphi_0'(x) + 2\int_0^1 s\varphi_0'(sx)\varphi_0((1-s)x)\,ds,$$

$$\varphi_{tt}(x, t_0) = -\varphi_0^2(0)\varphi_0(x) - 3\varphi_0(0)\int_0^1 \varphi_0(sx)\varphi_0((1-s)x)\,ds \qquad (2.4.3)$$

$$+2\int_0^1\int_0^1 \varphi_0((1-s)x)\varphi_0(ss'x)\varphi_0(s(1-s')x)\varphi_0((1-s)x)\,ds\,ds'.$$

### 2.4.1.1 Admitted Lie Group

The generator of the admitted Lie group is sought in the form

$$X = \xi(x, t, \varphi)\partial_x + \eta(x, t, \varphi)\partial_t + \zeta(x, t, \varphi)\partial_\varphi.$$

The determining equation for (2.4.1) is

$$D_t\psi(x, t) + \psi(0, t)\varphi(x, t) + \psi(x, t)\varphi(0, t)$$

$$-2\int_0^1 \varphi(x(1-s)s, t)\psi(xs, t)\,ds = 0, \qquad (2.4.4)$$

where $\varphi(x, t)$ is an arbitrary solution of (2.4.1), $D_t$ is the total derivative with respect to $t$, and the function $\psi(x, t)$ is

$$\psi(x, t) = \zeta(x, t, \varphi(x, t)) - \xi(x, t, \varphi(x, t))\varphi_x(x, t) - \eta(x, t, \varphi(x, t))\varphi_t(x, t).$$

In the determining equation (2.4.4) the derivatives $\varphi_t$, $\varphi_{xt}$ and $\varphi_{tt}$ are defined by formulae (2.4.3).

The method of solving the determining equation (2.4.4) consists of in studying the properties of the functions $\xi(x, t, \varphi)$, $\eta(x, t, \varphi)$ and $\zeta(x, t, \varphi)$. These properties are obtained by sequentially considering the determining equation on a particular class of solutions of (2.4.1). This class of solutions is defined by the initial conditions

$$\varphi_0(x) = bx^n \qquad (2.4.5)$$

at the given (arbitrary) time $t = t_0$. Here $n$ is a positive integer. The determining equation is considered for any arbitrary initial time $t_0$.

During solving the determining equation we use the following properties. Multiplying any solution of (2.4.1) by $e^{\lambda x}$, one maps it into a solution of the same equation (2.4.1). Taking into account the $\beta$-function [39]

$$B(m+1, n+1) = \int_0^1 s^m (1-s)^n \, ds = \frac{m! n!}{(m+n+1)!}$$

one uses the notations

$$P_n = \frac{(n!)^2}{(2n+1)!}, \quad Q_n = 2 P_n \frac{(2n)! n!}{(3n+1)!}.$$

Notice that

$$2 P_{n+1} = P_n \frac{1}{1 + \frac{1}{n+1}}$$

and

$$\lim_{n \to \infty} P_n = 0, \quad \lim_{n \to \infty} Q_n = 0, \quad \lim_{n \to \infty} \frac{Q_n}{P_n} = 0.$$

Assume that the coefficients of the infinitesimal generator $X$ are represented by the formal Taylor series with respect to $\varphi$:

$$\xi(x, t, \varphi) = \sum_{l \geq 0} q_l(x, t) \varphi^l,$$

$$\eta(x, t, \varphi) = \sum_{l \geq 0} r_l(x, t) \varphi^l, \quad \zeta(x, t, \varphi) = \sum_{l \geq 0} p_l(x, t) \varphi^l.$$

Equation (2.4.4) is studied by setting $n = 0, 1, 2, \ldots$, and varying the parameter $b$.

If $n = 0$, then the determining equation (2.4.4) becomes

$$\hat{\zeta}(x, t) + b(\hat{\zeta}(0, t) + \hat{\zeta}(x, t)) - 2b \int_0^1 \hat{\zeta}(xs, t) \, ds = 0.$$

From this equation one obtains

$$\frac{\partial p_0}{\partial t} = 0, \quad \frac{\partial p_{l+1}}{\partial t}(x, t) + p_l(x, t) + p_l(0, t) - 2 \int_0^1 p_l(xs, t) \, ds = 0 \quad (2.4.6)$$

$$(l = 0, 1, \ldots).$$

Here and below $\hat{\zeta}$, $\hat{\xi}$ and $\hat{\eta}$ are the coefficients of the operator $X$ evaluated for the initial data (2.4.5).

If $n \geq 1$ in (2.4.5) one finds that

$$\varphi_t(x, t_0) = P_n b^2 x^{2n}, \quad \varphi_x(x, t_0) = nbx^{n-1},$$

$$\varphi_{tt}(x, t_0) = Q_n b^3 x^{3n}, \quad \varphi_{tx}(x, t_0) = 2n P_n b^2 x^{2n-1}.$$

The determining equation (2.4.4) becomes

$$\hat{\zeta}_t + b\left(-nx^{n-1}\hat{\xi}_t + x^n\hat{\zeta}(0, t) - 2x^n \int_0^1 (1-s)^n \hat{\zeta}(xs, t)\, ds\right)$$

$$+ b^2\left(-P_n x^{2n}\hat{\eta}_t + P_n x^{2n}\hat{\zeta}_\varphi - 2n P_n x^{2n-1}\hat{\xi} - \delta_{n1}\hat{\xi}(0, t)\right.$$

$$\left. + 2nx^{2n-1}\int_0^1 (1-s)^n s^{n-1}\hat{\xi}(xs, t)\, ds\right)$$

$$+ b^3\left(-n P_n x^{2n-1}\hat{\xi}_\varphi - Q_n x^{3n}\hat{\eta} + 2 P_n x^{3n}\int_0^1 (1-s)^n s^{2n}\hat{\eta}(xs, t)\, ds\right)$$

$$- b^4\left(P_n^2 x^{4n}\hat{\eta}_\varphi\right) = 0. \tag{2.4.7}$$

Using the arbitrariness of the value $b$, the equation (2.4.7) can be split into a series of equations by equating to zero the coefficients of $b^k$ $(k = 0, 1, \ldots)$ in the left-hand side of (2.4.7).

For $k = 0$ the corresponding coefficient in the left-hand side of (2.4.7) vanishes because of the first equation of (2.4.6).

For $k = 1$, the equation (2.4.7) yields:

$$x\left(-p_0(x, t) + 2\int_0^1 (1 - (1-s)^n)p_0(xs, t)\, ds\right) - n\frac{\partial q_0(x, t)}{\partial t} = 0.$$

By virtue of arbitrariness of $n$, one finds

$$p_0(x, t) = 0, \qquad \frac{\partial q_0(x, t)}{\partial t} = 0.$$

These relations provide that $\hat{\zeta}(0, t) = 0$.

For $k = 2$ one obtains the equation

$$x\left(-p_1(x, t) - p_1(0, t) + 2\int_0^1 (1 - (1-s)^n s^n)p_1(xs, t)\, ds\right.$$

$$\left. + P_n\left(p_1(x, t) - \frac{\partial r_0(x, t)}{\partial t}\right)\right) - n\frac{\partial q_1(x, t)}{\partial t} - 2n P_n q_0(x, t)$$

$$+ 2n\int_0^1 (1-s)^n s^{n-1}q_0(xs, t)\, ds = 0.$$

Consecutively dividing by $n$, $P_n$ and letting $n \to \infty$, one obtains

$$p_1(x, t) = c_0 + c_1 x, \qquad \frac{\partial q_1(x, t)}{\partial t} = 0, \qquad q_0(x, t) = c_2 x, \qquad \frac{\partial r_0(x, t)}{\partial t} = -c_0,$$

where $c_0, c_1, c_2$ are arbitrary constants.

For $k = 3$, one has

$$x^{n+1}\left(-p_2(x,t) - p_2(0,t) + 2\int_0^1 (1 - (1-s)^n s^{2n})p_2(xs,t)\,ds\right.$$

$$- P_n\frac{\partial r_1(x,t)}{\partial t} + 2P_n p_2(x,t) + 2P_n\int_0^1 (1-s)^n s^{2n} r_0(xs,t)\,ds - Q_n r_0(x,t)\Big)$$

$$+ x^n\left(-n\frac{\partial q_2(x,t)}{\partial t} - 2n P_n q_1(x,t) + 2n\int_0^1 (1-s)^n s^{2n-1}q_1(xs,t)\,ds\right)$$

$$- n P_n q_1(x,t) = 0.$$

Similar to the previous case ($k = 2$) one finds

$$q_1(x,t) = 0, \qquad \frac{\partial q_2(x,t)}{\partial t} = 0, \qquad p_2(x,t) = 0,$$

$$\frac{\partial r_1(x,t)}{\partial t} = 0, \qquad r_0(x,t) = -c_0 t + c_3,$$

where $c_3$ is an arbitrary constant.

For $k = 4 + \alpha$ ($\alpha = 0, 1, \ldots$), the equation (2.4.7) yields

$$x^{n+1}\left(\frac{\partial p_{\alpha+4}(x,t)}{\partial t} - 2\int_0^1 (1-s)^n s^{(3\alpha)n} p_{3+\alpha}(xs,t)\,ds\right.$$

$$+ (3+\alpha)P_n p_{3+\alpha}(x,t) - P_n\frac{\partial r_{2+\alpha}(x,t)}{\partial t} - (\alpha+1)P_n^2 r_{\alpha+1}(x,t)$$

$$+ 2P_n\int_0^1 (1-s)^n s^{(3+\alpha)n} r_{\alpha+1}(xs,t)\,ds - Q_n r_{\alpha+1}(x,t)\Big)$$

$$+ nx^n\left(-\frac{\partial q_{\alpha+3}(x,t)}{\partial t} - 2P_n q_{\alpha+2}(x,t)\right.$$

$$+ 2\int_0^1 (1-s)^n s^{(\alpha+3)n-1}q_{\alpha+2}(xs,t)\,ds\Big)$$

$$- n(\alpha+2)P_n q_{\alpha+2}(x,t) = 0.$$

From this equation one obtains

$$p_{\alpha+3}(x,t) = 0, \qquad q_{\alpha+2}(x,t) = 0, \qquad r_{\alpha+1}(x,t) = 0 \quad (\alpha = 0, 1, \ldots).$$

Thus, from the above equations, one finds

$$\xi = c_2 x, \qquad \eta = c_3 - c_0 t, \qquad \zeta = (c_1 x + c_0)\varphi \tag{2.4.8}$$

with the arbitrary constants $c_0, c_1, c_2, c_3$. Formulae (2.4.8) are the necessary conditions for the coefficients of the generator $X$ to satisfy the determining equation (2.4.4). One can directly check that they also satisfy the determining equation (2.4.4). Thus, the calculations provide the unique solution of the determining equation (2.4.4).

Because of the uniqueness of the obtained solution of the determining equation (2.4.4) one finds a constructive proof of the next statement.

**Theorem 2.4.1** *The four-dimensional Lie algebra $L^4 = \{X_1, X_2, X_3, X_4\}$ spanned by the generators*

$$X_1 = \partial_t, \quad X_2 = x\varphi\partial_\varphi, \quad X_3 = x\partial_x, \quad X_4 = \varphi\partial_\varphi - t\partial_t \qquad (2.4.9)$$

*defines the complete Lie group $G^4$ admitted by (2.4.1).*

### 2.4.1.2  Invariant Solutions

For constructing an invariant solution one has to choose a subalgebra. Since any subalgebra is equivalent to one of the representatives of an optimal system of admitted subalgebras, it is sufficient to study invariant solutions corresponding to the optimal system of subalgebras. Choosing a subalgebra from the optimal system of subalgebras, finding invariants of the subalgebra, and assuming dependence between these invariants, one obtains the representation of an invariant solution. Substituting this representation into (2.4.1) one gets the reduced equations: for the invariant solutions the original equation is reduced to the equation for a function with a single independent variable.

The optimal system of one-dimensional subalgebras of $L^4$ consists of the subalgebras

$$X_1, \quad X_4 + cX_3, \quad X_2 - X_1, \quad X_4 \pm X_2, \quad X_1 + X_3, \qquad (2.4.10)$$

where $c$ is an arbitrary constant. The corresponding representations of the invariant solutions are the following.

The invariants of the subalgebra $\{X_1\}$ are $\varphi$ and $x$. Hence, an invariant solution has the representation $\varphi = g(x)$, where the function $g$ has to satisfy the equation

$$g(x)g(0) - \int_0^1 g(xs)g(x(1-s))\,ds = 0. \qquad (2.4.11)$$

The Maxwell solution $\varphi = pe^{\lambda x}$ is an invariant solution with respect to this subalgebra. Let a solution of (2.4.11) be represented through the formal series $g(x) = \sum_{j\geq 0} a_j x^j$. For the coefficients of the formal series one obtains

$$a_0\left(1 - \frac{2}{(k+1)!}\right)a_k = \sum_{j=1}^{k-1} \frac{j!(k-j)!}{(k+1)!} a_j a_{k-j} \quad (k = 2, 3, \ldots).$$

Noticing that the value $a_0 = 0$ leads to the trivial case $g = 0$, so that one has to assume $a_0 \neq 0$. Because (2.4.11) admits scaling of the function $g$, one can set $a_0 = 1$. Since the multiplication by the function $e^{\lambda x}$ transforms any solution of (2.4.11) into another solution, one also can set $a_1 = 0$. Hence, all other coefficients vanish, $a_j = 0$ $(j = 2, 3, \ldots)$. Thus, the general solution of (2.4.11) is $g = e^{\lambda x}$. This means the uniqueness of the absolute Maxwell distribution as was mentioned in the above section.

In the case of the subalgebra $\{X_4 + cX_3\}$ the representation of an invariant solution is $\varphi = t^{-1}g(y)$, where $y = xt^c$, and the function $g$ has to satisfy the equation

$$cyg'(y) - g(y) + g(y)g(0) - \int_0^1 g(ys)g(y(1-s))\,ds = 0. \qquad (2.4.12)$$

Assuming that a solution is represented through the formal series $g(y) = \sum_{j \geq 0} a_j y^j$, one obtains the equations for the coefficients

$$a_0 = 0, \quad (c-1)a_1 = 0, \quad (ck-1)a_k = \sum_{j=0}^{k} \frac{j!(k-j)!}{(k+1)!} a_j a_{k-j} \quad (k = 2, 3, \ldots).$$

The case where $ck \neq 1$ for all $k$ $(k = 1, 2, \ldots)$ leads to the trivial solution $g = 0$ of (2.4.12). If $c = \alpha^{-1}$ where $\alpha$ is integer, then $a_k = 0$ $(k = 1, 2, \ldots, \alpha - 1)$, the coefficient $a_\alpha$ is arbitrary, and for other coefficients $a_k$ $(k = \alpha + 1, \alpha + 2, \ldots)$ one obtains the recurrence formula

$$(\alpha^{-1}k - 1)a_k = \sum_{j=1}^{k-1} \frac{j!(k-j)!}{(k+1)!} a_j a_{k-j}.$$

The representation of an invariant solution of the subalgebra $\{X_2 - X_1\}$ is $\varphi = e^{-xt}g(x)$, where the function $g$ satisfies the equation

$$-xg(x) + g(x)g(0) - \int_0^1 g(xs)g(x(1-s))\,ds = 0.$$

If one assumes that a solution can be represented through the formal series $g(x) = \sum_{j \geq 0} a_j x^j$, the first two terms of the series, obtained after substitution, are

$$a_0 = 0, \quad a_1(6 + a_1) = 0.$$

The case $a_1 = 0$ leads to the trivial solution $g = 0$. If $a_1 \neq 0$, then the other coefficients are defined by the recurrent formula

$$\left(1 - \frac{6}{k(k+1)}\right)a_{k-1} = -\sum_{j=1}^{k-2} \frac{j!(k-j)!}{(k+1)!} a_j a_{k-j} \quad (k = 3, 4, \ldots).$$

An invariant solution of the subalgebra $\{X_1 + X_3\}$ has the form $\varphi = g(y)$, where $y = xe^{-t}$. The function $g$ has to satisfy the equation

$$-yg'(y) + g(y)g(0) - \int_0^1 g(ys)g(y(1-s))\,ds = 0. \qquad (2.4.13)$$

The solution of this equation $g = 6e^y(1 - y)$ is known as the BKW-solution [3, 42].[18] This solution was obtained by assuming that the series $g(y) = e^y \sum_{j \geq 0} a_j y^j$ can be terminated. In fact, substituting the function $g(y) = e^y \sum_{j \geq 0} a_j y^j$ into (2.4.13) for the coefficients $a_k$ one obtains the equations

$$a_0 + a_1 = 0, \quad 2(a_0 - 6)a_2 = a_1(6 + a_1), \quad 6(a_0 - 6)a_3 = a_2(12 + a_1),$$

$$\left(a_0(1 - \frac{2}{(k+1)}) - k\right)a_k = a_{k-1}\left(1 + \frac{1}{k(k+1)}a_1\right) + \frac{2}{k(k^2-1)}a_{k-2}a_2$$

$$+ \sum_{j=2}^{k-2} \frac{j!(k-j)!}{(k+1)!}a_j a_{k-j} \quad (k = 4, 5, \ldots).$$

$$(2.4.14)$$

One can check that the choice $a_0 = 6$, $a_1 = -6$, and $a_k = 0$ $(k = 2, 3, \ldots)$ satisfies (2.4.14).

A representation of an invariant solution of the subalgebra $\{X_4 \pm X_2\}$ is $\varphi = t^{-(1\pm x)}g(x)$, where the function $g$ has to satisfy the equation

$$(1 \pm x)g(x) - g(x)g(0) + \int_0^1 g(xs)g(x(1-s))\,ds = 0.$$

### 2.4.2 Equations of One-Dimensional Viscoelastic Continuum Motion

One of models describing the one-dimensional motion of a viscoelastic continuum is based on the equations [60]

$$v_t = \sigma_x, \quad e_t = v_x, \quad \sigma + \int_0^t K(t, \tau)\sigma(x, \tau)\,d\tau = \varphi(e), \qquad (2.4.15)$$

where the time $t$ and the distance $x$ are the independent variables, the stress $\sigma$, the velocity $v$, and the strain $e$ are the dependent variables. The Volterra integral equation in the system (2.4.15) describes a dependence of the stress $\sigma$ on the strain $e$,

---

[18]This solution is usually considered as invariant solution with respect to the subalgebra $\{X_2 - X_3 + c^{-1}X_1\}$ which is similar to $\{X_1 + X_3\}$.

$K(t, \tau)$ is a kernel of heredity, $\varphi(e)$ is a known function. It is assumed that $K \neq 0$ and $\varphi'(e) \neq 0$.

Let the infinitesimal generator of a Lie group admitted by (2.4.15) be

$$X = \zeta^e \partial_e + \zeta^v \partial_v + \zeta^\sigma \partial_\sigma + \xi^x \partial_x + \xi^t \partial_t$$

with the coefficients depending on $(t, x, v, e, \sigma)$. The determining equations are

$$\left(D_t \widehat{\zeta^v} - D_x \widehat{\zeta^\sigma}\right)_{|(S)} = 0, \quad \left(D_t \widehat{\zeta^e} - D_x \widehat{\zeta^v}\right)_{|(S)} = 0, \tag{2.4.16}$$

$$\left(\varphi' \widehat{\zeta^e} - \widehat{\zeta^\sigma} - \int_0^t K(t, \tau) \widehat{\zeta^\sigma}(x, \tau) d\tau\right)_{|(S)} = 0, \tag{2.4.17}$$

where

$$\widehat{\zeta^e} = \zeta^e - \xi^x e_x - \xi^t e_t, \quad \widehat{\zeta^v} = \zeta^v - \xi^x v_x - \xi^t v_t, \quad \widehat{\zeta^\sigma} = \zeta^\sigma - \xi^x \sigma_x - \xi^t \sigma_t$$

with the functions $e(x, t)$, $v(x, t)$, $\sigma(x, t)$ satisfying (2.4.15) substituted in them. The complete set of solutions of the determining equations is sought under the assumption that there exists a solution of the Cauchy problem[19]

$$e(x_o, t) = e_0(t), \quad v(x_o, t) = v_0(t), \quad \sigma(x_o, t) = \sigma_0(t)$$

with arbitrary sufficiently smooth functions $e_0(t), v_0(t), \sigma_0(t)$.

Derivatives of the functions $e(x, t), v(x, t), \sigma(x, t)$ at the point $x = x_o$ can be found from (2.4.15):

$$v_t = v_o', \quad \sigma_t = \sigma_o', \quad \sigma_x = v_o', \quad v_x = e_t = \frac{g_1}{\varphi'}, \quad e_x = \frac{g_2}{\varphi'}, \tag{2.4.18}$$

where

$$g_1 = \sigma_o' + K(t, t)\sigma_o + \int_0^t K_t(t, \tau)\sigma_o(\tau) d\tau, \quad g_2 = v_o' + \int_0^t K(t, \tau)v_o'(\tau) d\tau.$$

Substituting the derivatives $v_t, \sigma_t, \sigma_x, v_x, e_t, e_x$ into the determining equations (2.4.16), considered at the point $x_o$, one obtains

$$v_o'\left(\zeta_v^v - \eta_t - \zeta_\sigma^\sigma + \xi_x - \eta_e g_1\right) + (v_o')^2(-\eta_v + \xi_\sigma) + g_2(-\zeta_e^\sigma + \sigma_o'\eta_e) + v_o' g_2 \xi_e$$
$$+ \zeta_\sigma^v \sigma_o' + \zeta_t^v - \zeta_x^\sigma + \sigma_o'\eta_x + g_1\left(\zeta_e^v - \zeta_v^\sigma + \sigma_o'\eta_v - \xi_\sigma \sigma_o' - \xi_t - \xi_e g_1\right) = 0,$$
$$v_o'(\zeta_v^e - \zeta_\sigma^\sigma + \eta_x + \xi_\sigma g_1) + (v_o')^2 \eta_\sigma + g_2(-\xi_\sigma \sigma_o' - \zeta_e^\sigma - \xi_t) + v_o' g_2(-\xi_v + \eta_e)$$
$$+ \zeta_\sigma^e \sigma_o' + \zeta_t^e - \zeta_x^v + g_1(\zeta_e^e - \eta_\sigma \sigma_o' - \eta_t - \zeta_v^v + \xi_x + g_1(-\eta_e + \xi_v)) = 0.$$

These equations can be split with respect to $v_o, v_o', v_o' + \int_0^t K(t, \tau)v_o'(\tau) d\tau$. In fact, setting the function $v_o(t)$ such that

$$v_o(t) = a_1 + a_2(t - t_o) + a_3 \frac{(t - t_o)^{n+1}}{(n + 1)} \quad (n \geq 1),$$

---

[19]These conditions are boundary conditions, rather than initial conditions.

one finds at the time $t = t_o$:

$$v_o(t_o) = a_1, \quad v_o'(t_o) = a_2,$$

$$v_o'(t_o) + \int_0^{t_o} K(t_o, \tau) v_o'(\tau) d\tau \tag{2.4.19}$$

$$= a_2 \left( 1 + \int_0^{t_o} K(t_o, \tau) d\tau \right) + a_3 \int_0^{t_o} K(t_o, \tau)(\tau - t_o)^n d\tau.$$

Since the set of the functions $(t - t_o)^n$ $(n \geq 0)$ is complete in the space $L_2(0, t_o]$, and $t_o$ is such that $K(t_o, \tau) \neq 0$, there exists $n$ for which $\int_0^{t_o} K(t_o, \tau)(\tau - t_o)^n d\tau \neq 0$. Hence, for the given values $v_o(t_o), v_o'(t_o), \int_0^{t_o} K(t_o, \tau) v_o'(\tau) d\tau$ one can solve (2.4.19) with respect to the coefficients $a_1, a_2, a_3$. This means that the values $v_o, v_o', v_o' + \int_0^t K(t, \tau) v_o'(\tau) d\tau$ are arbitrary and one can split the determining equations with respect to them. Splitting the determining equations, one finds

$$\xi_v = \xi_e = \xi_\sigma = 0, \quad \eta_v = \eta_e = \eta_\sigma = 0, \quad \zeta_e^v = -\xi_t, \tag{2.4.20}$$

$$\xi_x - \eta_t = \zeta_\sigma^\sigma - \zeta_v^v, \quad \zeta_e^\sigma = 0, \quad \zeta_v^e - \zeta_\sigma^e = -\eta_x,$$

$$(\zeta_\sigma^v + \eta_x)\sigma_o' + \zeta_t^v - \zeta_x^\sigma = g_1(2\xi_t + \zeta_v^\sigma)), \tag{2.4.21}$$

$$\zeta_\sigma^e \sigma_o' + \zeta_t^e - \zeta_x^v = g_1(\eta_t + \zeta_v^v - \xi_x - \zeta_e^e).$$

Equations (2.4.21) also can be split with respect to

$$\sigma_o(t_o), \quad \sigma_o'(t_o), \quad e(t_o), \quad \sigma_o'(t_o) + K(t_o, t_o)\sigma_o(t_o) + \int_0^{t_o} K_t(t_o, \tau)\sigma_o(\tau) d\tau.$$

In fact, let

$$\sigma_o(\tau) = a_1 + a_2(\tau - t_o) + (t_o - \tau)^2 (a_3 \psi_1(\tau) + a_4 \psi_2(\tau)).$$

If the determinant

$$\Delta = \left( \int_0^{t_o} K(t_o, \tau)(t_o - \tau)^2 \psi_1(\tau) d\tau \right) \left( \int_0^{t_o} K_t(t_o, \tau)(t_o - \tau)^2 \psi_2(\tau) d\tau \right)$$

$$- \left( \int_0^{t_o} K(t_o, \tau)(t_o - \tau)^2 \psi_2(\tau) d\tau \right) \left( \int_0^{t_o} K_t(t_o, \tau)(t_o - \tau)^2 \psi_1(\tau) d\tau \right)$$

is equal to zero for all functions $\psi_1, \psi_2 \in L_2[0, t_o]$, then by virtue of $K(t, \tau) \neq 0$ one obtains that there exists a function $f(t)$ such that

$$K_t(t, \tau) = f(t) K(t, \tau). \tag{2.4.22}$$

The general solution of this equation in some neighborhood of the point $t = t_o$ has the form

$$K(t, \tau) = h(t) g(\tau), \tag{2.4.23}$$

where $f(t) = h'(t)/h(t)$. The kernels of the type[20] (2.4.23) are excluded from the study, because for these kernels system of equations (2.4.15) is reduced to a system of differential equations. Thus, for nondegenerate kernels, (2.4.21) can be split with respect to the considered values:

$$\zeta_\sigma^v + \eta_x = 0, \qquad \zeta_t^v - \zeta_x^\sigma = 0, \, 2\xi_t + \zeta_v^\sigma = 0, \qquad \zeta_\sigma^e = 0,$$

$$\zeta_t^e - \zeta_x^v = 0, \qquad \eta_t + \zeta_v^v - \xi_x - \zeta_e^e = 0. \tag{2.4.24}$$

For the case $z = -\infty$ one also obtains (2.4.24).

Integrating (2.4.20), (2.4.24), one finds

$$\xi = t(c_1 x + c_2) + c_3 x^2 + c_5 x + c_6, \qquad \eta = x(c_3 t + c_4) + c_1 t^2 + c_7 t + c_8,$$

$$\zeta^v = -e(c_1 x + c_2) - \sigma(c_3 t + c_4) - v(2c_1 t + 2c_3 x + c_5 - c_9) + \lambda_{xt},$$

$$\zeta^\sigma = -\sigma(3c_1 t + c_3 x + c_7 - c_9) - 2v(c_1 x + c_2) + \lambda_{tt},$$

$$\zeta^e = -e(c_1 t + 3c_3 x + 2c_5 - c_7 - c_9) - 2v(c_3 t + c_4) + \lambda_{xx}. \tag{2.4.25}$$

Here $c_i$ $(i = 1, 2, \ldots, 9)$ are arbitrary constants, and $\lambda(x, t)$ is an arbitrary function of two arguments.

For studying the remaining determining equations (2.4.17) it is convenient to write

$$z_0 = \zeta^\sigma + 2v\xi_t = -\sigma(3c_1 t + c_3 x + c_7 - c_9) + \lambda_{tt},$$

$$z_1 = \zeta^e + 2v\eta_x = -e(c_1 t + 3c_3 x + 2c_5 - c_7 - c_9) + \lambda_{xx}. \tag{2.4.26}$$

Substituting (2.4.18) into (2.4.17) and evaluating some integrals by parts, one obtains

$$\varphi' z_1 - z_0 - \int_0^t K(t, \tau) z_0(\tau) d\tau + 2v_0(\xi_t - \varphi' \eta_x) + v_0(0) K(t, 0)(\xi(t) - \xi(0))$$

$$+ \int_0^t v_0(\tau) \left( (\xi(t) - \xi(\tau)) K_\tau(t, \tau) + \xi_t(\tau) K(t, \tau) \right) d\tau - K(t, 0) \eta(0) \sigma_0(0)$$

$$- \int_0^t \sigma_0(\tau) \left( K_\tau(t, \tau) \eta(\tau) + K_t(t, \tau) \eta(t) + K(t, \tau) \eta_t(\tau) \right) d\tau = 0. \tag{2.4.27}$$

Because of the arbitrariness of the function $v_0(t)$, from the last equation one finds

$$K(t, 0)(\xi(t) - \xi(0)) = 0, \tag{2.4.28}$$

$$\xi_t - \varphi' \eta_x = 0, \tag{2.4.29}$$

$$(\xi(t) - \xi(\tau)) K_\tau(t, \tau) + \xi_t(\tau) K(t, \tau) = 0, \tag{2.4.30}$$

---

[20]They are called degenerate kernels.

$$\varphi' z_1 - z_o - \int_0^t K(t, \tau) z_o(\tau) \, d\tau - K(t, 0) \eta(0) \sigma_o(0)$$

$$- \int_0^t \sigma_o(\tau) \big( K_\tau(t, \tau) \eta(\tau) + K_t(t, \tau) \eta(t)$$

$$+ K(t, \tau) \eta_t(\tau) \big) \, d\tau = 0. \tag{2.4.31}$$

Substituting (2.4.25) into (2.4.29) and splitting them with respect to $x$, one obtains

$$c_1 = 0, \quad c_3 = 0, \tag{2.4.32}$$

$$c_2 = \varphi' c_4. \tag{2.4.33}$$

Equations (2.4.28)–(2.4.31) become

$$c_4 K(t, 0) = 0, \tag{2.4.34}$$

$$c_4 \big( (t - \tau) K_\tau(t, \tau) + K(t, \tau) \big) = 0, \tag{2.4.35}$$

$$\varphi' (\lambda_{xx} + e_o(c_7 + c_9 - 2c_5)) + c_7 \sigma_o - \lambda_{tt}$$

$$- \int_0^t K(t, \tau) \lambda_{tt}(\tau) \, d\tau - c_9 \varphi(e_o) - c_8 K(t, 0) \sigma_o(0)$$

$$- \int_0^t \sigma_o(\tau)(c_4 x + c_7 \tau + c_8) K_\tau(t, \tau)$$

$$+ (c_4 x + c_7 \tau + c_8) K_t(t, \tau) \, d\tau = 0. \tag{2.4.36}$$

If there exist functions $\psi_i(\tau) = (t - \tau)^{n_i}$ $(i = 1, 2)$ such that the determinant

$$\Delta_1 = \left( \int_0^t z_3(t, \tau, x) \psi_1(\tau) \tau(t - \tau) \, d\tau \right) \left( \int_0^t K(t, \tau) \tau(t - \tau) \psi_2(\tau) \, d\tau \right)$$

$$- \left( \int_0^t z_3(t, \tau, x) \psi_2(\tau) \tau(t - \tau) \, d\tau \right) \left( \int_0^t K(t, \tau) \tau(t - \tau) \psi_1(\tau) \, d\tau \right)$$

is not equal to zero, then choosing the function $\sigma_o(\tau)$ one can obtain contradictory relations. Hence, $\Delta_1 = 0$ for all functions $\psi_i(\tau)$. Here $z_3(t, \tau, x) = (c_4 x + c_7 \tau + c_8) K_\tau + (c_4 x + c_7 t + c_8) K_t$. Because $K(t, \tau) \neq 0$ and the system of the functions $(t - \tau)^n$ is complete in $L_2[0, t]$, there exists a function $f_1(t, x)$ such that

$$z_3(t, \tau, x) = f_1(t, x) K(t, \tau). \tag{2.4.37}$$

Substituting (2.4.37) into (2.4.36), using (2.4.16), and splitting with respect to $\sigma_o(0)$, $\sigma_o(t)$ and $e_o(t)$, one obtains

$$c_7 + f_1 = 0, \tag{2.4.38}$$

$$c_8 K(t, 0) = 0, \tag{2.4.39}$$

$$\varphi'(\lambda_{xx} + e_o(c_7 + c_9 - 2c_5)) + \varphi(c_7 - c_9) - \lambda_{tt}$$

$$- \int_0^t K(t, \tau)\lambda_{tt}(\tau)\, d\tau = 0. \tag{2.4.40}$$

Splitting (2.4.37) with respect to $x$, and because of (2.4.38), one finds

$$c_4(K_\tau + K_t) = 0, \tag{2.4.41}$$

$$(c_7 t + c_8)K_t + (c_7 \tau + c_8)K_\tau = -c_7 K. \tag{2.4.42}$$

Regarding (2.4.34), (2.4.35) and (2.4.41), one obtains

$$c_4 = 0. \tag{2.4.43}$$

If $c_7^2 + c_8^2 \neq 0$, then from (2.4.39), (2.4.42), one finds that $c_8 = 0$ and $K = (c_7 t)^{-1} R(\tau/t)$. The kernels of this type are excluded from the study, because they have a singularity at the time $t = 0$. Hence,

$$c_7 = 0, \quad c_8 = 0, \tag{2.4.44}$$

and the group classification of (2.4.15), (2.4.16) is reduced to the study of (2.4.40).

From (2.4.40) it follows that the kernel of the admitted Lie groups is given by the generators

$$X_1 = \partial_x, \quad X_2 = \partial_v. \tag{2.4.45}$$

Extensions of the kernel (2.4.45) are obtained for specific functions $\varphi(e)$.

If $\varphi'' \neq 0$, then the classifying equations are

$$\varphi'(c_{10} + e(c_9 - 2c_5)) - c_9\varphi = c_{11}, \tag{2.4.46}$$

$$\lambda_{tt} + \int_0^t K(t, \tau)\lambda_{tt}(\tau)\, d\tau = c_{11}, \tag{2.4.47}$$

where $c_{10}, c_{11}$ are arbitrary constants. Hence, the extension of the kernel of admitted Lie groups occurs for the following cases:

(a) If $\varphi = \alpha + \beta \ln(a + ce)$, then the additional generator is

$$Y_1 = -cx/2\partial_x + cv/2\partial_v + (a + ce)\partial_e + \beta c\mu(t)\partial_\sigma.$$

(b) If $\varphi = \alpha(a + ce)^\beta + \gamma$ $(\beta \neq 1)$, then system of equations (2.4.15) admits the generator

$$Y_2 = (\beta - 1)cx\partial_x + (\beta + 1)cv\partial_v + 2(ce + a)\partial_e + 2\beta c(\sigma - \gamma\mu(t))\partial_\sigma.$$

(c) If $\varphi = \alpha + \exp(\gamma e)$ $(\gamma \neq 0)$, then there is the additional generator

$$Y_3 = \gamma x\partial_x + \gamma v\partial_v + 2\gamma(\sigma - \alpha\mu(t))\partial_\sigma + 2\partial_e.$$

(d) If the function $\varphi(e)$ is linear $\varphi = Ee + E_1$, then along with the generators $X_1, X_2$ system (2.4.15), (2.4.16) also admits the generators

$$Y_4 = v\partial_v + \sigma\partial_\sigma + e\partial_e, \quad Y_\lambda = \lambda_{xt}\partial_v + \lambda_{tt}\partial_\sigma + \lambda_{xx}\partial_e.$$

Here $\alpha, \beta, \gamma, a, c$ are constant, the function $\mu(t)$ is an arbitrary solution of the equation

$$\mu(t) + \int\limits_0^t K(t,\tau)\mu(\tau)\,d\tau = 1,$$

and the function $\lambda(x, t)$ is a solution of the equation

$$E\lambda_{xx} = \lambda_{tt} + \int\limits_0^t K(t,\tau)\lambda_{tt}(\tau)\,d\tau.$$

**Remark 2.4.1** This approach was also used for other models of elasticity in [57, 63].

# References

1. Abraham-Shrauner, B.: Exact, time-dependent solutions of the one-dimensional Vlasov–Maxwell equations. Phys. Fluids **27**(1), 197–202 (1984)
2. Balesku, R.: Statistical Mechanics of Charged Particle. Wiley, London (1963)
3. Bobylev, A.V.: The method of Fourier transform in the theory of the Boltzmann equation for Maxwell molecules. Dokl. Akad. Nauk SSSR **225**, 1041–1044 (1975)
4. Bobylev, A.V.: On exact solutions of the Boltzmann equation. Dokl. AS USSR **225**(6), 1296–1299 (1975)
5. Bobylev, A.V.: A class of invariant solutions of the Boltzmann equation. Sov. Phys. Dokl. **21**(11), 632–634 (1976)
6. Bobylev, A.V.: Exact solutions of the Boltzmann equation. Sov. Phys. Dokl. **21**(12), 822–824 (1976)
7. Bobylev, A.V.: On one class of invariant solutions of the Boltzmann equation. Dokl. AS USSR **231**(3), 571–574 (1976)
8. Bobylev, A.V.: On exact solutions of the nonlinear Boltzmann equation and its models. In: Struminsky V. (ed.) Molecular Gasdynamics, pp. 50–54. Nauka, Moscow (1982)
9. Bobylev, A.V.: Exact solutions of the nonlinear Boltzmann equation and relaxation theory of the Maxwellian gas. Theor. Math. Phys. **60**(2), 280–310 (1984)
10. Boltzmann, L.: Further studies on the thermal equilibrium among gas-molecules. Collected Works **1**, 275–370 (1872)
11. Buckwar, E., Luchko, Y.: Invariance of a partial differential equation of fractional order under the Lie group of scaling transformations. J. Math. Anal. Appl. **227**, 71–97 (1998)
12. Bunimovich, A.I., Krasnoslobodtsev, A.V.: Invariant-group solutions of kinetic equations. Meh. Zidkosti Gas. (4), 135–140 (1982)
13. Bunimovich, A.I., Krasnoslobodtsev, A.V.: On some invariant transformations of kinetic equations. Vestn. Moscow State Univ., Ser. 1., Mat. Meh. (4), 69–72 (1983)
14. Carleman, T.: Problemes Mathematiques Dans la Theorie Cinetique des Gas. Almqvist & Wiksell, Uppsala (1957)

15. Cercignani, C.: Exact solutions of the Boltzmann equation. In: Modern Group Analysis: Advanced Analytical and Computational Methods in Mathematical Physics, pp. 125–136. Kluwer Academic, Dordrecht (1993)
16. Chapman, S., Cowling, T.G.: The Mathematical Theory of Non-uniform Gases. Cambridge University Press, Cambridge (1952)
17. Chesnokov, A.A.: Characteristic properties and exact solutions of the kinetic equation of bubbly liquid. J. Appl. Mech. Tech. Phys. **44**(3), 336–343 (2003)
18. Chetverikov, V.N., Kudryavtsev, A.G.: A method for computing symmetries and conservation laws of integro-differential equations. Acta Appl. Math. **41**, 45–56 (1995)
19. Chetverikov, V.N., Kudryavtsev, A.G.: Modelling integro-differential equations and a method for computing their symmetries and conservation laws. Amer. Math. Soc. Transl. **167**, 1–22 (1995)
20. Cohen, A.: An Introduction to the Lie Theory of One-Parameter Groups, with Applications to the Solutions of Differential Equations. Health, New York (1911)
21. Cornille, H.: Oscillating Maxwellians. J. Phys. A, Math. Gen. **18**, L839–L844 (1985)
22. Dorodnitsyn, V.A.: Group Properties of Difference Equations. Fizmatlit, Moscow (2001) (in Russian)
23. Ernst, M.H.: Exact solution of the non-linear Boltzmann equation for Maxwell models. Phys. Lett. A **69**(6), 390–392 (1979)
24. Ernst, M.H.: Exact solutions of the nonlinear Boltzmann equation. J. Stat. Phys. **34**(5/6), 1001–1017 (1984)
25. Ferziger, J.H., Kaper, H.G.: Mathematical Theory of Transport Processes in Gases. North-Holland, Amsterdam (1972)
26. Gazizov, R.K., Kasatkin, A.A., Lukashchuk, A.Y.: Symmetries and group-invariant solutions of nonlinear fractional differential equations. Phys. Scr. **227** (2009). Topical articles: Fractional Differentiation and Its Applications (FDA 08) (Ankara, 5–7 November 2008)
27. Grad, H.: On the kinetic theory of rarefied gases. Commun. Pure Appl. Math. **II**, 331–407 (1949)
28. Grigoriev, Y.N., Meleshko, S.V.: Investigation of invariant solutions of the Boltzmann kinetic equation and its models. Preprint of Institute of Theoretical and Applied Mechanics (1986)
29. Grigoriev, Y.N., Meleshko, S.V.: Group analysis of the integro-differential Boltzmann equation. Dokl. AS USSR **297**(2), 323–327 (1987)
30. Grigoriev, Y.N., Meleshko, S.V.: Group analysis of kinetic equations. Russ. J. Numer. Anal. Math. Model. **9**(5), 425–447 (1995)
31. Grigoriev, Y.N., Meleshko, S.V., Sattayatham, P.: Classification of invariant solutions of the Boltzmann equation. J. Phys. A, Math. Gen. **32**, 337–343 (1999)
32. Grigoryev, Y.N.: A class of exact solutions of one nonlinear kinetic equation. Dyn. Contin. (26), 30–43 (1976) (in Russian)
33. Grigoryev, Y.N., Mikhalitsyn, A.N.: A spectral method of solving Boltzmann's kinetic equation numerically. U.S.S.R. Comput. Math. Math. Phys. **23**(6), 105–111 (1985)
34. Ibragimov, N.H.: Transformation Groups Applied to Mathematical Physics. Nauka, Moscow (1983). English translation: Reidel, Dordrecht (1985)
35. Ibragimov, N.H. (ed.): CRC Handbook of Lie Group Analysis of Differential Equations, vols. 1, 2, 3. CRC Press, Boca Raton (1994, 1995, 1996)
36. Ibragimov, N.H.: Elementary Lie Group Analysis and Ordinary Differential Equations. Wiley, Chichester (1999)
37. Ibragimov, N.H., Kovalev, V.F., Pustovalov, V.V.: Symmetries of integro-differential equations: a survey of methods illustrated by the Benney equations. Nonlinear Dyn. **28**(2), 135–153 (2002)
38. Kac, M.: Foundation of kinetic theory. In: Proc. 3rd Berkeley Sympos. Math. Stat. and Prob., vol. 8, pp. 171–197 (1956)
39. Korn, G.A., Korn, T.M.: Mathematical Handbook. McGraw–Hill, New York (1968)
40. Kovalev, V.F., Krivenko, S.V., Pustovalov, V.V.: Group symmetry of the kinetic equations of the collisionless plasma. JETP Lett. **55**(4), 256–259 (1992)

41. Krasnoslobodtsev, A.V.: Gasdynamic and kinetic analogies in the theory of vertical nonuniform shallow water. In: Proceedings of Institute of General Physics of USSR Ac Sci
42. Krook, M., Wu, T.T.: Formation of Maxwellian tails. Phys. Rev. Lett. **36**(19), 1107–1109 (1976)
43. Krook, M., Wu, T.T.: Exact solutions of the Boltzmann equation. J. Phys. Fluids **20**(10), 1589–1595 (1977)
44. Kupershmidt, B., Manin, Y.: Long wave equation with free boundary. i Conservation laws and solutions. Funct. Anal. Appl. **11**(3), 188–197 (1977)
45. Landau, L., Teller, E.: Theory of sound dispersion. Phys. Z. Sowjetunion **34**(10), 34–43 (1936)
46. Lebowitz, J., Montroll, E.: Nonequilibrium Phenomena I. The Boltzmann Equation. North-Holland, Amsterdam (1983)
47. Lie, S., Scheffers, G.: Lectures on Differential Equations with Known Infinitesimal Transformations. Teubner, Leipzig (1891)
48. Maxwell, J.C.: On the dynamical theory of gases. Philos. Trans. R. Soc. Lond. **157**, 49–88 (1867)
49. Meleshko, S.V.: Methods for Constructing Exact Solutions of Partial Differential Equations. Springer, New York (2005)
50. Morgenstern, D.: Analytical studies related to the Maxwell–Boltzmann equation. J. Ration. Mech. Anal. **4**(4), 533–555 (1955)
51. Nikolskii, A.A.: The simplest exact solutions of the Boltzmann equation of a rarefied gas motion. Dokl. AS USSR **151**(2), 299–301 (1963)
52. Nikolskii, A.A.: Three dimensional homogeneous expansion–contraction of a rarefied gas with power interaction functions. Dokl. AS USSR **151**(3), 522–524 (1963)
53. Nikolskii, A.A.: Homogeneous motion of displacement of monatomic rarefied gas. Inz. J. **5**(4), 752–755 (1965)
54. Nonenmacher, T.F.: Application of the similarity method to the nonlinear Boltzmann equation. J. Appl. Math. Phys. (ZAMP) **35**(9), 680–691 (1984)
55. Olver, P.J.: Applications of Lie Groups to Differential Equations. Springer, New York (1986)
56. Ovsiannikov, L.V.: Group Analysis of Differential Equations. Nauka, Moscow (1978). English translation by Ames, W.F. (ed.), published by Academic Press, New York (1982)
57. Özer T.: Symmetry group classification of two-dimensional elastodynamics problem in nonlocal elasticity. Int. J. Eng. Sci. **41**(18), 2193–2211 (2003)
58. Polya, G.: Mathematics and Plausible Reasoning. Princeton University Press, Princeton (1954)
59. Polyachenko, V.L., Fridman, A.M.: Equilibrium and Stability of Gravitating Systems. Nauka, Moscow (1976)
60. Rabotnov, Y.N.: Elements of Hereditary Solid Mechanics. Nauka, Moscow (1977)
61. Sedov, L.I.: Similarity and Dimensional Methods in Mechanics. Academic Press, New York (1959)
62. Seiler, W.M.: Involution and symmetry reductions. Math. Comput. Model. **25**(8/9), 63–73 (1997)
63. Senashov, S.I.: Group classification of viscoelastic bar equation. Model. Meh. **4**(21)(1), 69–72 (1990)
64. Taranov, V.B.: On symmetry of one-dimensional high frequency motions of collisionless plasma. J. Tech. Phys. **46**(6), 1271–1277 (1976)
65. Taranov, V.B.: Symmetry extensions and their physical reasons in the kinetic and hydrodynamic plasma models. SIGMA **4**, 006 (2008) (7 pages)
66. Tenti, G., Hui, W.H.: Some classes of exact solutions of the nonlinear Boltzmann equation. J. Math. Phys. **19**(4), 774–779 (1978)
67. Teshukov, V.M.: Hyperbolicity of the long-wave equations. Dokl. Akad. Nauk SSSR **284**(3), 555–562 (1985)
68. Teshukov, V.M.: Simple waves on a shear free-boundary flow of an ideal incompressible fluid. J. Appl. Mech. Tech. Phys. **38**(2), 211–218 (1997)

69. Teshukov, V.M.: Spatial simple waves on a shear flow. J. Appl. Mech. Tech. Phys. **43**(5), 661–670 (2002)
70. Teshukov, V.M.: Spatial stationary long waves in shear flows. J. Appl. Mech. Tech. Phys. **45**(2), 172–180 (2004)
71. Voloshchuk, V.M.: Kinetic Coagulation Theory. Gidrometeoizdat, Leningrad (1984)
72. Whittaker, E.T., Watson, J.N.: A Course of Modern Analysis. I. Cambridge University Press, Cambridge (1927)

# Chapter 3
# The Boltzmann Kinetic Equation and Various Models

This chapter deals with applications of the group analysis method to the full Boltzmann kinetic equation and some similar equations. Calculations of the 11-parameter Lie group $G^{11}$ admitted by the full Boltzmann equation with arbitrary intermolecular potential and its extensions for power potentials are presented. The found isomorphism of these Lie groups with the Lie groups admitted by the ideal gas dynamics Euler equations allowed one to obtain an optimal system of admitted subalgebras and to classify all invariant solutions of the full Boltzmann equation. For equations similar to the full Boltzmann equation complete admitted Lie groups are derived by solving determining equations. The corresponding optimal systems of admitted subalgebras are directly calculated and representations of all invariant solutions are obtained in explicit forms.

## 3.1 Studies of Invariant Solutions of the Boltzmann Equation

The Boltzmann kinetic equation is a basis of the classical kinetic theory of rarefied gases and had served as the standard in developing of other statistical kinetic theories. Starting with its appearance in the papers of D. Maxwell [30] and L. Boltzmann [8], the mathematical theory of this equation was studied by many researchers. A particular interest in the study of the Boltzmann equation was always related with searching for exact (invariant) solutions directly associated with the fundamental properties of the equation. After the studies of the class of the local Maxwellians [8, 11, 30] new classes of invariant solutions were constructed in 1960s in [32–34]. The decade later the BKW-solution was almost simultaneously derived in [1] and in [28]. Contrary to the Maxwellians, the Boltzmann collision integral is not zero for this solution. The discovery of the BKW-solution stimulated a great splash of studies of invariant solutions of different kinetic equations. However, the progress of that time was really restricted to obtaining BKW-type solutions for different simplified models of the Boltzmann equation.[1]

---

[1] See the review [15].

Y.N. Grigoriev et al., *Symmetries of Integro-Differential Equations*,
Lecture Notes in Physics 806,
DOI 10.1007/978-90-481-3797-8_3, © Springer Science+Business Media B.V. 2010

The positive consequence of these years consisted in the comprehension that further progress of the studies is only possible with successive use of the group analysis method. The first steps in this direction were done in [9, 10]. Applying the moment method the admitted Lie group $G^{11}$ (see the next section) was originally calculated in these papers for the model Bhatnagar–Gross–Krook kinetic equation with a simple relaxational collision integral. Then it was directly verified that the found infinitesimals are also admitted by the full Boltzmann equation. Using the "scaling conjecture" method, the authors [17] derived the admitted Lie group $G^{11}$ immediately for the full Boltzmann equation. Extensions of $G^{11}$ for special intermolecular potentials were also obtained. As a result an approach for applying the group analysis method to integro-differential equations was proposed [17, 18]. In particular, the methods of constructing and solving determining equations for an admitted Lie group were worked out (see Chap. 2). Using the proposed method, the admitted Lie groups for some kinetic equations with a small number of independent variables were obtained in [19, 20, 22]. The uniqueness of the general solution of determining equations allows one to consider these results as constructive proofs of the completeness of the obtained Lie groups. In papers [4, 7] a Lie algebra $L_{11}(X)$ was announced as a complete Lie group admitted by the full Boltzmann equation. But calculations were practically carried out by ad hoc approach, which was equivalent to "scaling conjecture". Only in the recent publication [5] a proof of $L^{11}$ completeness was really derived by solving the corresponding determining equations. A group classification of solutions of the full Boltzmann equation invariant with respect to the Lie group $G^{11}$ and its extensions was carried out in [23]. This classification allows one to separate the set of invariant solutions into non-intersecting essentially different classes. The representations of the invariant solutions for different classes and some reductions of the full Boltzmann equation to factor-equations were obtained.

## 3.2  Introduction to the Boltzmann Equation

The Boltzmann kinetic equation allows one to consider rarefied gas flows on a molecular level. These flows are realized in the wide range of scales from astrophysical up to microscopical: the so-called "jets" and turbulent "piles" in deep space, flows around satellites and spaceships during their landing in atmospheres of planets, flows in vacuum chemical reactors, as well as flows of aerosols of micron scales in ecology problems, flows in micro electrical machine systems, scattering of ultrasound waves to name only a few.

The Boltzmann equation is an integro-differential equation that describes the evolution of rarefied gas in terms of a molecular distribution function. The distribution function in general case

$$f = f(\mathbf{x}, \mathbf{v}, t), \qquad f : \mathbb{R}^3 \times \mathbb{R}^3 \times \mathbb{R}^+ \longrightarrow \mathbb{R}^+, \qquad (3.2.1)$$

depends on seven independent variables and gives the probability distribution of molecules in the phase space, more precisely

$$f(\mathbf{x}, \mathbf{v}, t)\, d\mathbf{x}\, d\mathbf{v} \qquad\qquad (3.2.2)$$

gives the expected number of molecules in the element volume $d\mathbf{x}\,d\mathbf{v}$ centered at the phase space point $(\mathbf{x}, \mathbf{v})$, at time $t$.

Without external forces, the Boltzmann equation for a monatomic gas can be written as

$$\frac{\partial f}{\partial t} + \mathbf{v} \cdot \nabla_{\mathbf{x}} f = J(f, f), \quad \mathbf{x}, \mathbf{v} \in \mathbb{R}^3. \tag{3.2.3}$$

The term $J(f, f)$ is the so-called collision operator which is defined by the formula

$$J(f, f)(\mathbf{x}, \mathbf{v}, t) = \int\limits_{\mathbb{R}^3} \int\limits_{S^2} B(|\mathbf{v} - \mathbf{v_1}|, \theta)[f(\mathbf{x}, \mathbf{v}', t) f(\mathbf{x}, \mathbf{v_1}', t)$$
$$- f(\mathbf{x}, \mathbf{v}, t) f(\mathbf{x}, \mathbf{v_1}, t)]\, d\mathbf{n}\, d\mathbf{v_1},$$

where $\mathbf{v}'$ and $\mathbf{v_1}'$ are the velocities after a collision of two particles that had velocities $\mathbf{v}$ and $\mathbf{v_1}$ before the collision. The deflection angle $\theta$ is the angle between $\mathbf{v} - \mathbf{v_1}$ and $\mathbf{v}' - \mathbf{v_1}'$. Velocities in the collision satisfy the microscopic momentum and energy conservation laws,

$$\mathbf{v}' + \mathbf{v_1}' = \mathbf{v} + \mathbf{v_1}, \qquad |\mathbf{v}'|^2 + |\mathbf{v_1}'|^2 = |\mathbf{v}|^2 + |\mathbf{v_1}|^2. \tag{3.2.4}$$

The postcollision velocities can be obtained by solving algebraic equations (3.2.4) and are parameterized by

$$\mathbf{v}' = \frac{1}{2}(\mathbf{v} + \mathbf{v_1} + |\mathbf{v} - \mathbf{v_1}|\mathbf{n}), \quad \mathbf{v_1}' = \frac{1}{2}(\mathbf{v} + \mathbf{v_1} - |\mathbf{v} - \mathbf{v_1}|\mathbf{n}),$$

where $\mathbf{n}$ is a unit vector varying on the sphere

$$S^2 = \{\mathbf{n} \in \mathbb{R}^3, \ |\mathbf{n}| = 1\}.$$

The scattering function $B$ has the form

$$B(|\mathbf{v} - \mathbf{v_1}|, \theta) = |\mathbf{v} - \mathbf{v_1}|\sigma(|\mathbf{v} - \mathbf{v_1}|, \cos(\theta)) \tag{3.2.5}$$

where $\cos(\theta) = \frac{(\mathbf{v} - \mathbf{v_1}, \mathbf{n})}{|\mathbf{v} - \mathbf{v_1}|}$. The function $\sigma : \mathbb{R}^+ \times [-1, 1] \to \mathbb{R}^+$ is the differential cross-section and $\theta$ is the scattering angle. The scattering function $B$ characterizes the details of the binary interactions, and depends on the physical properties of gas. Some types of collision scattering function frequently considered in the Boltzmann equation are the following.

For inverse power intermolecular potentials $U(r) \sim r^{-(\nu-1)}$ ($\nu > 2$) the scattering function acquires the form

$$B(|\mathbf{v} - \mathbf{v_1}|, \theta) = b_\gamma(\cos(\theta))|\mathbf{v} - \mathbf{v_1}|^\gamma, \quad \gamma = \frac{\nu - 5}{\nu - 1} \quad (\nu > 2), \tag{3.2.6}$$

where $b_\gamma(\cos(\theta))$ is a known function.

The special case $\gamma = 0$ corresponds to the Maxwell molecules with

$$B(|\mathbf{v} - \mathbf{v_1}|, \theta) = b_0(\cos(\theta)). \tag{3.2.7}$$

The collision scattering function $B(|\mathbf{v} - \mathbf{v_1}|, \theta)$ here does not depend on the relative velocity $|\mathbf{v} - \mathbf{v_1}|$.

The other particular case is a hard spheres model for $\gamma = 1$

$$B(|\mathbf{v} - \mathbf{v_1}|, \theta) = \frac{d^2}{4}|\mathbf{v} - \mathbf{v_1}|, \qquad (3.2.8)$$

where $d$ denotes the diameter of the particles.

The Boltzmann collision operator has the following fundamental properties of the conservation of mass, momentum, and energy:

$$\int_{\mathbb{R}^3} Q(f, f)d\mathbf{v} = 0, \qquad (3.2.9)$$

$$\int_{\mathbb{R}^3} \mathbf{v}Q(f, f)d\mathbf{v} = 0, \qquad (3.2.10)$$

$$\int_{\mathbb{R}^3} |\mathbf{v}|^2 Q(f, f)d\mathbf{v} = 0, \qquad (3.2.11)$$

which are directly related with its symmetries.

## 3.3 Group Analysis of the Full Boltzmann Equation

In this section using the "scaling conjecture" the calculations of the Lie algebras admitted by full Boltzmann equation and its Fourier transform are presented. The isomorphisms of Lie groups (algebras) admitted by the full Boltzmann equation and Euler gas dynamics (EGD)-system is set up. The proven isomorphism allows one to apply the optimal systems of subalgebras already obtained for the EGD-system [14, 16, 36] for classifications of invariant solutions of the full Boltzmann equation. The representations of essentially different H-solutions of the spatially inhomogeneous Boltzmann equation with one and two independent invariant variables are derived in explicit form.

### 3.3.1 Admitted Lie Algebras

The full Boltzmann equation (3.2.3) is rewritten as follows

$$\frac{\partial f}{\partial t} + \mathbf{v}\frac{\partial f}{\partial \mathbf{x}} = J(f, f) \qquad (3.3.1)$$

where the collision integral is

$$J(f, f) = \int d\mathbf{w}\,d\mathbf{n}B\left(g, \frac{\mathbf{gn}}{g}\right)[f(\mathbf{v}^*)f(\mathbf{w}^*) - f(\mathbf{v})f(\mathbf{w})],$$

$$\qquad (3.3.2)$$

$$\mathbf{v}^* = \frac{1}{2}(\mathbf{v} + \mathbf{w} + g\mathbf{n}), \quad \mathbf{w}^* = \frac{1}{2}(\mathbf{v} + \mathbf{w} - g\mathbf{n}), \quad \mathbf{g} = \mathbf{v} - \mathbf{w}, \quad g = |\mathbf{g}|, \quad |\mathbf{n}| = 1.$$

Let us assume that the admitted Lie group $G(T_a)$ of point transformations $T_a$ of the Boltzmann equation (3.3.1) has the form[2]

$$\bar{f} = \lambda(\bar{t}, \bar{\mathbf{x}}; a)f, \quad t = \tau(\bar{t}, \bar{\mathbf{x}}; a), \quad \mathbf{x} = \mathbf{h}(\bar{t}, \bar{\mathbf{x}}; a), \quad \mathbf{v} = A(\bar{t}, \bar{\mathbf{x}}; a)\bar{\mathbf{v}} + \mathbf{b}(\bar{t}, \bar{\mathbf{x}}; a),$$

$$(3.3.3)$$

where $a$ is the group parameter, $A(\bar{t}, \bar{\mathbf{x}}; a)$ is a $3 \times 3$ matrix such that for any vector $\mathbf{v}$ and any unit vector $\mathbf{n}$ the following property is satisfied

$$|A\mathbf{v}|^{-1}(A\mathbf{v}, \mathbf{n}) = |\mathbf{v}|^{-1}(\mathbf{v}, \mathbf{n}). \tag{3.3.4}$$

This property is satisfied if and only if the matrix $A$ has the form

$$A = qG,$$

where $G(\bar{t}, \bar{\mathbf{x}}; a)$ is an orthogonal matrix and $q(\bar{t}, \bar{\mathbf{x}}; a) > 0$ is a scalar function.

The functions (3.3.3) have necessarily to satisfy the main group superposition property

$$T_b T_a = T_{a+b}, \tag{3.3.5}$$

and the identity property for the group parameter $a = 0$:

$$\lambda(\bar{t}, \bar{\mathbf{x}}; 0) = 1, \quad \tau(\bar{t}, \bar{\mathbf{x}}; 0) = \bar{t}, \quad \mathbf{h}(\bar{t}, \bar{\mathbf{x}}; 0) = \bar{\mathbf{x}},$$

$$A(\bar{t}, \bar{\mathbf{x}}; 0) = I, \quad \mathbf{b}(\bar{t}, \bar{\mathbf{x}}; 0) = 0. \tag{3.3.6}$$

Transformations (3.3.3) map a function $f(t, \mathbf{x}, \mathbf{v})$ into the function

$$\bar{f}(\bar{t}, \bar{\mathbf{x}}, \bar{\mathbf{v}}; a) = \lambda(\bar{t}, \bar{\mathbf{x}}; a)f(\tau(\bar{t}, \bar{\mathbf{x}}; a), \mathbf{h}(\bar{t}, \bar{\mathbf{x}}; a), q(\bar{t}, \bar{\mathbf{x}}; a)G(\bar{t}, \bar{\mathbf{x}}; a)\bar{\mathbf{v}}$$

$$+ \mathbf{b}(\bar{t}, \bar{\mathbf{x}}; a)). \tag{3.3.7}$$

Using (3.3.7), one can check that the nonlinear collision integral has the scaling property:

$$J(\bar{f}, \bar{f}) = \frac{\lambda^2}{q^{\gamma+3}} J(f, f). \tag{3.3.8}$$

Substituting the derivatives of the function $\bar{f}(\bar{t}, \bar{\mathbf{x}}, \bar{\mathbf{v}}; a)$:

$$\frac{\partial \bar{f}}{\partial \bar{t}} = f\frac{\partial \lambda}{\partial \bar{t}} + \lambda\left(f_t\frac{\partial \tau}{\partial \bar{t}} + f_{x_i}\frac{\partial h_i}{\partial \bar{t}} + f_{v_i}\left(\frac{\partial q}{\partial \bar{t}}G_{ij}\bar{v}_j + q\frac{\partial G_{ij}}{\partial \bar{t}}\bar{v}_j + \frac{\partial b_i}{\partial \bar{t}}\right)\right),$$

$$\frac{\partial \bar{f}}{\partial \bar{x}_k} = f\frac{\partial \lambda}{\partial \bar{x}_k} + \lambda\left(f_t\frac{\partial \tau}{\partial \bar{x}_k} + f_{x_i}\frac{\partial h_i}{\partial \bar{x}_k} + f_{v_i}\left(\frac{\partial q}{\partial \bar{x}_k}G_{ij}\bar{v}_j + q\frac{\partial G_{ij}}{\partial \bar{x}_k}\bar{v}_j + \frac{\partial b_i}{\partial \bar{x}_k}\right)\right),$$

into the equation

$$\frac{\partial \bar{f}}{\partial \bar{t}} + \bar{v}_k\frac{\partial \bar{f}}{\partial \bar{x}_k} = J(\bar{f}, \bar{f}),$$

---

[2]The text follows [17], see also Chap. 2.

and using the property that

$$J(f, f) = \frac{\partial f}{\partial t} + v_k \frac{\partial f}{\partial x_k},$$

one obtains

$$f \left( \frac{\partial \lambda}{\partial \bar{t}} + \bar{v}_k \frac{\partial \lambda}{\partial \bar{x}_k} \right) + \lambda f_t \left( \frac{\partial \tau}{\partial \bar{t}} + \bar{v}_k \frac{\partial \tau}{\partial \bar{x}_k} - \frac{\lambda}{q^{\gamma+3}} \right)$$

$$+ \lambda f_{x_i} \left( \frac{\partial h_i}{\partial \bar{t}} + \bar{v}_k \frac{\partial h_i}{\partial \bar{x}_k} - \frac{\lambda}{q^{\gamma+3}} (q G_{ij} \bar{v}_j + b_i) \right)$$

$$+ \lambda f_{v_i} \left( \frac{\partial b_i}{\partial \bar{t}} + \bar{v}_k \frac{\partial b_i}{\partial \bar{x}_k} + \left( \frac{\partial q}{\partial \bar{t}} + \bar{v}_k \frac{\partial q}{\partial \bar{x}_k} \right) G_{ij} \bar{v}_j \right.$$

$$\left. + q \left( \frac{\partial G_{ij}}{\partial \bar{t}} + \bar{v}_k \frac{\partial G_{ij}}{\partial \bar{x}_k} \right) \bar{v}_j \right) = 0.$$

By virtue of the arbitrariness of the function $f(t, \mathbf{x}, \mathbf{v})$ the last equation[3] can be split with respect to its derivatives:

$$\frac{\partial \lambda}{\partial \bar{t}} + \bar{v}_j \frac{\partial \lambda}{\partial \bar{x}_j} = 0, \quad \frac{\partial \tau}{\partial \bar{t}} + \bar{v}_j \frac{\partial \tau}{\partial \bar{x}_j} = \frac{\lambda}{q^{\gamma+3}},$$

$$\frac{\partial h_i}{\partial \bar{t}} + \bar{v}_j \frac{\partial h_i}{\partial \bar{x}_j} = \frac{\lambda}{q^{\gamma+3}} (q G_{ij} \bar{v}_j + b_i), \tag{3.3.9}$$

$$\frac{\partial b_i}{\partial \bar{t}} + \bar{v}_j \frac{\partial b_i}{\partial \bar{x}_j} + \left( \frac{\partial q}{\partial \bar{t}} + \bar{v}_k \frac{\partial q}{\partial \bar{x}_k} \right) G_{ij} \bar{v}_j + q \left( \frac{\partial B_{ij}}{\partial \bar{t}} + \bar{v}_k \frac{\partial G_{ij}}{\partial \bar{x}_k} \right) \bar{v}_j = 0$$

$$(i = 1, 2, 3).$$

Since the functions $\lambda$, $\tau$, $h_i$, $q$, $G_{ij}$ and $b_i$ do not depend on $\bar{\mathbf{v}}$, the equations (3.3.9) can be once again split with respect to $\bar{\mathbf{v}}$

$$\frac{\partial \lambda}{\partial \bar{t}} = 0, \quad \frac{\partial \lambda}{\partial \bar{x}_k} = 0, \quad \frac{\partial b_i}{\partial \bar{t}} = 0,$$

$$\frac{\partial b_i}{\partial \bar{x}_j} + \frac{\partial (q G_{ij})}{\partial \bar{t}} = 0, \quad \frac{\partial (q G_{ij})}{\partial \bar{x}_k} = 0, \tag{3.3.10}$$

$$\frac{\partial \tau}{\partial \bar{t}} = \frac{\lambda}{q^{\gamma+3}}, \quad \frac{\partial \tau}{\partial \bar{x}_j} = 0, \tag{3.3.11}$$

$$\frac{\partial h_i}{\partial \bar{t}} = \frac{\lambda}{q^{\gamma+3}} b_i, \quad \frac{\partial h_i}{\partial \bar{x}_j} = \frac{\lambda}{q^{\gamma+2}} G_{ij} \tag{3.3.12}$$

$$(i, j, k = 1, 2, 3).$$

The general solution of (3.3.10) is

$$\lambda = \lambda(a), \quad b_i = Q_{ij}(a)\bar{x}_j + p_i(a), \quad q G_{ij} = -\bar{t} Q_{ij}(a) + q_{ij}(a) \quad (i = 1, 2, 3).$$

Since $G$ is an orthogonal matrix, one obtains

$$q^2 \delta_{ik} = Q_{ij} Q_{kj} \bar{t}^2 - \bar{t}(q_{ij} Q_{kj} + Q_{ij} q_{kj}) + q_{ij} q_{kj} \quad (i, k = 1, 2, 3).$$

---

[3]The theory of existence and uniqueness of a local solution of the Boltzmann equation can be found in [12, 13, 27].

This also gives that the function $q$ has the form

$$q^2 = f_2 \bar{t}^2 + f_1 \bar{t} + f_0, \qquad (3.3.13)$$

where

$$f_2 = Q_{i_o j} Q_{i_o j}, \quad f_1 = -(q_{i_o j} Q_{i_o j} + Q_{i_o j} q_{i_o j}),$$
$$f_0 = q_{i_o j} q_{i_o j} \quad (i_o = 1, 2, 3). \qquad (3.3.14)$$

Here is no summation with respect to $i_o$.

Equations (3.3.12) become

$$\frac{\partial h_i}{\partial \bar{t}} = \frac{\lambda}{q^{\gamma+3}} (Q_{ij} \bar{x}_j + p_i),$$
$$\frac{\partial h_i}{\partial \bar{x}_k} = \frac{\lambda}{q^{\gamma+3}} (-\bar{t} Q_{ik} + q_{ik}) \quad (i, k = 1, 2, 3). \qquad (3.3.15)$$

Equating the mixed derivatives $\frac{\partial}{\partial \bar{x}_k}(\frac{\partial h_i}{\partial \bar{t}}) = \frac{\partial}{\partial \bar{t}}(\frac{\partial h_i}{\partial \bar{x}_k})$, one finds

$$2q\, Q_{ik} = (\gamma + 3) \frac{\partial q}{\partial \bar{t}} (\bar{t} Q_{ik} - q_{ik}) \quad (i, k = 1, 2, 3). \qquad (3.3.16)$$

Multiplying these equations by $q$, using (3.3.13), and splitting with respect to $\bar{t}$, one obtains

$$f_2(\gamma + 1) Q_{ik} = 0 \quad (i, k = 1, 2, 3). \qquad (3.3.17)$$

If $f_2 = 0$, then by virtue of the definition (3.3.14) one gets that $Q_{ik} = 0$ ($i, k = 1, 2, 3$). Hence (3.3.17) separates the study in two cases: (a) $Q_{ij} = 0, \forall i, j$; (b) there exists $i$ and $j$ such that $Q_{ij} \neq 0$.

Assume that $Q_{ik} = 0, \forall i, j$. Integrating (3.3.15), the transformations (3.3.3) become

$$\bar{f} = \lambda f, \quad \bar{t} = \frac{\lambda}{q^{\gamma+3}} \bar{t} + \tau_o,$$
$$x_i = \frac{\lambda}{q^{\gamma+3}} (\bar{t} p_i + \bar{x}_j q_{ij} + s_i), \quad v_i = q_{ij} \bar{v}_j + p_i. \qquad (3.3.18)$$

Here $\lambda(a)$, $\tau_o(a)$, $p_i(a)$, $q(a)$, $s_i(a)$ and $q_{ij}(a)$ are found from the property to compose a Lie group: $T_0$ is the identical transformation and the conditions (3.3.5) hold. The Lie group corresponding to the transformations (3.3.18) is defined by the infinitesimal generators:

$$X_1 = \partial_x, \ X_2 = \partial_y, \ X_3 = \partial_z, \ X_4 = t\partial_x + \partial_u, \ X_5 = t\partial_y + \partial_v, \ X_6 = t\partial_z + \partial_w,$$
$$X_7 = y\partial_z - z\partial_y + v\partial_w - w\partial_v, \quad X_8 = z\partial_x - x\partial_z + w\partial_u - u\partial_w,$$
$$X_9 = x\partial_y - y\partial_x + u\partial_v - v\partial_u, \quad X_{10} = \partial_t, \quad X_{11} = t\partial_t + x\partial_x + y\partial_y + z\partial_z - f\partial_f,$$
$$X_{12} = t\partial_t - u\partial_u - v\partial_v - w\partial_w + (\gamma + 2)f\partial_f.$$

Here we used the individual notations for the space variables and velocities:

$$x = x_1, \quad y = x_2, \quad z = x_3, \quad u = v_1, \quad v = v_2, \quad w = v_3.$$

The generators $X_1, X_2, X_3, X_{10}$ correspond to shifts with respect to the space variables and time; the generators $X_4, X_5, X_6$ correspond to the Galilean transformations; the generators $X_7, X_8, X_9$ correspond to rotations; the generator $X_{11}$ corresponds to a scaling transformation. One can check that the Lie group of transformations $G^{11}$ corresponding to the Lie algebra $L_{11}(X)$ with the basis $\{X_1, X_2, X_3, \ldots, X_{11}\}$ is also admitted by the Boltzmann equation with an arbitrary scattering function $B$.

Assuming that $Q_{i_o j_o} \neq 0$ for some $i_o$ and $j_o$, one has $\gamma = -1$. Solving (3.3.16) with $i = i_o$ and $j = j_o$, one finds $q = \alpha(a)\bar{t} + \beta(a)$, where $\alpha = Q_{i_o j_o}$. Equations (3.3.16) give $q_{ij} = -Q_{ij}\beta/\alpha$. Integrating (3.3.15), one obtains

$$\bar{f} = \lambda f, \ t = -\frac{\lambda}{\alpha(\alpha\bar{t} + \beta)} + \tau_o, \ x_i = -\frac{\lambda}{(\alpha\bar{t} + \beta)}(\tilde{Q}_{ij}\bar{x}_j + \tilde{p}_i) + s_i,$$

$$v_i = (\alpha\bar{t} + \beta)\tilde{Q}_{ij}\bar{v}_j + \alpha(\tilde{Q}_{ij}\bar{x}_j + \tilde{p}_i), \tag{3.3.19}$$

where $Q_{ij} = \alpha\tilde{Q}_{ij}$, $p_i = \alpha\tilde{p}_i$ $(i, j = 1, 2, 3)$, and the matrix $\tilde{Q}$ is an orthogonal matrix. The Lie group corresponding to transformations (3.3.19) is defined by the Lie algebra $L_{13}(X)$ with the basis of generators $X_1, X_2, \ldots, X_{12}$ and

$$X_{13} = t^2 \partial_t + t\mathbf{x}\partial_{\mathbf{x}} + (\mathbf{x} - t\mathbf{v})\partial_{\mathbf{v}}.$$

The generator $X_{13}$ determines projective transformations.

The Lie algebra $L_{11}(X)$ with its extensions given by the generators $X_{12}, X_{13}$ defines a particular *group classification* of the Boltzmann equation (3.3.1) with respect to specifications of the scattering function $B$ as *a parameter element* (in terminology of [35]).

**Remark 3.3.1** Since transformations (3.3.18) and (3.3.19) map a solution of the Boltzmann equation into a solution, they attract attention of scientists independently of forming a Lie group. Such type of transformations were also considered in [10].

**Remark 3.3.2** In [9] the generator $X_{12}$ was presented with reference to A.A. Nikol'skii (as a private communication). The relation between $X_{13}$ and point transformations of the Boltzmann equation for $\gamma = -1$ found by Nikol'skii [33] was pointed out in [10].

Similar approach can be applied to the Fourier representation of the full Boltzmann equation with a power intermolecular potential. This equation is written as follows [25]

$$\frac{\partial \varphi}{\partial t} + i \frac{\partial^2 \varphi}{\partial x_j \partial w_j} = C(\gamma) \int d\mathbf{w}_1 \, d\mathbf{n} \Phi(\mathbf{n})\varphi(\mathbf{w}/2 + \mathbf{w}_1)\varphi(\mathbf{w}/2 - \mathbf{w}_1)$$

$$\times \left( |\mathbf{w}_1 - |\mathbf{w}|\mathbf{n}/2|^{-(\gamma+3)} - |\mathbf{w}_1 - \mathbf{w}/2|^{-(\gamma+3)} \right). \tag{3.3.20}$$

Here

$$\varphi(t, \mathbf{x}, \mathbf{w}) = \int d\mathbf{v} e^{-i\mathbf{w}\mathbf{v}} f(t, \mathbf{x}, \mathbf{v})$$

is the Fourier transform of the distribution function $f(t, \mathbf{x}, \mathbf{v})$,

$$C(\gamma) = \frac{2^\gamma \pi^{-3/2} \Gamma((\gamma + 3)/2)}{\Gamma(-\gamma/2)},$$

and $\Gamma(z)$ is the Euler gamma-function. The result of calculations is the admitted Lie group $G^9$ corresponding to the Lie algebra $L_9(Z)$ with the basis generators

$$Z_1 = \partial_x, \; Z_2 = \partial_y, \; Z_3 = \partial_z,$$

$$Z_4 = y\partial_z - z\partial_y + v\partial_w - w\partial_v, \; Z_5 = z\partial_x - x\partial_z + w\partial_u - u\partial_w,$$

$$Z_6 = x\partial_y - y\partial_x + u\partial_v - v\partial_u, \; Z_7 = \partial_t, \; Z_8 = t\partial_t + x\partial_x + y\partial_y + z\partial_z - \varphi\partial_\varphi,$$

$$Z_9 = t\partial_t + u\partial_u + v\partial_v + w\partial_w + (\gamma - 1)\varphi\partial_\varphi.$$

**Remark 3.3.3** For a homogeneous case there are the additional generators

$$Z_{10} = u\varphi\partial_\varphi, \quad Z_{11} = v\varphi\partial_\varphi, \quad Z_{12} = w\varphi\partial_\varphi.$$

Moreover, for the (pseudo) Maxwellian molecules in the homogeneous case there is one more extension of the admitted Lie algebra

$$Z_{13} = (u^2 + v^2 + w^2)\varphi\partial_\varphi.$$

Further a classification of all solutions invariant with respect to the Lie group $G^{11}$ of the Boltzmann equation (3.3.1) (shortly, H-solutions) is presented.

## 3.3.2 Isomorphism of Algebras

The classification separates the set of H-solutions into equivalent (similar) classes. Any two H-solutions $f_1$ and $f_2$ are elements of the same equivalent class if there exists a transformation $T_a \in G$ that $f_2 = T_a f_1$. Otherwise $f_1$, $f_2$ belong to different classes and they are called essentially different H-solutions. A list of all essentially different H-solutions (one representative from each class) composes an optimal system of invariant solutions that defines the sought-for classification. An optimal system of invariant solutions is related with an optimal system of subalgebras $\Theta_L$ of the Lie algebra $L$ [35]. The general algorithms for constructing such systems are known in the theory of group analysis.[4] Optimal systems for some kinetic equations with a small number of independent variables were obtained in [9, 18, 20]. For a low dimension of a Lie algebra $L$ calculations of $\Theta_L$ are sufficiently simple. The higher the dimension of an algebra $L$, the greater the difficulty in the construction of an optimal system. A realization of an optimal system of subalgebras for large dimensional algebras such as $L_{11}$ requires extremely long and tedious calculations.

---

[4]See also Chap. 1.

For example, despite the fact that the group classification of the gas dynamics equations (EGD-system) was obtained in the 1970s [35], optimal systems of admitted subalgebras were only calculated almost two decades later [14, 16, 36].

The Lie algebra $L_{11}(X)$ admitted by the Boltzmann equation (3.3.1) and its extensions $L_{12}(X)$ and $L_{13}(X)$ possess a remarkable feature that allows avoiding tedious calculations. The lucky fact for the full Boltzmann equation is that there is an isomorphism of the Lie algebra $L_{11}(X)$ and the Lie algebra admitted by EGD-system of equations with the general state equation [23]. There are also isomorphisms of the extensions of $L_{11}(X)$ up to the $L_{12}(X)$ and $L_{13}(X)$ for specified intermolecular potentials and the Lie algebras studied in the gas dynamics for a polytropic gas. The isomorphisms allow one to solve the problem of classification of invariant H-solutions of the full Boltzmann equation by using optimal systems of subalgebras known for the EGD-system of equations.

**Theorem 3.3.1** *The Lie algebra $L_{11}(X)$ admitted by the full Boltzmann equation is isomorphic to the Lie algebra $L_{11}(Y)$ admitted by the EGD-system.*

*Proof* The EGD-system is written as

$$\frac{d\rho}{dt} + \rho\nabla\mathbf{v} = 0, \qquad \rho\frac{d\mathbf{v}}{dt} + \nabla p = 0,$$

$$\frac{dp}{dt} + A(\rho, p)\nabla\mathbf{v} = 0, \tag{3.3.21}$$

where $\rho$, $p$ are the density and the pressure of a gas, $\nabla$ is the nabla operator, $\frac{d}{dt} = \frac{\partial}{\partial t} + \mathbf{v}\nabla$. As above $t \in R^1_+$, $\mathbf{x} = (x, y, z) \in R^3_{\mathbf{x}}$, $\mathbf{v} = (u, v, w) \in R^3_{\mathbf{v}}$, but the vector $\mathbf{v}$ in (3.3.21) is the vector of gas macroscopic velocity. The function $A(\rho, p)$ is determined by the state equation of a gas.

For an arbitrary state equation system (3.3.21) admits the 11-dimensional Lie algebra $L_{11}(Y)$ with the basis generators [35]:

$$Y_1 = \partial_x, \ Y_2 = \partial_y, \ Y_3 = \partial_z, \ Y_4 = t\partial_x + \partial_u, \ Y_5 = t\partial_y + \partial_v, \ Y_6 = t\partial_z + \partial_w,$$

$$Y_7 = y\partial_z - z\partial_y + v\partial_w - w\partial_v, \ Y_8 = z\partial_x - x\partial_z + w\partial_u - u\partial_w,$$

$$Y_9 = x\partial_y - y\partial_x + u\partial_v - v\partial_u, \quad Y_{10} = \partial_t, \quad Y_{11} = t\partial_t + x\partial_x + y\partial_y + z\partial_z.$$

Let $Q(Y) = X$ be a linear transformation of $L_{11}(Y)$ onto $L_{11}(X)$, defined by $Q(Y_k) = X_k$ $(k = 1, 2, \ldots, 11)$. It is obvious that $Q$ conserves the commutators[5]

$$Q([Y_k, Y_j]) = [Q(Y_k), Q(Y_j)] \quad (j, k = 1, 2, \ldots, 11), \tag{3.3.22}$$

where $[A, B] = AB - BA$. Hence, the Lie algebras $L_{11}(X)$ and $L_{11}(Y)$ are isomorphic, and $Q$ is an isomorphism.                                                                          □

It is known [35] that any construction of the optimal system of subalgebras of a given Lie algebra is completely defined by the table of commutators of basis gen-

---

[5]See Appendix A.

erators. From (3.3.22) it follows that the tables of commutators of both algebras $L_{11}(X)$ and $L_{11}(Y)$ coincide. Hence, by virtue of the proven isomorphism of the Lie algebras $L_{11}(X)$ and $L_{11}(Y)$, their optimal systems of subalgebras are also isomorphic. It is similar verified that the Lie algebra $L_{12}(X)$ is isomorphic to the Lie algebra $L_{12}(Y)$.

Thus, the following statement is proven.

**Corollary 3.3.1** *For classifying and constructing essentially different H-solutions of the Boltzmann equation (3.3.1) one can use the optimal system of subalgebras constructed for the EGD-system (3.3.21).*

**Remark 3.3.4** If the function $A(p, \rho) = \kappa\rho$ (which corresponds to a polytropic gas), then there is an extension of the admitted Lie algebra from $L_{11}(Y)$ to the Lie algebra $L_{13}(Y)$ with the following additional two basis generators:

$$Y_{12} = t\partial_t - u\partial_u - v\partial_v - w\partial_w + 2\rho\partial_\rho, \quad Y_{13} = \rho\partial_\rho + p\partial_p.$$

The operator $Y_{13}$ composes a center of the Lie algebra $L_{13}(Y) = L_{12}(Y) \oplus \{Y_{13}\}$: $[Y_i, Y_{13}] = 0$ ($i = 1, 2, \ldots, 12$). The optimal system of subalgebras of $L_{12}(Y)$ is the subset of an optimal system of subalgebras for the Lie algebra $L_{13}(Y)$ without $Y_{13}$. The optimal system of subalgebras of $L_{13}(Y)$ was constructed in [16].

In the case of a monatomic gas $\kappa = (n + 2)/n$ ($n$ is the dimension of a flow) the EGD-system (3.3.21) admits one more generator:

$$Y_{14} = t^2\partial_t + t(x\partial_x + y\partial_y + z\partial_z)$$
$$+ (x - tu)\partial_u + (y - tv)\partial_v$$
$$+ (z - tw)\partial_w - nt\rho\partial_\rho - (n + 2)tp\partial_p.$$

A relation between the generators $X_{13}$ and $Y_{14}$ was also noted in [6].

### 3.3.3 Invariant Solutions of the Full Boltzmann Equation

Further study is restricted to the Lie algebra $L_{11}(X)$ admitted by the Boltzmann equation (3.3.1) with an arbitrary scattering function $B$. An application of the optimal system of subalgebras of $L_{11}(X)$ for constructing invariant solutions of the full Boltzmann equation and EGD-system differs. This is related due to the differences in the sets of the independent and dependent variables. For example, H-solutions of the Boltzmann equation (3.3.1) with one and two independent variables can only be obtained for subalgebras with dimension more than five. In this case using the optimal system of subalgebras,[6] one finds 11 different classes of invariant solutions with one independent variable and 38 with two independent variables. Their functional expressions are presented in Table 3.1 and Table 3.2.

---

[6]The part of the optimal system of six and seven dimensional subalgebras of [36] is presented in Appendix A.

**Table 3.1** Representations of H-solutions with one independent invariant variable

| No. | $f$ | | No. | $f$ | |
|---|---|---|---|---|---|
| 1 | $e^{\varepsilon\theta}g(q)$ | 6,3 | 7 | $g(u)$ | 6,11 |
| 2 | $t^{-1}g(W^2 + (V - rt^{-1})^2)$ | 6,4 | 8 | $g(u - t)$ | 6,23 |
| 3 | $t^{-1}g(q)$ | 6,5 | 9 | $e^{\varepsilon\theta}g(w), \varepsilon \neq 0$ | 6,18 |
| 4 | $t^{-1}g(u - xt^{-1})$ | 6,7 | 10 | $t^{-1}g(q/t)$ | 7,2 |
| 5 | $t^{-1}g(u - \varepsilon \ln t)$ | 6,20 | 11 | $x^{-1}g(u)$ | 7,3 |
| 6 | $g(t)$ | 6,14 | | | |

In Table 3.1 and Table 3.2 $\alpha, \beta, \varepsilon$ are arbitrary constants. In the second column representations of H-solutions are given. In the last column pair of indices $m, i$ means that the representation of an invariant solution is obtained with respect to the subalgebra $m, i$, where $m$ is the dimension of the corresponding subalgebra and $i$ is its number in Table 6 of [36]. The capital S means that the given representation should be considered in the spherical coordinate system $(x, \varphi, \theta, u, V, W)$, where:

$$x = r \sin\theta \cos\varphi, \quad y = r \sin\theta \sin\varphi, \quad z = r \cos\theta,$$

$$u = U \sin\theta \cos\varphi + V \cos\theta \cos\varphi - W \sin\varphi,$$

$$v = U \sin\theta \sin\varphi + V \cos\theta \sin\varphi + W \cos\varphi,$$

$$w = U \cos\theta - V \sin\theta, \quad Q = \sqrt{u^2 + v^2 + w^2} = \sqrt{U^2 + V^2 + W^2}.$$

The capital C corresponds to the cylindrical coordinate system $(x, r, \theta, u, V, W)$, where:

$$y = r \cos\theta, \quad z = r \sin\theta, \quad v = V \cos\theta - W \sin\theta, \quad w = V \sin\theta + W \cos\theta,$$

$$q = \sqrt{v^2 + w^2} = \sqrt{V^2 + W^2}.$$

Other representations are considered in the Cartesian coordinate system.

It should be noted that for many subalgebras from Table 3.1 and Table 3.2 H-solutions either do not exist or do not have a physical meaning. For some of solutions this statement is easily verified. In the general case it is necessary to substitute a representation of a solution into the Boltzmann equation (3.3.1) and to study the corresponding factor equation. But in difference to the EGD-system (3.3.21) obtaining the factor equation for the Boltzmann equation with the complicated collision integral (3.3.2) is sufficiently difficult. As an example the factor equation for H-solution (number 38 in Table 3.2) derived in [37] is presented here:

$$\frac{\partial f(t, Q)}{\partial t} = 8\frac{\pi^2\sigma^2}{Q} \int\limits_0^Q \int\limits_{\sqrt{Q^2 - P^2}}^Q f(t, P)f(t, R)\sqrt{P^2 + R^2 - Q^2}\, PR dP dR$$

**Table 3.2** Representations of H-solutions with two independent invariant variables

| No. | $f$ | | No. | $f$ | |
|---|---|---|---|---|---|
| 1 | $r^{-1}g(U,q)$ | S 5,1 | 17 | $g(t,u)$ | 5,17 |
| 2 | $r^{-1}g(V,W)$ | C 5,2 | 18 | $g(x-t^2/2, u-t)$ | 5,18 |
| 3 | $x^{-1}g(u,q)$ | 5,3 | 19 | $g(x,u)$ | 5,19 |
| 4 | $e^{-\alpha\theta}g(u-\beta\theta,q)$ | C 5,4 | 20 | $g(q, \arcsin(v/q)+t)$ | 5,20 |
| 5 | $t^{-1}g(u-x/t, (v-y/t)^2$ | 5,5 | 21 | $x^{-1}g(u, w-\beta\ln x)$ | 5,21 |
| | $\quad +(w-z/t)^2)$ | | 22 | $e^{-\beta u}g(v,w)$ | 5,22 |
| 6 | $t^{-1}g(u-x/t, q)$ | 5,6 | 23 | $t^{-1}g(v-y/t, w-z/t)$ | 5,24 |
| 7 | $t^{-1}g(u-\beta\ln t$ | 5,7 | 24 | $t^{-1}g(u-x/t,$ | 5,25 |
| | $\quad +\alpha\arcsin((v-y/t)/q),q)$ | | | $\quad v-\alpha^{-1}(x-\beta\ln t))$ | |
| | $q=\sqrt{(v-y/t)^2+(w-z/t)^2}$ | | 25 | $t^{-1}g(x/t-\beta\ln t,$ | 5,26 |
| 8 | $t^{-1}g(u-\beta\ln t$ | 5,8 | | $\quad u-\beta\ln t)$ | |
| | $\quad +\alpha\arcsin(v/q),q)$ | | 26 | $t^{-1}g(u-x/t, v-\beta\ln t)$ | 5,27 |
| 9 | $g(t,q),$ | 5,9 | 27 | $t^{-1}g(u-\beta\ln t, v)$ | 5,28 |
| | $q=\sqrt{(v-y/t)^2+(w-z/t)^2}$ | $\alpha=0$ | 28 | $t^{-1}g(v,w)$ | 5,29 |
| 10 | $t^{-1}g(q, \arcsin((v-y/t)/q)$ | 5,9 | 29 | $g(u-t, v-\alpha^{-1}(x-t^2/2))$ | 5,30 |
| | $\quad +\alpha^{-1}\ln t),$ | $\alpha\neq0$ | 30 | $g(x-t^2/2, u-t)$ | 5,31 |
| | $q=\sqrt{(v-y/t)^2+(w-z/t)^2}$ | | 31 | $g(u,v-x)$ | 5,32 |
| 11 | $t^{-1}g(x/t, u-\alpha^{-1}\beta\ln t)$ | 5,10 | 32 | $g(x,u)$ | 5,33 |
| 12 | $t^{-1}g(\arcsin(v/q)-\alpha^{-1}\ln t, q)$ | 5,11 | 33 | $g(u-t,v)$ | 5,34 |
| 13 | $g(u-\alpha\arcsin(v/q),q)$ | 5,12 | 34 | $g(t, u-x/t)$ | 5,35 |
| 14 | $g(t,u-x/t)$ | 5,13 | 35 | $g(t, w+ut-x)$ | 5,36 |
| 15 | $g(t,q)$ | 5,15 | 36 | $g(t,u)$ | 5,37 |
| 16 | $g(t,q),$ | 5,16 | 37 | $g(t, (\mathbf{u}-\mathbf{x}/t)^2)$ | 6,9 |
| | $q=(v-(yt+z)/(1+t^2))^2$ | | 38 | $g(t,Q)$ | 6,10 |
| | $\quad +(w+(y-zt)/(1+t^2))^2$ | | | | |

$$+8\pi^2\sigma^2\left(\int_Q^\infty f(t,P)P\,dP\right)^2$$

$$+16\frac{\pi^2\sigma^2}{Q}\int_0^Q f(t,P)P^2\,dP\int_Q^\infty f(t,P)P\,dP$$

$$-\frac{2}{3}\frac{\pi^2\sigma^2}{Q}f(t,Q)\int_0^\infty f(t,P)[(Q+P)^3-|Q-P|^3]P\,dP.$$

**Table 3.3**  Optimal system of subalgebras of $L_9(Z)$

| $r = 9$ | | | $r = 6$ | | |
|---|---|---|---|---|---|
| 1 | 1,2,3; 4,5,6; 7,8,9 | $=9,1$ | 1 | 2,3; 4; 7,8,9 | $=6,1$ |
| $r = 8$ | | | 2 | 4,5,6; 7,8,9 | $=6,2$ |
| 1 | 1,2,3; 4,5,6; 8,9 | $=8,1$ | 3 | 1,2,3; 4; 8,9 | $=6,3$ |
| 2 | 1,2,3; 4,5,6; $7,9+\alpha 8$ | 9,1 | 4 | 1,2,3; 4; $7,9+\alpha 8$ | 7,1 |
| 3 | 1,2,3; 4,5,6; 7,8 | 9,1 | 5 | 1,2,3; 4; 7,8 | 7,1 |
| $r = 7$ | | | 6 | 1,2,3; 4,5,6 | 9,1 |
| 1 | 1,2,3; 4; 7,8,9 | $=7,1$ | | | |
| 2 | 1,2,3; 4,5,6; $9+\alpha 8$ $(\alpha \neq -1)$ | 8,1 | | | |
| 3 | 1,2,3; 4,5,6; $9-8$ | 9,1 | | | |
| 4 | 1,2,3; 4,5,6; $9-8+\mu 7$ | $8,2^{-1}$ | | | |
| 5 | 1,2,3; 4,5,6; 8 | 8,1 | | | |
| 6 | 1,2,3; 4,5,6; 7 | 9,1 | | | |

### 3.3.4  Classification of Invariant Solutions of the Fourier Transformation of the Full Boltzmann Equation

A self-normalized optimal system of subalgebras of the Lie algebra $L_9(Z)$ admitted by (3.3.20) is calculated here. The two-step algorithm developed in [36] is used. This algorithm allows one to reduce the problem of the construction of an optimal system of subalgebras of $L_9$ to a classification of subalgebras with less dimensions.

At the first step the original Lie algebra $L_9(Z)$ was presented as $L_9(Z) = J^1 \oplus N^1$, where $J^1 = \{Z_1, Z_2, Z_3, Z_4, Z_5, Z_6, Z_7\}$ is an ideal and $N^1 = \{Z_8, Z_9\}$ is a subalgebra. Then the optimal system for the assigned subalgebra $N^1$ is constructed. The latter is written in the form $N^1 = J^2 \oplus N^2$ with the ideal $J^2 = \{Z_7\}$ and the subalgebra $N^2 = \{Z_1, Z_2, Z_3, Z_4, Z_5, Z_6\}$. Further the process is repeated: the subalgebra $N^2$ is decomposed $N^2 = J^3 \oplus N^3$ where again $J^3 = \{Z_1, Z_2, Z_3\}$ is an ideal and $N^3 = \{Z_4, Z_5, Z_6\}$ is a subalgebra.

The final result of a part of an optimal system of subalgebras of $L_9$ is presented in Table 3.3. Subalgebra-representatives are notated by a pair of numbers $r, i$, where $r$ is the dimension and $i$ is the number of the subalgebra of the dimension $r$ which is given in the first column. In the second column a basis of the subalgebra is presented in the symbolical form: corresponding numbers of basis generators are only presented. The restrictions for parameters are also given in this column. Notice that $\mu \neq 0$ in all subalgebras and the absence of restrictions for the parameters means that they are arbitrary real numbers. In the third column the normalizer of the subalgebra is presented. The sign $=$ means that the subalgebra is self-normalized. The upper index $^{-1}$ points out that the normalizer is contained in the set of subalgebras and it is obtained for the parameter equal to $-1$.

**Table 3.4** Representations of H-solutions of (3.3.20) with a single independent invariant variable

| No. | $\varphi$ | No. | $\varphi$ |
|---|---|---|---|
| 1 | $t^{\gamma\beta-1}\psi(Qt^{-\beta})$ | 6 | $u^{\gamma-1}x^{-1}\psi(uq^{-1})$ |
| 2 | $Q^\gamma\psi(t)$ | 7 | $k^{\gamma-1}r^{-\gamma}\psi(rQk^{-1})$ |
| 3 | $Q^\gamma\psi(Qe^{-t})$ | 8 | $u^\gamma t^{-1}\psi(uq^{-1})$ |
| 4 | $t^{-1}\psi(Q)$ | 9 | $u^{-\alpha}\psi(uq^{-1})$ |
| 5 | $\psi(Q)$ | | |

**Table 3.5** Representations of H-solutions of (3.3.20) with two independent invariant variables

| No. | $\varphi$ | No. | $\varphi$ |
|---|---|---|---|
| 1 | $\psi(t,Q)$ | 11 | $x^{-1}\psi(u,q)$ |
| 2 | $u^{\gamma-1}r^{-1}\psi((vy+wz)(ru)^{-1},(-vz+wy)(ru)^{-1})$ | 12 | $(x-t)^{-1}\psi(u,q)$ |
| 3 | $u^{\gamma-1+\delta}e^{-\beta\theta}\psi(xu^\delta e^{-\beta\theta},uq^{-1})$ | 13 | $t^{\gamma\beta-1}\psi(ut^{-\beta},qt^{-\beta})$ |
| 4 | $x^{-1}e^{(\gamma-\beta)\theta}\psi(qe^{-\beta\theta},ue^{-\beta\theta})$ | 14 | $u^\gamma\psi(t,uq^{-1})$ |
| 5 | $x^{-1}u^{(\gamma-1)}\psi(\theta,uq^{-1})$ | 15 | $e^{\gamma t}\psi(ue^{-t},qe^{-t})$ |
| 6 | $x^{-1}u^{(\gamma-1)}\psi(xut^{-1},uq^{-1})$ | 16 | $t^{-1}\psi(u,q)$ |
| 7 | $x^{-1}u^{(\gamma-1)}\psi(ux^\beta,uq^{-1})$ | 17 | $\psi(u,q)$ |
| 8 | $e^{(\gamma-1)\beta(x-\theta)}\psi(ue^{\beta(\theta-x)},qe^{\beta(\theta-x)})$ | 18 | $t^{\gamma-1}r^{-\gamma}\psi(RUt^{-1},Rqt^{-1})$ |
| 9 | $e^{(\gamma-1)x}\psi(ue^{-x},qe^{-x})$ | 19 | $U^{\gamma-1-\alpha}\psi(U^\alpha R^{-1},Uq^{-1})$ |
| 10 | $u^{\gamma-1}\psi(x,uq^{-1})$ | 20 | $R^{-1}\psi(U,q)$ |

Representations of invariant solutions of (3.3.20) are obtained by finding corresponding invariants. These representations are given in Table 3.4 and Table 3.5. There

$$Q=\sqrt{u^2+v^2+w^2}, \quad \beta\neq0, \quad q=\sqrt{v^2+w^2},$$
$$r=\sqrt{y^2+z^2}, \quad k=ux+vy+w$$

and in the spherical coordinate system (S):

$$q=\sqrt{V^2+W^2}, \quad R=\sqrt{x^2+y^2+z^2}.$$

The parameters $\alpha$ and $\beta$ are arbitrary constants.

**Remark 3.3.5** The well-known BKW-solution for is in the class 7,4 with $\gamma=0$.

## 3.4 Complete Group Analysis of Some Kinetic Equations

In the group analysis of integro-differential equations the most difficulties are related with solving determining equations. However, for kinetic equations with a small number of independent variables the proposed regular method allows one to overcome these difficulties effectively (see examples in Chap. 2). In such cases one can derive not only a complete admitted Lie group but also an optimal system of sub-algebras, and a description of the complete set of invariant solutions can be obtained. In order words, comprehensive group analysis is successively applied to these equations.

### 3.4.1 The Boltzmann Kinetic Equation with an Approximate Asymptotic Collision Integral

This equation for isotropic scattering model has the form

$$\Phi(f) \equiv x^\alpha f_t(x,t) + \theta(f)f(x,t) - \int_0^1 f(sx,t)f((1-s)x,t)\,ds = 0.$$

(3.4.1)

where $f(t,x)$ is the molecular distribution function with respect to the molecular energies $x = v^2/2$, $\theta(f)$ is some functional of $f$. Equation (3.4.1) approximately describes the spatially homogeneous relaxation of a monatomic gas for high molecular energies with the power interaction potential $U = k/r^{\nu-1}$. The exponent $\alpha$ is defined as $(5-\nu)(\nu-1)^{-1}/2$. The meaning of (3.4.1) is that it gives an approximate intermediate asymptotic of the Boltzmann kinetic equation at $x \to \infty$ and large but finite values of time $t$. This equation is interesting in the context of calculating the kinetic processes passing on the high energy tail of the distribution function [26, 28, 31]. In spite of its simplicity, (3.4.1) has the main features of the full Boltzmann equation. In particular, for $\nu = 5$ and $\theta(f) = f(0,t)$ the equation (3.4.1) becomes the Fourier transform of the Boltzmann equation of the Maxwell isotropic molecular model. Results of its group analysis were presented in Chap. 2. For $\nu = 7/3$ and $\theta(f) = f(0,t) = 1$ another invariant solution of the BKW-type in the explicit form was obtained in [28]. Using the regular method outlined in Chap. 2, the equation (3.4.1) was completely studied in [17, 26]. The principal results of the group analysis are formulated as follows.

**Theorem 3.4.1** *The four-dimensional Lie algebra* $L_4 = \{X_1, X_2, X_3, X_4\}$ *spanned by the generators*

$$X_1 = \partial_t, \quad X_2 = xf\partial_f, \quad X_3 = \alpha f\partial_f + x\partial_x, \quad X_4 = f\partial_f - t\partial_t, \qquad (3.4.2)$$

*defines a complete Lie group* $G^4$ *admitted by (3.4.1). The optimal system of one-dimensional subalgebras admitted by (3.4.1) consists of the subalgebras:*

$$\{X_1\}, \{X_2\}, \{X_3\}, \{X_4 + \beta X_3\}, \{X_2 + X_1\}, \{X_3 + X_1\}, \{X_4 \pm X_2\},$$

(3.4.3)

*where* $\beta$ *is an arbitrary constant.*

**Remark 3.4.1** The proof of the theorem was obtained for the functional $\theta$ satisfying the conditions

$$\theta(\beta f(x,t)) = \beta\theta(f(x,t)), \qquad \theta(e^{\alpha x} f(x,t)) = \theta(f(x,t)),$$

where $\beta > 0$ is an arbitrary constant. For example, $\theta(f(x,t)) = f(0,t)$ or $\theta(f(x,t)) = 0$.

The optimal system of subalgebras (3.4.3) allows one to specify all classes of H-solutions of (3.4.1) which are essentially different with respect to the Lie group $G^4$. Representations of these solutions are

$$X_1 : f = g(y), \quad y = x,$$
$$X_3 : f = x^\alpha g(y), \quad y = t,$$
$$X_4 + \beta X_3 : f = \frac{x^\alpha}{t} g(y), \quad y = xt^\beta,$$
$$X_2 + X_1 : f = e^{xt} g(y), \quad y = x,$$
$$X_3 + X_1 : f = x^\alpha g(y), \quad y = xe^{-t},$$
$$X_4 \pm X_2 : f = t^{-(1\pm x)} g(y), \quad y = x.$$

The corresponding factor-equations are obtained by substituting the representation of an invariant solution into (3.4.1).

One class of solutions among the H-solutions defined by the group $G^4$ is the most attractive. Two of these solutions were derived in explicit form for $\alpha = 0, 1$ in [1, 28]. For arbitrary real $\alpha$ and natural numbers $\nu \in [2, \infty)$ the class of BKW-solutions is specified by the infinitesimal generator

$$X = c_1 X_2 + X_3 - \alpha X_4. \tag{3.4.4}$$

The corresponding H-solutions are of the form

$$f = e^{-c_1 x} F(y), \quad y = x(\alpha ct)^{-1/\alpha}. \tag{3.4.5}$$

**Remark 3.4.2** For the isotropic Maxwell scattering model at $\alpha = 0$ the term $c^{-1} X_1$ is used instead of the last term in (3.4.4) and in this case the BKW-solution has the form

$$f = e^{-c_1 x} F(y), \quad y = xe^{ct}.$$

Substituting (3.4.5) into (3.4.1) one gets

$$cy^{1+\alpha} \frac{dF}{dy} = -F(y) + \int_0^1 ds\, F(sy) F((1-s)y). \tag{3.4.6}$$

In [17, 26] the BKW-solutions was searched for in the form

$$F(y) = e^{-y} \left( 1 + \sum_{\{p(n)\}} a_n y^{p(n)} / \Gamma(p(n) + 1) \right), \tag{3.4.7}$$

where $\{p(n)\}$ is the finite subset of an ordered countably additive of rational numbers and $\Gamma(x)$ is the Euler gamma-function. The following statement was proven.

**Theorem 3.4.2** *The BKW-solutions of* (3.4.6), (3.4.1) *in the form of finite series* (3.4.7) *exist only for* $\alpha = 0; 1$.

### 3.4.2 The Smolukhovsky Kinetic Equation

The Smolukhovsky kinetic equation of homogeneous coagulations is usually considered in the form

$$\frac{\partial f(t, v)}{\partial t} = \frac{1}{2} \int_0^v d v_1 \beta(v - v_1, v_1) f(t, v - v_1) f(t, v_1)$$

$$- f(t, v) \int_0^\infty d v_1 \beta(v, v_1) f(t, v_1), \qquad (3.4.8)$$

where $f(t, v)$ is the distribution function of coagulating particles with respect to their volumes, the coagulation kernel $\beta(v, v_1)$ is a symmetric nonnegative function $\beta(v, v_1) = \beta(v_1, v) \geq 0$ [38].

The Cauchy problem for this equation is stated with the following initial data

$$f(0, v) = f_0(v).$$

Equation (3.4.8) occurs in the kinetic theory of dispersive systems such as atmospheric aerosols, colloidal solutions, suspensions, protoplanet space matter. This equation has many features which make it similar to the Boltzmann equation.

Multiple references to H-solutions of (3.4.8) with the "parabolic" coagulation kernel $\beta(v, v_1) = c(h + v)(h + v_1)$ exist, where $c$ and $h$ are constant. Using a change of the variables, (3.4.8) with this kernel is reduced to the "spectrum age" equation

$$\frac{\partial \theta(x, \tau)}{\partial \tau} = \int_0^x dx_1 (h + x - x_1)(h + x_1) \theta(x - x_1, \tau) \theta(x_1, \tau). \qquad (3.4.9)$$

However, in this case the complete admitted Lie group of (3.4.9) (also (3.4.8)) had not been known, hence there was no description of the complete set of its H-solutions.

There are other kernels which describe interaction of coagulating particles more adequately [38]:

$$\beta(v, v_1) = c(v^a v_1^b + v^b v_1^a). \qquad (3.4.10)$$

In [20], using a change of the variables similar to the "spectrum age" variables, (3.4.8) for particular values of the parameters $a$ and $b$ was reduced to the following universal form

$$\frac{\partial u(x, \tau)}{\partial \tau} = x^\gamma \int_0^1 s^{k_1} (1 - s)^{k_2} u(x(1 - s), \tau) u(xs, \tau) ds. \qquad (3.4.11)$$

Notice that (3.4.11) has a structure similar to (3.4.1). The group analysis of (3.4.1) allows one to state the following theorem.

**Theorem 3.4.3** *The four-dimensional Lie algebra $L_4 = \{X_1, X_2, X_3, X_4\}$ spanned by the generators*

$$X_1 = \partial_\tau, \quad X_2 = xu\partial_u, \quad X_3 = x\partial_x - \gamma u\partial_u, \quad X_4 = u\partial_u - \tau\partial_\tau, \qquad (3.4.12)$$

*defines the complete Lie group $G^4$ admitted by (3.4.11). The optimal system of one-parameter subalgebras admitted by (3.4.11) consists of the subalgebras*

$$\{X_1\}, \{X_2\}, \{X_3\}, \{X_4 + \beta X_3\}, \{X_2 + X_1\}, \{X_3 + X_1\}, \{X_4 \pm X_2\},$$

$$(3.4.13)$$

*where $\beta$ is an arbitrary constant.*

The set of H-solutions of (3.4.11) essentially different with respect to the Lie group $G^4$ is exhausted by the solutions with the following representations

$$\begin{aligned}
X_1 &: u = g(y), \quad y = x, \\
X_3 &: u = x^{-\gamma} g(y), \quad y = t, \\
X_4 + \beta X_3 &: u = x^{-(\gamma+1)/\beta} g(y), \quad y = xt^\beta, \\
X_2 + X_1 &: u = e^{\pm xt} g(y), \quad y = x, \\
X_3 + X_1 &: u = x^{-\gamma} g(y), \quad y = xe^{\pm t}, \\
X_4 \pm X_2 &: u = t^{1 \pm x} g(y), \quad y = x.
\end{aligned}$$

### 3.4.3 A System of the Space Homogeneous Boltzmann Kinetic Equations

The system of the Boltzmann equations describing evolution of an $N$-component gas mixture to an equilibrium has the form

$$\frac{\partial f_\alpha(\mathbf{v}_\alpha, t)}{\partial t} = \frac{1}{4\pi} \sum_{\beta=1}^{N} \sigma_{\alpha\beta} \iint d\mathbf{w}_\beta \, d\Omega [f_\alpha(\mathbf{v}'_\alpha) f_\beta(\mathbf{w}'_\beta) - f_\alpha(\mathbf{v}_\alpha) f_\beta(\mathbf{w}_\beta)],$$

$$(3.4.14)$$

where $\mathbf{v}_\alpha, \mathbf{w}_\beta \in \mathbb{R}^3$, $v_\alpha = |\mathbf{v}_\alpha|$, $f_\alpha(\mathbf{v}_\alpha, t)$ is the velocity distribution function of molecules of $\alpha$-species, $m_\alpha$ is the molecular mass, $\Omega$ is a unit vector, and $d\Omega$ is the area element on the unit sphere,

$$\mathbf{v}'_\alpha = \mathbf{R}_{\alpha\beta} + \mu_{\alpha\beta} g_{\alpha\beta} \Omega, \quad \mathbf{w}'_\beta = \mathbf{R}_{\alpha\beta} - \mu_{\alpha\beta} g_{\alpha\beta} \Omega, \quad \mathbf{R}_{\alpha\beta} = \mu_{\alpha\beta} \mathbf{v}_\alpha + \mu_{\beta\alpha} \mathbf{w}_\beta,$$

$$\mathbf{g}_{\alpha\beta} = \mathbf{v}_\alpha - \mathbf{w}_\beta, \quad \mu_{\alpha\beta} = m_\alpha/(m_\alpha + m_\beta) \quad (\alpha, \beta = 1, 2, \ldots, N).$$

The scattering of molecules here is described by the isotropic Maxwell model [15] with the scattering functions $B_{\alpha\beta}(\Omega, g_{\alpha\beta}) = \sigma_{\alpha\beta}/4\pi$. It is worth to notice that the results of the present subsection hold true for other Maxwell models of interactions where $B_{\alpha\beta}$ is independent of the relative velocity of the scattering particles.

The initial value data for (3.4.14) are set as

$$f_\alpha(\mathbf{v}_\alpha, 0) = f_\alpha^{(0)}(\mathbf{v}_\alpha) \quad (\alpha = 1, 2, \dots, N).$$

Solutions of (3.4.14) satisfy the conservation laws of the component concentrations and the total energy of the heat motion of molecules. In the chosen system of the dimensionless variables, the conservation laws are

$$n_\alpha(t) = \int d\mathbf{v}_\alpha f_\alpha(\mathbf{v}_\alpha, t) = n_\alpha(0), \quad \sum_{\alpha=1}^{N} n_\alpha = 1, \tag{3.4.15}$$

$$E(t) = \sum_{\alpha=1}^{N} \int d\mathbf{v}_\alpha v_\alpha^2 f_\alpha(\mathbf{v}_\alpha, t) = E(0) = 3. \tag{3.4.16}$$

As $t \to \infty$ solutions of (3.4.14) converge asymptotically to the stationary Maxwellian distributions

$$f_{\alpha,M}(\mathbf{v}_\alpha) = \frac{n_\alpha}{(2\pi)^{3/2}} \exp(-v_\alpha^2/2).$$

For further calculations it is convenient to use along with (3.4.14) the Fourier transform of the Boltzmann equations constructed in [3]. Let the direct and inverse Fourier transforms be determined by the formulae

$$\begin{aligned}
\varphi_\alpha(\mathbf{k}_\alpha, t) &= \int d\mathbf{v}_\alpha e^{i\mathbf{k}_\alpha \mathbf{v}_\alpha} f_\alpha(\mathbf{v}_\alpha, t), \\
f_\alpha(\mathbf{v}_\alpha) &= \frac{1}{(2\pi)^3} \int d\mathbf{k}_\alpha e^{i\mathbf{k}_\alpha \mathbf{v}_\alpha} \varphi_\alpha(\mathbf{k}_\alpha, t).
\end{aligned} \tag{3.4.17}$$

The consideration is restricted here to the study of solutions of (3.4.14) which are isotropic in the velocity space. In this case the distribution functions only dependent of the modulus of the molecular velocity and Fourier transforms of (3.4.14) are

$$\frac{\partial \varphi_\alpha(x_\alpha, t)}{\partial t} = \sum_{\beta=1}^{N} \sigma_{\alpha\beta} \int_0^1 ds [\varphi_\alpha(x_\alpha(1 - \varepsilon_{\alpha\beta}s))\varphi_\beta(x_\alpha \varepsilon_{\alpha\beta}s)$$
$$- \varphi_\alpha(x_\alpha)\varphi_\beta(0)], \tag{3.4.18}$$

where $x_\alpha = k_\alpha^2/2$, $\varepsilon_{\alpha\beta} = 4\mu_{\alpha\beta}\mu_{\beta\alpha}$ and $(\alpha = 1, \dots, N)$. The conservation laws (3.4.16) and (3.4.17) become

$$\varphi_\alpha(0, t) = n_\alpha \quad (\alpha = 1, 2, \dots, N); \quad \sum_{\alpha=1}^{N} \varphi_{\alpha,x_\alpha}(0, t) = -1. \tag{3.4.19}$$

The subscripts after comma mean differentiation with respect to the corresponding variables. The Fourier transforms of the Maxwellian distribution functions are $\varphi_{\alpha,M} = n_\alpha \exp(-x_\alpha)$.

An infinitesimal generator of the Lie group admitted by system (3.4.18) is sought for in the form [21]

$$X = \xi^t(\mathbf{u})\partial_t + \sum_{\alpha=1}^{N}(\xi^\alpha(\mathbf{u})\partial_{x_\alpha} + \zeta^\alpha(\mathbf{u})\partial_{\varphi_\alpha}), \qquad (3.4.20)$$

where $\mathbf{u} = (t, x_1, \ldots, x_N, \varphi_1, \ldots, \varphi_N)$.

**Theorem 3.4.4** *The complete Lie group of point transformations admitted by system (3.4.18) is the four-parameter group $G^4$ determined by the Lie algebra $L_4$ with the basis generators*

$$X_1 = \partial_t, \quad X_2 = \sum_{\alpha=1}^{N} x_\alpha \varphi_\alpha \partial_{\varphi_\alpha},$$

$$X_3 = \sum_{\alpha=1}^{N} x_\alpha \partial_{x_\alpha}, \quad X_4 = \sum_{\alpha=1}^{N} \varphi_\alpha \partial_{\varphi_\alpha} - t\partial_t. \qquad (3.4.21)$$

*Proof* The dependence $\varphi_\alpha = \varphi_\alpha(x_\alpha, t)$ $(\alpha = 1, 2, \ldots, N)$, in terms of its differential representation is written down as

$$\frac{\partial \varphi_\alpha}{\partial x_\beta} = 0 \quad (\alpha \neq \beta; \ \beta, \alpha = 1, 2, \ldots, N). \qquad (3.4.22)$$

The part of the determining equations related with (3.4.22) is the result of the action of the prolonged operator[7] $\bar{X}$ on relations (3.4.22):

$$\bar{X}\left(\frac{\partial \varphi_\alpha}{\partial x_\beta}\right) = D_\beta \psi_\alpha = 0, \quad \alpha \neq \beta \quad (\alpha, \beta = 1, 2, \ldots, N). \qquad (3.4.23)$$

Here $D_\alpha$ is the operator of the total differentiation with respect to $x_\alpha$ and

$$\psi_\alpha = \zeta^\alpha - \frac{\partial \varphi_\alpha}{\partial x_\alpha}\xi^\alpha - \frac{\partial \varphi_\alpha}{\partial t}\xi^t.$$

The determining equations (3.4.23) become

$$\frac{\partial \zeta^\alpha}{\partial \varphi_\beta}\varphi_{\beta,x_\beta} + \frac{\partial \zeta^\alpha}{\partial x_\beta} - \varphi_{\alpha,t}\left(\frac{\partial \xi^t}{\partial \varphi_\beta}\varphi_{\beta,x_\beta} + \frac{\partial \xi^t}{\partial x_\beta}\right)$$

$$- \varphi_{\alpha,x_\alpha}\left(\frac{\partial \xi^\alpha}{\partial \varphi_\beta}\varphi_{\beta,x_\beta} + \frac{\partial \xi^\alpha}{\partial x_\beta}\right) = 0 \qquad (3.4.24)$$

$$(\alpha \neq \beta; \ \alpha, \beta = 1, 2, \ldots, N).$$

Recall that the determining equations have to be satisfied for any solution of (3.4.18). Consider solutions corresponding to the initial data at $t = t_0$:

$$\varphi_\alpha(x_\alpha, t_0) = a_\alpha x_\alpha^{k_\alpha} \quad (\alpha = 1, 2, \ldots, N), \qquad (3.4.25)$$

---

[7]The operator $X$ is considered as the equivalent canonical Lie–Bäcklund operator.

where $t_0$ is an arbitrary moment of time and $a_\alpha$ are arbitrary constants. For this set of solutions one can find all derivatives involved in (3.4.24) at $t = t_0$. Notice that for finding the derivatives $\varphi_{\alpha,t}$ one has to use (3.4.18). Substituting the derivatives into (3.4.24), and varying the values $k_\alpha$ and $a_\alpha$, one can find the general solution of the determining equations (3.4.24).

For example, for $k_\alpha = 0$ $(\alpha = 1, 2, \ldots, N)$ and by virtue of the arbitrariness of the choice of $a_\alpha$, one can conclude that $\frac{\partial \zeta^\alpha}{\partial x_\beta} = 0$ $(\alpha \neq \beta; \alpha, \beta = 1, 2, \ldots, N)$ or $\zeta^\alpha = \zeta^\alpha(\varphi_1, \ldots, \varphi_N, x_\alpha, t)$.

Setting $k_\alpha = 0$ and $k_\beta \geq 1$ for $\beta \neq \alpha$, and splitting them with respect to $a_\alpha$, one has

$$\frac{\partial \xi^t}{\partial \varphi_\beta} = 0, \quad \frac{\partial \xi^t}{\partial x_\beta} = 0, \quad \frac{\partial \zeta^\alpha}{\partial \varphi_\beta} = 0, \quad \frac{\partial \zeta^\alpha}{\partial x_\beta} = 0 \quad (\alpha \neq \beta; \; \alpha, \beta = 1, 2, \ldots, N).$$

Finally (3.4.24) become

$$\varphi_{\alpha, x_\alpha} \left( \frac{\partial \xi^\alpha}{\partial \varphi_\beta} \varphi_{\beta, x_\beta} + \frac{\partial \xi^\alpha}{\partial x_\beta} \right) = 0.$$

The last set of equations gives

$$\frac{\partial \xi^\alpha}{\partial \varphi_\beta} = 0, \quad \frac{\partial \xi^\alpha}{\partial x_\beta} = 0 \quad (\alpha \neq \beta; \; \alpha, \beta = 1, 2, \ldots, N).$$

Thus, one obtains that the coefficients of the infinitesimal generator $X$ have the form

$$\xi^t = \xi^t(t), \quad \zeta^\alpha = \zeta^\alpha(\varphi_\alpha, x_\alpha, t), \quad \xi_\alpha = \xi_\alpha(\varphi_\alpha, x_\alpha, t) \quad (\alpha = 1, 2, \ldots, N).$$

$$(3.4.26)$$

The second part of the determining equations is obtained by applying the operator $\bar{X}$ to (3.4.18). Taking into account (3.4.26) these equations are

$$D_t \psi_\alpha + \sum_{\beta=1}^{N} \sigma_{\alpha\beta} \left\{ \psi_\alpha \varphi_\beta(0) - \int_0^1 ds \psi_\alpha(x_\alpha(1 - \varepsilon_{\alpha\beta} s)) \varphi_\beta(x_\alpha \varepsilon_{\alpha\beta} s) \right.$$

$$\left. - \int_0^1 ds \varphi_\alpha(x_\alpha(1 - \varepsilon_{\alpha\beta} s)) \psi_\beta(x_\alpha \varepsilon_{\alpha\beta} s) \right\} = 0$$

$$(3.4.27)$$

$$(\alpha = 1, 2, \ldots, N),$$

where $D_t$ is the operator of the total differentiation with respect to $t$.

For solving (3.4.27), one can use the same approach as in the case of the single Boltzmann equation considered in Chap. 2. Equations (3.4.27) are considered on the solutions of (3.4.18) corresponding to initial data (3.4.25).

First, setting in (3.4.25) all $k_\alpha = 0$, the equations (3.4.27) take the form

$$\frac{\partial \zeta^\alpha}{\partial t} + \sum_{\beta=1}^{N} \sigma_{\alpha\beta} (a_\beta \zeta^\alpha + a_\alpha \zeta^\beta(0))$$

$$- \sum_{\beta=1}^{N} \sigma_{\alpha\beta} \left( a_\beta \int_0^1 \zeta^\alpha (x_\alpha (1 - \varepsilon_{\alpha\beta} s)) \, ds + a_\alpha \int_0^1 \zeta^\beta (x_\alpha \varepsilon_{\alpha\beta} s) \, ds \right) = 0 \qquad (3.4.28)$$

$(\alpha = 1, 2, \ldots, N)$.

The determining equations are solved expanding the coefficients

$$\zeta^\alpha(\varphi_\alpha, x_\alpha, t) = \sum_{n=0}^{\infty} g_\alpha^{(n)}(x_\alpha, t)(\varphi_\alpha)^n,$$

$$\xi^\alpha(\varphi_\alpha, x_\alpha, t) = \sum_{n=0}^{\infty} h_\alpha^{(n)}(x_\alpha, t)(\varphi_\alpha)^n, \qquad (3.4.29)$$

and splitting them with respect to $a_\alpha$. Indeed, considering in (3.4.28) the terms with $(a_\alpha)^0$, one finds

$$\frac{\partial g_\alpha^{(0)}}{\partial t} + \sum_{\beta=1, \beta \neq \alpha}^{N} \sigma_{\alpha\beta} a_\beta \left( g_\alpha^{(0)} - \int_0^1 g_\alpha^{(0)}(x_\alpha(1 - \varepsilon_{\alpha\beta} s)) \, ds \right) = 0 \quad (\alpha = 1, 2, \ldots, N).$$

Since the values $a_\alpha$ are arbitrary, the last equations imply that

$$\frac{\partial g_\alpha^{(0)}}{\partial t} = 0, \quad g_\alpha^{(0)} - \int_0^1 g_\alpha^{(0)}(x_\alpha(1 - \varepsilon_{\alpha\beta} s)) \, ds = 0 \quad (\alpha = 1, 2, \ldots, N).$$

Obviously, that constant values of $g_\alpha^{(0)}$ satisfy these equations.

The coefficients with $(a_\alpha)^1$ are

$$\frac{\partial g_\alpha^{(1)}}{\partial t} + \sum_{\beta=1, \beta \neq \alpha}^{N} \sigma_{\alpha\beta} a_\beta \left( g_\alpha^{(1)} - a_\beta \int_0^1 g_\alpha^{(1)}(x_\alpha(1 - \varepsilon_{\alpha\beta} s), t) \, ds \right)$$

$$+ \sum_{n=0}^{\infty} a_\beta^n \left( g_\alpha^{(n)}(0, t) - \int_0^1 g_\beta^{(n)}(x_\alpha \varepsilon_{\alpha\beta} s, t) \, ds \right) = 0$$

$(\alpha = 1, \ldots, N)$.

Splitting the last equations with respect to $(a_\beta)^n$, one has

$$g_\alpha^{(1)} = c_0 x_\alpha + c_\alpha, \quad g_\alpha^{(n)} = g_\alpha^{(n)}(t), \quad n \geq 2,$$

where $c_\alpha$ $(\alpha = 0, 1, 2, \ldots, N)$ are arbitrary constants.

Notice that the results of [20] hint one to assume that (3.4.18) are invariant with respect to the Lie group corresponding to (3.4.21). Correctness of this assumption is verified immediately. In particular, this allows one to subtract the operator $c_0 X_2$

from $X$. Hence, one can assume that $c_0 = 0$. Then (3.4.28) implies that all coefficients $g_\alpha^{(n)}$ ($n \geq 0$) are constant.

To proceed we set the powers $k_i > 1$ and $k_\alpha = 0$ ($\alpha \neq i$) in (3.4.25), substitute (3.4.29) into (3.4.27), and consider again the coefficients of $(a_\alpha)^n$, $n \geq 0$. For example, for $\alpha \neq i$ and $n = 0$, one finds $g_\alpha^{(0)} = 0$. For $\alpha = i$ one obtains

$$k_i h_i^{(0)} \varepsilon_{\alpha i} + \frac{d\xi^t}{dt} x_i \varepsilon_{\alpha i} + \sum_{n=0}^{\infty} \frac{(k_i + 1)}{(n+1)k_i + 1} a_i^n (x_i \varepsilon_{\alpha 1})^{nk_i + 1} g_i^{(n+1)}$$

$$- k_i(k_i + 1) \sum_{n=0}^{\infty} a_i^n (x_i \varepsilon_{\alpha i})^{nk_i} \int_0^1 s^{(n+1)k_i - 1} h_\alpha^{(1)} (x_i \varepsilon_{\alpha i} s) \, ds = 0, \quad (3.4.30)$$

where there is no summation with respect to $i$ and the term with $a_i (x_\alpha \varepsilon_{\alpha i})^{k_i - 1}$ is excluded.

The coefficient of $(a_i)^0$ in (3.4.30) is

$$k_i h_i^{(0)} \varepsilon_{\alpha i} + \left( \frac{d\xi^t}{dt} + g_i^{(1)} \right) x_i \varepsilon_{\alpha i} - k_i (k_i + 1) \int_0^1 s^{k_i - 1} h_i^{(0)} (x_i \varepsilon_{\alpha i} s) \, ds = 0.$$

By the arbitrariness of $k_i$, one can conclude that $h_i^{(0)} = p_i(t)x_i$ and $g_i^{(1)} + d\xi^t/dt = 0$. Hence, $g_i^{(1)} = c$, where $c$ is an arbitrary constant. This means that $\xi^t = -ct + c_0$. Since (3.4.18) admits the operators $X_1$ and $X_4$ and factoring over them, without loss of generality one can assume that $\xi^t = 0$ and $g_i^{(1)} = 0$ ($i = 1, \ldots, N$).

Since (3.4.18) admits the operators $X_1$, $X_4$ and factoring over them, then one can assume without loss of generality that $\xi^t = 0$ and $g_i^{(1)} = 0$ ($i = 1, \ldots, N$).

The comparison of the coefficients of the powers of $(a_i)^n$ ($n > 0$) in (3.4.30) yields the relations

$$\frac{1}{nk_i + 1} g_i^{(n)} x_i \varepsilon_{\alpha i} + k_i \int_0^1 s^{nk_i - 1} h_i^{(n-1)} (x_i \varepsilon_{\alpha i} s) \, ds = 0.$$

By the arbitrariness of $k_i$, the last equations imply that $g_i^{(n)} = 0$ and $h_i^{(n-1)} = 0$, $n > 1$. Since $i$ is also arbitrary, one obtains $g_\alpha^{(n)} = 0$ ($\alpha = 1, 2, \ldots, N$).

Thus, factoring over the admitted generators $X_1$, $X_2$, and $X_4$, one comes to the relations

$$\zeta^\alpha = 0, \quad \xi^t = 0, \quad \xi^\alpha = p_\alpha(t)x_\alpha \quad (\alpha = 1, 2, \ldots, N).$$

Taking these relations into account, (3.4.30) become

$$(p_i - p_\alpha) \int_0^1 s \frac{\partial \varphi_i}{\partial x_i} (x_i \varepsilon_{\alpha 1} s) \, ds = 0.$$

This means that $p_i = p(t)$ ($i = 1, \ldots, N$). Substituting these expressions into (3.4.27), one finds that $dp/dt = 0$. As the result, one concludes that system (3.4.18) only admits the generators (3.4.21). $\qquad\qquad\qquad\qquad\qquad\qquad\square$

**Theorem 3.4.5** *The classes of essentially different with respect to the Lie group $G^4$ invariant solutions of* (3.4.18) *are determined by the optimal system of one-dimensional subalgebras*

$$\{X_1\}, \quad \{X_4 + \beta X_3\}, \quad \{X_2 - X_1\}, \quad \{X_4 \pm X_2\}, \quad \{X_1 + X_3\}, \qquad (3.4.31)$$

*where $\beta$ is an arbitrary constant.*

Representations of invariant solutions corresponding to the subalgebras of the optimal system are

$$X_1 : \varphi_\alpha = g(y_\alpha), \quad y_\alpha = x_\alpha,$$

$$X_4 + aX_3 : \varphi_\alpha = g(y_\alpha), \quad y_\alpha = x_\alpha t^a,$$

$$X_2 - X_1 : \varphi_\alpha = e^{-x_\alpha t} g(y_\alpha), \quad y_\alpha = x_\alpha,$$

$$X_3 + X_1 : \varphi_\alpha = g(y_\alpha), \quad y_\alpha = x_\alpha e^{-t},$$

$$X_4 \pm X_2 : \varphi_\alpha = t^{-(1 \pm x_\alpha)} g(y_\alpha), \quad y_\alpha = x_\alpha$$

$$(\alpha = 1, 2, \ldots, N).$$

As in the case of a one-component gas special interest is in finding the BKW-type solutions [2, 28] in elementary functions.[8] The class of BKW-solutions is defined by the subalgebra $\{X_1 + X_3\}$. Usually the BKW-solutions are considered with respect to the subalgebra $\{X_2 - X_3 + c^{-1}X_1\}$ that is similar to the subalgebra $\{X_1 + X_3\}$ with respect to automorphisms of the Lie algebra $L_4$. Using the subalgebra $\{X_2 - X_3 + c^{-1}X_1\}$ we look for BKW-solutions of the following form

$$\varphi_\alpha(x_\alpha, t) = \exp(y_\alpha - x_\alpha)\Phi_\alpha(y_\alpha), \quad y = \theta_0 e^{ct} x_\alpha. \qquad (3.4.32)$$

Substituting (3.4.32) into (3.4.18), one comes to the factor-equations for the functions $\Phi_\alpha(y_\alpha)$:

$$cy_\alpha \left( \frac{d\Phi_\alpha(y_\alpha)}{dy_\alpha} + \Phi_\alpha(y_\alpha) \right)$$

$$= \sum_{\beta=1}^{N} \sigma_{\alpha\beta} \int_0^1 ds [\Phi_\alpha((1 - \varepsilon_{\alpha\beta}s)y_\alpha)\Phi_\beta(\varepsilon_{\alpha\beta}sy_\alpha) - \Phi_\alpha(y_\alpha)\Phi_\beta(0)].$$

$$(3.4.33)$$

A solution of (3.4.33) is sought for in the form of the power series

$$\Phi_\alpha(y_\alpha) = n_\alpha \left( 1 + \sum_{k=1}^{\infty} b_k^{(\alpha)} \frac{y_\alpha^k}{k!} \right). \qquad (3.4.34)$$

---

[8]The previous results [29] obtained in this direction were far from being complete.

Here the mass conservation law (3.4.19) is taken into account: $\varphi_\alpha(0, t) = \Phi_\alpha(0) = n_\alpha$. The energy conservation law imposes the following constraint on the first coefficients of the series (3.4.34):

$$\sum_{\alpha=1}^{N} n_\alpha b_1^{(\alpha)} = -1.$$

(3.4.35)

The study of the system for the coefficients

$$\mathbf{b}_k = \left(b_k^{(1)}, \ldots, b_k^{(N)}\right)$$

of the power series (3.4.34) allows one to prove [21] the following.[9]

**Theorem 3.4.6**  1. *The BKW-solutions (3.4.14) have the form*

$$f_\alpha(v_\alpha, t) = \frac{n_\alpha}{[2\pi T(t)]^{3/2}} \left[1 + \frac{1 - T(t)}{T(t)} \left(\frac{v_\alpha^2}{2T(t)} - \frac{3}{2}\right)\right] \exp\left(-\frac{v_\alpha^2}{2T(t)}\right),$$

$$T(t) = 1 - \theta_0 e^{ct}, \quad \theta_0 \in [0, 2/5] \quad (\alpha = 1, 2, \ldots, N),$$

(3.4.36)

*similar to the BKW-solutions for a one-component gas [2, 28] if and only if the parameters of the gas mixture obey the constraints*

$$-\frac{1}{6}\left[n_\alpha \sigma_{\alpha\alpha} + \sum_{(\beta)}' n_\beta \sigma_{\alpha\beta}(3 - 2\varepsilon_{\alpha\beta})\varepsilon_{\alpha\beta}\right] = c, \quad \alpha = 1, \ldots, N.$$

*and*

$$c = -\frac{1}{6N} \sum_{\alpha=1}^{N}\left[n_\alpha \sigma_{\alpha\alpha} + \sum_{(\beta)}' n_\beta \sigma_{\alpha\beta}(3 - 2\varepsilon_{\alpha\beta})\varepsilon_{\alpha\beta}\right].$$

2. *For the components with indices $\alpha = 1, \ldots, l, l \leq r$, the BKW-solutions have the form of the nonstationary Maxwellian distribution functions*

$$f_\alpha(v_\alpha, t) = \frac{n_\alpha}{[2\pi T(t)]^{3/2}} \exp\left(-\frac{v_\alpha^2}{2T(t)}\right);$$

(3.4.37)

*whereas for the components with indices $\alpha = l+1, \ldots, r$,*

$$f_\alpha(v_\alpha, t) = \frac{n_\alpha}{[2\pi T(t)]^{3/2}}\left[1 + q_\alpha(t)b^{(\alpha)}\right]\exp\left(-\frac{v_\alpha^2}{2T(t)}\right);$$

(3.4.38)

*and for the components with indices $\alpha = r+1, \ldots, N$,*

$$f_\alpha(v_\alpha, t) = \frac{n_\alpha}{[2\pi T(t)]^{3/2}}[1 + q q_\alpha(t)]\exp\left(-\frac{v_\alpha^2}{2T(t)}\right),$$

(3.4.39)

---

[9]The spectral properties of the linearized system corresponding to (3.4.18) are substantially used in the proof.

*where*

$$q = \frac{(1 - \sum_{\alpha=l+1}^{r} b^{(\alpha)} n_\alpha)}{\sum_{\alpha=r+1}^{N} n_\alpha}, \quad q_\alpha(t) = \frac{1 - T(t)}{T(t)} \left( \frac{v_\alpha^2}{2T(t)} - \frac{3}{2} \right),$$

$b^{(\alpha)}$ $(\alpha = l + 1, \ldots, r)$ *are arbitrarily-chosen coefficients such that all* $b^{(\alpha)}$ *are strictly greater than zero and* $\sum_{\alpha=l+1}^{r} b^{(\alpha)} n_\alpha < 1$, *if and only if the conditions*

$$\frac{1}{2} \sigma_{\alpha\beta} \varepsilon_{\alpha\beta} = -c, \quad \alpha < \beta, \quad \alpha = 1, \ldots, r, \ \beta = 2, \ldots, N,$$

*are satisfied and the parameters of the mixture obey the extra constraints*

$$c = -\frac{1}{6} \left[ n_\alpha \sigma_{\alpha\alpha} b_1^{(\alpha)} + \sum_{(\beta)}' n_\beta \sigma_{\alpha\beta} b_1^{(\beta)} (3 - 2\varepsilon_{\alpha\beta}) \varepsilon_{\alpha\beta} \right]$$

$$(\alpha = 1, 2, \ldots, N), \tag{3.4.40}$$

$$b_1^{(\alpha)} = -b^{(\alpha)} \quad (\alpha = l + 1, \ldots, r),$$

$$b_1^{(\alpha)} = -q \quad (\alpha = r + 1, \ldots, N).$$

*In this case, the parameter* $\theta_0$ *in* (3.4.37)–(3.4.39) *is chosen within the interval*

$$0 \le \theta_0 \le \min[\theta_{01}, \theta_{02}], \tag{3.4.41}$$

*where*

$$\theta_{01} = \frac{2}{2 + 3b_M}, \quad b_M = \max_{(\alpha)} b^{(\alpha)}, \quad \theta_{02} = 2(2 + 3q).$$

**Remark 3.4.3** More complete formulation of this theorem with elements of the proof one can find in [22].

The found BKW-solutions (3.4.38)–(3.4.39) with arbitrary coefficients $b^{(\alpha)}$ substantially extend the set of the BKW-solutions in elementary functions. The nonstationary Maxwellian distribution functions (3.4.37) are of special interest. They were never considered before in the framework of the BKW-class.

The presented BKW-modes allow one to model in an explicit form many physically interesting kinetic processes even for $N = 2$. For example, to investigate the relaxation in mixture with disparate species masses such as Lorentz gas or Rayleigh gas. It is possible to study the behavior of disequilibrium impurity in Maxwellian bath or vice versa of Maxwellian impurity on the non-equilibrium background, etc.

## 3.4.4 Homogeneous Relaxation of a Binary Model Gas

In conclusion the BKW-solutions of the system of kinetic equations which is a low-dimensional mathematical model of the system of the Boltzmann equations for a binary gas mixture are presented.

In order to give some manageable results for constructing and testing numerical algorithms, a system of two equations with the simplest model of molecular interaction is considered [24]. Despite of its simplicity, the system reproduces the main mathematical features of the kinetics for a multi-component mixture. This system is written as

$$\frac{\partial f_i(t, v_i)}{\partial t} = \int_{-\infty}^{\infty} dw_i \int_{-\pi}^{\pi} d\theta g_{ii}(\theta)[f_i(v_i')f_i(w_i') - f_i(v_i)f_i(w_i)]$$

$$+ \int_{-\infty}^{\infty} dw_j \int_{-\pi}^{\pi} d\theta g_{ij}(\theta)[f_i(v_i')f_i(w_j') - f_i(v_i)f_j(w_j)] \quad (3.4.42)$$

$$(i = 1, 2).$$

Here $f_i(t, v_i)$ is the distribution function of the $i$-th component, $t \in \mathbb{R}_+^1$, $v_i \in \mathbb{R}^1$, and $g_{ij}(\theta) = g_{ji}(\theta)$ and $g_{ii}(\theta)$ are nonnegative scattering functions subject to the normalization conditions

$$\int_{-\pi}^{\pi} g_{ii}(\theta) d\theta = \int_{-\pi}^{\pi} g_{ij}(\theta) d\theta = 1.$$

A "collision" of model molecules $(v_i, w_j) \rightarrow (v_i', w_j')$ is described by the orthogonal rotation in the velocity plane

$$(v_i', w_j')' = A(v_i, w_i)', \quad A = \begin{pmatrix} \cos\theta & -\sin\theta \\ \sin\theta & \cos\theta \end{pmatrix}.$$

During collisions, the elementary conservation laws of the energy and the number of particles hold: $v_i^2 + w_j^2 = v_i'^2 + w_j'^2$. The spatially homogeneous equilibrium solutions of (3.4.42) reached as $t \rightarrow \infty$ have the form

$$f_{i,0} = \frac{n_i}{\sqrt{2\pi}} e^{-v_i^2/2}, \quad i = 1, 2,$$

where $n_i$ is the particle concentration of the $i$-th species.

For system (3.4.42) the conservation laws of concentrations and energy hold:

$$\sum_{i=1}^{2} \int_{-\infty}^{\infty} f_i(t, v_i) dv_i = \sum_{i=1}^{2} \int_{-\infty}^{\infty} f_{i,0}(v_i) dv_i = \sum_{i=1}^{2} n_i = 1,$$

$$\sum_{i=1}^{2} \int_{-\infty}^{\infty} v_i^2 f_i(t, v_i) dv_i = \sum_{i=1}^{2} \int_{-\infty}^{\infty} v_i^2 f_{i,0}(v_i) dv_i = 1.$$

The direct and inverse Fourier transforms are defined as follows

$$\varphi_i(k_i) = \int_{-\infty}^{\infty} f_i(v_i) e^{-2\pi i k_i v_i} dv_i, \quad f_i(v_i) = \int_{-\infty}^{\infty} \varphi_i(k_i) e^{2\pi i k_i v_i} dk_i \quad (i = 1, 2).$$

The Fourier representation of system (3.4.42) has the form

$$\frac{\partial \varphi_i(t, k_i)}{\partial t} = \sum_{j=1}^{2} \int_{-\pi}^{\pi} d\theta g_{ij}(\theta)[\varphi_i(k_i \cos\theta)\varphi_j(k_i \sin\theta) - \varphi_i(k_i)\varphi_j(0)] \quad (i = 1, 2).$$

$$(3.4.43)$$

The Fourier transforms of the equilibrium solutions are

$$\varphi_{i,0}(k_i) = n_i e^{-2\pi^2 k_i^2}.$$

In terms of the Fourier transforms the conservation law of particle concentrations takes the form

$$\sum_{i=1}^{2} \int_{-\infty}^{\infty} f_i(t, v_i)\, dv_i = \sum_{i=1}^{2} \varphi_i(t, 0) = 1, \quad \varphi_i(t, 0) = n_i.$$

The conservation law of energy becomes

$$\sum_{i=1}^{2} \int_{-\infty}^{\infty} v_i^2 f_i(t, v_i)\, dv_i = -\sum_{i=1}^{2} \varphi_i''(0)\frac{1}{4\pi^2} = 1.$$

System (3.4.43) admits the same complete Lie group of point transformations as the admitted Lie group of system (3.4.18) obtained above. The BKW-solutions associated with the operator $X_2 - X_3 + c^{-1}X_1$ are considered in the form

$$\varphi_i(k_i, t) = \exp[4\pi^2(y_i - x_i)]\Phi_i(y_i), \quad y_i = x_i\theta_0 e^{ct}, \quad x_i = \frac{k_i^2}{2}. \quad (3.4.44)$$

The substitution of (3.4.44) into (3.4.43) yields the system of the factor-equations for $\Phi_i(y_i)$:

$$cy_i\left(\frac{d\Phi_i}{dy_i} + 4\pi^2 y_i \Phi_i(y_i)\right)$$

$$= \sum_{i=1}^{2} \int_{-\pi}^{\pi} d\theta g_{ij}(\theta)[\Phi_i(y_i \cos\theta)\Phi_j(y_i \sin\theta) - \Phi_i(y_i)\Phi_j(0)].$$

Its solutions one seeks in the form of the power series

$$\Phi_i(y_i) = n_i\left(1 + \sum_{m=1}^{\infty} b_m^{(i)}\frac{y_i^m}{m!}\right).$$

Since $\varphi_i(0, t) = n_i$, the structure of these series is conformed with the conservation law of concentrations. The energy conservation law imposes the additional restriction on the first coefficients of the series

$$-\frac{1}{4\pi^2}\sum_{i=1}^{2} n_i b_1^{(i)} = 1.$$

The BKW-solutions of (3.4.42) in explicit form with the scattering model $g_{ij}(\theta) = v_{ij}/(2\pi)$ are described by the following theorem [24].

**Theorem 3.4.7** 1. *The BKW-solutions of* (3.4.42) *have the form*

$$f_i(v_i, t) = \frac{n_i}{\sqrt{2\pi(1-\theta(t))}} e^{-\frac{v_i^2}{2(1-\theta(t))}} \left[1 + \frac{\theta(t)}{2(1-\theta(t))}\left(\frac{v_i^2}{(1-\theta(t))} - 1\right)\right]$$
$$(i = 1, 2),$$

$$\theta = \theta_0 e^{ct}, \quad 0 < \theta_0 < \frac{2}{3},$$

*if and only if the parameters of the gas mixture satisfy the condition*

$$n_2\nu_{22} - n_1\nu_{11} = (n_2 - n_1)\nu_{12}$$

*and the parameter c is*

$$c = -\frac{1}{8}(n_i\nu_{ii} + n_j\nu_{ij}).$$

2. *If one puts* $c = \mu_2^{(2)} = -\frac{\nu_{12}}{2}$, *provided that* $\nu_{12} = \nu_{22}/8$, *then the BKW-solution takes the form*

$$f_1(v_1, t) = n_1 \frac{1}{\sqrt{2\pi(1-\theta)}} e^{-\frac{v_1^2}{2(1-\theta)}},$$

$$f_2(v_2, t) = n_2 \frac{1}{\sqrt{2\pi(1-\theta)}} e^{-\frac{v_2^2}{2(1-\theta)}} \left[1 + \frac{\theta}{2n_2(1-\theta)}\left(\frac{v_2^2}{1-\theta} - 1\right)\right],$$

*where*

$$\theta = \theta_0 e^{ct}, \quad 0 < \theta_0 \le \frac{2n_2}{1+2n_2}.$$

System (3.4.42) and its BKW-solutions in the explicit form give one a good material for testing different numerical methods which are intensively studied in recent years for the problems of the Boltzmann's kinetics of multi-component gas mixtures.

# References

1. Bobylev, A.V.: On exact solutions of the Boltzmann equation. Dokl. AS USSR **225**(6), 1296–1299 (1975)
2. Bobylev, A.V.: On one class of invariant solutions of the Boltzmann equation. Dokl. AS USSR **231**(3), 571–574 (1976)
3. Bobylev, A.V.: The structure of spatially uniform normal solutions of the nonlinear Boltzmann equation for a gas mixture. Dokl. Akad. Nauk SSSR **250**(2), 340–344 (1980)
4. Bobylev, A.V.: The Boltzmann equation and group transformations. Math. Models Methods Appl. Sci. **3**, 443–476 (1993)
5. Bobylev, A.V., Dorodnitsyn, V.A.: Symmetries of evolution equations with non-local operators and applications to the Boltzmann equation. Discrete Contin. Dyn. Syst. **24**(1), 35–57 (2009)
6. Bobylev, A.V., Ibragimov, N.H.: Relationships between the symmetry properties of the equations of gas kinetics and hydrodynamics. J. Math. Model. **1**(3), 100–109 (1989)

7. Bobylev, A.V., Caraffini, G.L., Spiga, G.: On group invariant solutions of the Boltzmann equation. J. Math. Phys. **37**(6), 2787–2795 (1996)

8. Boltzmann, L.: Further studies on the thermal equilibrium among gas-molecules. Collected Works **1**, 275–370 (1872)

9. Bunimovich, A.I., Krasnoslobodtsev, A.V.: Invariant-group solutions of kinetic equations. Meh. Zidkosti Gas. (4), 135–140 (1982)

10. Bunimovich, A.I., Krasnoslobodtsev, A.V.: On some invariant transformations of kinetic equations. Vestn. Moscow State Univ., Ser. 1., Mat. Meh. (4), 69–72 (1983)

11. Carleman, T.: Problemes Mathematiques Dans la Theorie Cinetique des Gas. Almqvist & Wiksell, Uppsala (1957)

12. Cercignani, C.: Theory and Application of the Boltzmann Equation. Chatto and Windus, London (1975)

13. Cercignani, C.: The Boltzmann Equation and Its Applications. Springer, New York (1988)

14. Cherevko, A.A.: Optimal system subalgebras for the Lie algebra of generators admitted by the gas dynamics equations with the state equation $p = f(s)\rho^{5/3}$. Preprint of Institute of Hydrodynamics SO RAS, Novosibirsk, Russia (1996)

15. Ernst, M.H.: Nonlinear model—Boltzmann equations and exact solutions. Phys. Rep. **78**, 1–171 (1981)

16. Golovin, S.V.: Optimal system of subalgebras for the Lie algebra of generators admitted by the polytropic gas dynamics equations. Preprint of Institute of Hydrodynamics SO RAS, Novosibirsk, Russia (1996)

17. Grigoriev, Y.N., Meleshko, S.V.: Investigation of invariant solutions of the Boltzmann kinetic equation and its models. Preprint of Institute of Theoretical and Applied Mechanics (1986)

18. Grigoriev, Y.N., Meleshko, S.V.: Group analysis of the integro-differential Boltzmann equation. Dokl. AS USSR **297**(2), 323–327 (1987)

19. Grigoriev, Y.N., Meleshko, S.V.: Group theoretical analysis of the kinetic Boltzmann equation and its models. Arch. Mech. **42**(6), 693–701 (1990)

20. Grigoriev, Y.N., Meleshko, S.V.: Group analysis of kinetic equations. Russ. J. Numer. Anal. Math. Model. **9**(5), 425–447 (1995)

21. Grigoriev, Y.N., Meleshko, S.V.: The full Lie group and invariant solutions of the system of Boltzmann equations of a multicomponent gas mixture. Sib. Math. J. **38**(3), 434–448 (1997)

22. Grigoriev, Y.N., Meleshko, S.V.: Bobylev–Krook–Wu modes for multicomponent gas mixtures. Phys. Rev. Lett. **81**(1), 93–95 (1998)

23. Grigoriev, Y.N., Meleshko, S.V., Sattayatham, P.: Classification of invariant solutions of the Boltzmann equation. J. Phys. A, Math. Gen. **32**, 337–343 (1999)

24. Grigoriev, Y.N., Omel'yanchuk, M.: Qualitative properties of a certain kinetic model of binary gas. Sib. Math. J. **46**(5), 813–825 (2005)

25. Grigoryev, Y.N., Mikhalitsyn, A.N.: A spectral method of solving Boltzmann's kinetic equation numerically. U.S.S.R. Comput. Math. Math. Phys. **23**(6), 105–111 (1985)

26. Grigoryev, Y.N., Mikhalitsyn, A.N.: Asymptotics of the Boltzmann kinetic equation for high-energy particles. Arch. Mech. **39**(4), 303–313 (1986)

27. Kogan, M.N.: Rarefied Gas Dynamics. Plenum Press, New York (1969)

28. Krook, M., Wu, T.T.: Formation of Maxwellian tails. Phys. Rev. Lett. **36**(19), 1107–1109 (1976)

29. Krook, M., Wu, T.T.: Exact solutions of the Boltzmann's equation for multicomponent systems. Phys. Rev. Lett. **38**(18), 991–993 (1977)

30. Maxwell, J.C.: On the dynamical theory of gases. Philos. Trans. R. Soc. Lond. **157**, 49–88 (1867)

31. Nakashev, N.K.: Properties of the solution of the Boltzmann equation at high energies of the translational molecular motion and their corollaries. Dokl. Acad. Sci. USSR **258**(1), 52–56 (1981)

32. Nikolskii, A.A.: The simplest exact solutions of the Boltzmann equation of a rarefied gas motion. Dokl. AS USSR **151**(2), 299–301 (1963)

33. Nikolskii, A.A.: Three dimensional homogeneous expansion–contraction of a rarefied gas with power interaction functions. Dokl. AS USSR **151**(3), 522–524 (1963)

34. Nikolskii, A.A.: Homogeneous motion of displacement of monatomic rarefied gas. Inz. J. **5**(4), 752–755 (1965)
35. Ovsiannikov, L.V.: Group Analysis of Differential Equations. Nauka, Moscow (1978). English translation by Ames, W.F. (ed.), published by Academic Press, New York (1982)
36. Ovsiannikov, L.V.: Program SUBMODELS. Gas dynamics. J. Appl. Math. Mech. **58**(4), 30–55 (1994)
37. Rykov, V.A.: Classification of invariant solutions of the Boltzmann equation. J. Appl. Math. Mech. **31**, 756–762 (1967)
38. Voloshchuk, V.M.: Kinetic Coagulation Theory. Gidrometeoizdat, Leningrad (1984)

# Chapter 4
# Plasma Kinetic Theory: Vlasov–Maxwell and Related Equations

## 4.1 Mathematical Model

Plasma is an ionized gas of charged particles. Plasma is distinguished from usual gases in the sense that plasma particles give rise to essential electromagnetic fields. Hence, the usual Boltzmann kinetic approach that takes into account only paired collisions should be supplemented by influence of electromagnetic fields generated in plasma on the motion of plasma particles. The plasma inhomogeneity caused by the electromagnetic field (i.e. inhomogeneous distribution of charged plasma particles) results in generation of induced charges and currents. The latter in turn creates the electromagnetic field, that anew modifies the motion of plasma particles. Therefore the correct description of a system of plasma particles should meet the condition of self-consistency.

The analysis of an infinite system of equations of motion for all plasma particles is conventionally replaced by studying a distribution function of coordinates and impulses of all plasma particles. The key point here is that plasma is a gas, thus all plasma particles move independently.[1] Therefore one can use one-particle distribution function $f^\alpha(t, r, p)$ that defines the probability to find a particle of $\alpha$ species with the impulse $p$ at time $t$ and point $r$. The conservation of probability yields

$$\frac{df^\alpha}{dt} \equiv f_t^\alpha + r_t f_r^\alpha + p_t f_p^\alpha = 0.$$

---

[1]Usually it is assumed for gas particles that the energy of their interaction is small compared to their kinetic energy. Up to the order of magnitude the latter can be estimated as $\kappa T$, where $T$ is the temperature and $\kappa$ is the Boltzmann constant. For charged plasma particles the energy of interaction is of the order of $e^2 N^{1/3}$, where $N^{-1/3}$ is the mean distance between particles, $e$ is a charge and $N$ is the number of particles in a unit volume. Hence the plasma demonstrates the gas property provided that

$$e^2 N^{1/3} \ll \kappa T.$$

This inequality holds for all real plasmas.

Y.N. Grigoriev et al., *Symmetries of Integro-Differential Equations*,
Lecture Notes in Physics 806,
DOI 10.1007/978-90-481-3797-8_4, © Springer Science+Business Media B.V. 2010

Noting that $r_t = v$ is a particle velocity and $p_t$ for charged particles is defined by the Lorentz force

$$e^\alpha \left\{ E + \frac{1}{c} [v \times B] \right\},$$

the equation for the distribution function for any plasma particle species takes the form:

$$f_t^\alpha + v f_r^\alpha + e^\alpha \left\{ E + \frac{1}{c} [v \times B] \right\} f_p^\alpha = 0. \tag{4.1.1}$$

The charge and current densities are defined via distribution function

$$\rho = \sum_\alpha e_\alpha m_\alpha^3 \int dv f^\alpha \gamma^5,$$

$$j = \sum_\alpha e_\alpha m_\alpha^3 \int dv f^\alpha \gamma^5 v, \quad \gamma = \frac{1}{\sqrt{1 - v^2/c^2}}, \tag{4.1.2}$$

where summation is taken over all plasma particle species. These charge and current densities enter the field equations and define electric and magnetic fields in plasma in a self-consistent manner. Equation (4.1.1) in view of the field equations

$$B_t + c \operatorname{rot} E = 0; \quad \operatorname{div} E = 4\pi \rho;$$
$$E_t - c \operatorname{rot} B + 4\pi j = 0; \quad \operatorname{div} B = 0, \tag{4.1.3}$$

are known as *kinetic equations with a self-consistent field*. The efficiency of this equation for description of plasma properties was first demonstrated by Vlasov [1]. At present Vlasov's kinetic equation with a self-consistent field is a basic equation in the theory of a collisionless[2] plasma (e.g., hot plasma used in the plasma fusion experiments).

Meanwhile in describing the evolution of distribution functions frequently it is more convenient to use not standard Vlasov equations (4.1.1) with Euler velocity $v$, but their hydrodynamic analogue [2–4] with Lagrangian velocity $w$. At transition to Lagrangian notations instead of the Euler velocity $v$ and the Euler momentum $p$ for each particle species two vector functions are introduced, the velocity $V^\alpha(t, r, q)$ and the momentum $P^\alpha(t, r, q)$, depending upon the Lagrangian momentum $q$ and Euler coordinates $r$ and time $t$, and related to $v$ and $p$ by the formulas

$$v = V^\alpha(q, r, t), \quad p = P^\alpha(q, r, t),$$
$$V^\alpha = c^2 P^\alpha (m^2 c^4 + c^2 (P^\alpha)^2)^{-1/2}. \tag{4.1.4}$$

---

[2]Equation (4.1.1) is approximate, as it neglects collisions of plasma particles. In view of particle collisions their motion becomes correlated. This effect leads to appearance of non-zero term in the right-hand side of (4.1.1), the so-called collision integral. However, the explicit form of the collision integral depends on particular conditions defined by the plasma properties in every concrete situation, and we will not discuss them here. In many particular problems collision effects can be neglected.

The change of variables (4.1.4) eliminates in the resulting equations for distribution functions for particle of each species the derivatives of these functions upon the Lagrangian momentum $q$. Hence, Lagrangian formulation of the kinetic description of plasma is fulfilled via the equations of hydrodynamic type for the density $N^\alpha(t, r, w)$ and the velocity $V^\alpha(t, r, w)$, which depend upon $t$, $r$ and $w$,

$$N_t^\alpha + \text{div}(N^\alpha V^\alpha) = 0,$$

$$V_t^\alpha + (V^\alpha \nabla)V^\alpha$$

$$= \frac{e_\alpha}{m_\alpha}\sqrt{1 - \left(\frac{V^\alpha}{c}\right)^2}\left\{E + \frac{1}{c}[V^\alpha \times B] - \frac{1}{c^2}V^\alpha(V^\alpha \cdot E)\right\}.$$

(4.1.5)

Here the index $\alpha$ indicates the plasma particle species with the charge $e_\alpha$ and mass $m_\alpha$ and the charge and current densities, $\rho$ and $j$, are in turn determined by the motion of plasma particles:

$$\rho = \sum_\alpha e_\alpha m_\alpha^3 \int dw N^\alpha \Gamma^5,$$

$$j = \sum_\alpha e_\alpha m_\alpha^3 \int dw N^\alpha V^\alpha \Gamma^5, \quad \Gamma = \frac{1}{\sqrt{1 - (w/c)^2}}.$$

(4.1.6)

It is typical, that (4.1.5) do not contain Lagrangian velocity $w$ (or Lagrangian momentum $q$) in explicit form. In order to find the dependence upon $q$ one should solve these equations with the "initial" conditions $V^\alpha = w$, $N^\alpha = N_0^\alpha(t_0, r, w)$ which hold for vanishing electric and magnetic fields $E = B = 0$ at some $t = t_0$. In particular, in homogeneous plasma "initial" conditions for the density $N^\alpha$ has the form $N_0^\alpha = n_{\alpha 0}f_0^\alpha(q)$, where the stationary and homogeneous function $f_0^\alpha(q)$ of Lagrangian momentum $q$ coincides with the function $f_0^\alpha(p)$ of Euler momentum $p$.

Given the density $N^\alpha(t, r, w)$ and the velocity $V^\alpha(t, r, w)$ that depend upon Lagrangian momentum the particles distribution function in Euler representation is restored with the help of the following relations (the index of particles species in these formulas is omitted)

$$N(t, r, q) = f\left(p = P(q, r, t), r, t\right)\det\left(\frac{\partial P_i}{\partial q_j}\right), \quad v = c^2 p(m^2 c^4 + c^2 p^2)^{-1/2},$$

$$w = c^2 q(m^2 c^4 + c^2 q^2)^{-1/2}, \quad V = c^2 P(m^2 c^4 + c^2 P^2)^{-1/2}.$$

(4.1.7)

The system of equations (4.1.4)–(4.1.7), (4.1.3) presents the Lagrangian formulation [2–4] of Vlasov–Maxwell equations, in which (4.1.6) appear as non-local material relations.

The search for particular solutions of the joint system of Vlasov–Maxwell equations (4.1.1)–(4.1.3) or its Lagrangian formulation (4.1.5)–(4.1.7) is very important both in theoretical treatment and practical applications. The group analysis of the system of Vlasov–Maxwell equations, which forms the essence of this chapter, offers a nice opportunity in constructing these solutions. The main obstacle in finding symmetry group for systems (4.1.1)–(4.1.3) and (4.1.5)–(4.1.7) with the help of a standard Lie algorithm is the non-locality of material relations (4.1.2) and (4.1.6). The first successful attempt in this field [5] deals with calculating the continuous point Lie group for the system (4.1.1)–(4.1.3) in the one-dimensional non-relativistic approximation of homogeneous electron plasma using the methods of moments. On the contrary we will follow a general algorithm [6–8] based on the direct method of calculation of symmetries, described in Chaps. 2 and 4.

This chapter is structured as follows. Section 4.2 introduces an approach for calculating symmetries of integro-differential equations used in this chapter. In Sect. 4.3.1 we describe in details the application of the general algorithm to the most simple one-dimensional non-relativistic model of one-component charged electron plasma that arises from (4.1.1)–(4.1.3) while treating only one particle species (electrons) and in one-dimensional plane geometry. We also neglect relativistic effects here. We consider this model since it is physically simple and informative from the group standpoint. The model has the same characteristic features as the complete three-dimensional system of kinetic equations for collisionless relativistic electron–ion plasma. The only difference is in the a smaller amount of calculations necessary for constructing and solving the group determining equations. For this reason, the next models are analyzed in less detail.

In the next sections we present the result of group analysis for the successively complicated systems that take into account other plasma species (Sects. 4.3.3, 4.3.4), relativistic effects (Sects. 4.3.2, 4.3.4), the presence of stationary or moving ion background (Sects. 4.3.5, 4.3.6). We also consider the so-called quasi-neutral approximation for plasma particles dynamics (Sect. 4.3.7). Symmetry of plasma kinetic equations in three dimensional geometry is analyzed for electron gas in Sect. 4.4.1 and for electron–ion plasma in Sect. 4.4.2. We also discuss the symmetry of plasma kinetic equations in Lagrangian variables (Sect. 4.5).

The special section is devoted to symmetry of Benney equations (Sect. 4.6). Here we apply both our algorithm and method of moments to demonstrate the incompleteness of the algorithm to describe all the admitted symmetries.

Section 4.7 is devoted mainly to particular problems that illustrate the efficiency of the symmetry approach to integro-differential equations to find solutions to various particular problems of interest. Here we especially draw attention to symmetries known in mathematical physics as *renormgroup symmetries*. Section 4.7 demonstrates the method of their construction as well as examples of applications.

## 4.2  Definition and Infinitesimal Test

To extend the classical Lie algorithm to integro-differential equations it appears necessary to resolve several problems. First, one should define the local one-parameter transformation group $G$ for the nonlocal (integro-differential) equations and formulate the invariance criteria that lead to determining equations, which appear also nonlocal. Secondly, and a procedure of solving nonlocal determining equations should be described.

### 4.2.1 Definition of Symmetry Group

Let an integro-differential equation under consideration be expressed as a zero equality for some functional (here we indicate only one argument for a function $f$), defined for $x_1 \leq x \leq x_2$,

$$F[f(x)] = 0, \qquad (4.2.8)$$

and let $G$ be a local one-parameter group that transforms $f$ to $\tilde{f}(x, a)$,

$$\tilde{f}(x, a) = f + a\varkappa + o(a), \quad \tilde{x} = x. \qquad (4.2.9)$$

Here we use the canonical group representation hence independent variables $x$ do not vary. Then the local group $G$ of point transformations (4.2.9) is called a symmetry group of integro-differential equations (4.2.8) iff for any $a$ the function $F$ does not vary [9] (see also Chap. 2),

$$F[\tilde{f}(x, a)] = 0. \qquad (4.2.10)$$

Differentiating (4.2.10) with respect to group parameter $a$ and assuming $a \to 0$ gives the invariance criterion in the infinitesimal form akin to (1.1.31) in Chap. 1. In view of the canonical form of transformations (4.2.9) the functional $F$ depends upon $a$ via $\tilde{f}$. Therefore to find the infinitesimal invariance criterion we should calculate the derivative $dF/da$.

### 4.2.2 Variational Derivative for Functionals

Let $f(x, a)$ be a differentiable function with respect to $a$, $f(x, a)$ and $\partial f(x, a)/\partial a$ continuous functions for $a \geq 0$, $x_1 \leq x \leq x_2$. The derivative $dF/da$ [10]

$$\frac{d}{da} F[f(x, a)] = \delta F\left[f(x, a); f_a'(x, a)\right], \qquad (4.2.11)$$

is given by variation of the functional $\delta F$, defined as a linear in $\delta f$ part of a difference

$$\delta F = F[f + \delta f] - F[f].$$

Let $F[f(x, a)]$ be a differentiable functional (recall that the functional $F$, defined on the interval $[x_1, x_2]$, is called a differentiable functional [10] if it has the first derivative in each point of this interval). Then the last formula is rewritten in the following form

$$\delta F = \int_{x_1}^{x_2} F'[f(x); q]\delta f(q)\, dq. \qquad (4.2.12)$$

Here the derivative $F'[f(x); y] = \delta F/\delta f(y)$ of the differentiable functional $F$ with respect to a function $f$ in the point $y$ is defined via the principal (linear) part of an increment of the functional as a limit (if it exists) (see [10]):

$$\frac{\delta F[f]}{\delta f(y)} = \lim_{\varepsilon \to 0} \frac{\{F[f + \delta f_\varepsilon] - F[f]\}}{\int_\Delta dy \delta f_\varepsilon(y)}; \quad y \in [x_1, x_2]. \tag{4.2.13}$$

In (4.2.13) the infinitesimal variation $\delta f_\varepsilon(y) \geq 0$ is a continuously differentiable function given on fixed interval $\Delta = [x_1, x_2]$ which differs from zero only in $\varepsilon$-vicinity of a point $y$, and the norm $\|\delta f_\varepsilon\|_{C^1} \to 0$ at $\varepsilon \to 0$.

**Example 4.2.1** Let $b(y)$ be a continuous function and $F[f]$ a linear functional

$$F[f] = \int_{x_1}^{x_2} b(y) f(y) \, dy.$$

By $\delta f_\varepsilon$ denote a variation that differs from zero only in $\varepsilon$-vicinity of a point $q$. Then using the mean value theorem

$$F[f + \delta f_\varepsilon] - F[f] = \int_{x_1}^{x_2} b(y) \delta f_\varepsilon \, dy = b(q) \int_{x_1}^{x_2} \delta f_\varepsilon \, dy,$$

we get the variation derivative

$$\frac{\delta F[f]}{\delta f(q)} = \lim_{\varepsilon \to 0} b(q) \frac{\int_\Delta \delta f_\varepsilon(y) \, dy}{\int_\Delta \delta f_\varepsilon(y) \, dy} = b(q). \tag{4.2.14}$$

Choosing $b(y) = 1/(\sqrt{2\pi}\sigma) \exp(-(y - y_0)^2/2\sigma^2)$ we obtain

$$\frac{\delta F[f]}{\delta f(q)} = \frac{1}{\sqrt{2\pi}\sigma} \exp\left(-\frac{(y - y_0)^2}{2\sigma^2}\right). \tag{4.2.15}$$

In the limit $\sigma \to 0$ we have $b(y) \to \delta(y - y_0)$, $F[f] \to f(y_0)$ and hence

$$\frac{\delta f(y_0)}{\delta f(q)} = \delta(y_0 - q). \tag{4.2.16}$$

### 4.2.3 Infinitesimal Criterion

According to Sect. 4.2.1 to write the infinitesimal criterion for the symmetry group for nonlocal equations one should differentiate (4.2.10) with respect to group parameter $a$ and assume $a \to 0$, i.e. calculate the limit of the derivative $dF/da$ for vanishing $a$. Combining (4.2.11), (4.2.12) and assuming $a \to 0$ in view of (4.2.9) we get

$$\left.\frac{dF[\tilde{f}]}{da}\right|_{a=0} = \int \varkappa(y) \frac{\delta F[f(x)]}{\delta f(y)} \, dy \equiv YF, \tag{4.2.17}$$

where we have introduced the generator $Y$ defined by its action on function $F$ as follows:

$$Y(F) = \int \varkappa(y) \frac{\delta F}{\delta f(y)} \, dy.$$

We will write this operator formally in the form

$$Y = \int \varkappa(y) \frac{\delta}{\delta f(y)} \, dy. \qquad (4.2.18)$$

Hence, the invariance criterion for $F$ with respect to the admitted group can be expressed in an infinitesimal form using the canonical group operator $Y$,

$$YF\big|_{F=0} = 0, \qquad (4.2.19)$$

which generalizes the action of a standard canonical group operator (see formula (1.5.7) in Chap. 1) not only on differential functions but on *functionals* as well using variational differentiation in the definition of $Y$ [7]. One can verify by direct calculation that the action of $Y$ on any differential function and its derivatives, e.g., $f$ and $f_x, \ldots$ produces the usual result: $Yf = \varkappa$, $Yf_x = D_x(\varkappa)$ and so on. Hence, if $F$ describe usual differential equations then formulas (4.2.19) lead to standard local determining equations, while for $F$ having the form of integro-differential equations formulas (4.2.19) can be treated as *nonlocal* determining equations as they depend both on local and nonlocal variables. As we treat local and nonlocal variables in determining equations as independent it is possible to separate these equations into local and nonlocal. The procedure of solving local determining equations is fulfilled in a standard way using Lie algorithm based on splitting the system of over-determined equations with respect to local variables and their derivatives. As a result we get expressions for coordinates of group generator that define the so-called group of *intermediate* symmetry [7]. In the similar manner the solution of nonlocal determining equations is fulfilled using the information borrowed from an intermediate symmetry and by "variational" splitting of nonlocal determining equations using the procedure of variational differentiation. Therefore, the algorithm of finding symmetries of nonlocal equations appears as an algorithmic procedure that consists of a sequence of several steps: (a) defining the set of local group variables, (b) constructing determining equations on basis of the infinitesimal criterion of invariance, that employs the generalization of the definition of the canonical operator, (c) separating determining equations into local and nonlocal, (d) solving local determining equations using a standard Lie algorithm, (e) solving nonlocal determining equations using the procedure of variational differentiation.

### 4.2.4 Prolongation on Nonlocal Variables

To complete we describe the procedure of prolongation of a symmetry group on nonlocal variables, say in the form of the integral relation

$$J(u) = \int \mathscr{F}(u(z)) \, dz. \qquad (4.2.20)$$

To fulfill this procedure one should first rewrite the operator, say $Y$, in a canonical form and then formally prolong this operator on the nonlocal variable $J$

$$Y + \varkappa^J \partial_J \equiv \varkappa \partial_u + \varkappa^J \partial_J. \qquad (4.2.21)$$

The integral relation between $\varkappa$ and $\varkappa^J$ is obtained by applying the generator (4.2.21) to the *definition* of $J$, i.e. to (4.2.20). Substituting the explicit expression for the coordinate $\varkappa$ of the known operator $Y$ and calculating integrals obtained gives the desired coordinate $\varkappa^J$,

$$\varkappa^J = \int \frac{\delta J(u)}{\delta u(z)} \varkappa(z) \, dz$$

$$\equiv \int \frac{\delta \mathscr{F}(u(z'))}{\delta u(z)} \varkappa(z) \, dz \, dz' = \int \mathscr{F}_u \varkappa(z) \, dz. \qquad (4.2.22)$$

## 4.3 Symmetry of Plasma Kinetic Equations in One-Dimensional Approximation

This section discuss the symmetry of Vlasov–Maxwell equations (4.1.1)–(4.1.3) for plane (one-dimensional) geometry. We start with the case of a one component non-relativistic electron plasma (electron gas) and proceed with a set of different models, including multi-component plasma, relativistic plasma, plasma with neu-tralizing moving and stationary ion background.

### 4.3.1 Non-relativistic Electron Gas

Consider the system of Vlasov–Maxwell equations (4.1.1)–(4.1.3) for charged elec-tron gas. In case of non-relativistic motion of electrons in the self-consistent electric field $E$ the one-dimensional Vlasov kinetic equation for the distribution function $f$ is written as follows:

$$f_t + vf_x + \frac{e}{m} Ef_v = 0. \qquad (4.3.1)$$

Here the potential field $E$ obeys the Poisson equation and the corresponding Maxwell equation

$$E_x = 4\pi\rho, \quad E_t + 4\pi j = 0, \qquad (4.3.2)$$

and charge density $\rho$ and current density $j$ are expressed as the integrals

$$\rho = em \int dv f, \quad j = em \int dv f v \qquad (4.3.3)$$

over electron velocities. Momentarily, we will assume that the charge $e$ and mass $m$ of the electron (parameters of the system) are invariants. The dependent variables $E$, $j$, and $\rho$ are functions of two arguments, time $t$ and coordinate $x$,

$$E = E(t, x), \quad j = j(t, x), \quad \rho = \rho(t, x), \qquad (4.3.4)$$

and the distribution function

$$f = f(t, x, v) \tag{4.3.5}$$

has three arguments, $t$, $x$ and electron velocity $v$. It follows from (4.3.4) that electric field intensity $E$, current density $j$, and charge density $\rho$ are independent of electron velocity $v$. Hence we have three additional differential constraints

$$E_v = 0, \quad j_v = 0, \quad \rho_v = 0, \tag{4.3.6}$$

which should be used in group analysis of the system (4.3.1)–(4.3.3) as well as compatibility condition for the field equations (4.3.2), known as continuity equation,

$$\rho_t - j_x = 0. \tag{4.3.7}$$

The coordinates $\xi$ and $\eta$ of the Lie point symmetry group generator

$$X = \xi^1 \frac{\partial}{\partial t} + \xi^2 \frac{\partial}{\partial x} + \xi^3 \frac{\partial}{\partial v} + \eta^1 \frac{\partial}{\partial f} + \eta^2 \frac{\partial}{\partial E} + \eta^3 \frac{\partial}{\partial j} + \eta^4 \frac{\partial}{\partial \rho}, \tag{4.3.8}$$

are considered as functions of the seven variables

$$t, \ x, \ v, \ f, \ E, \ J, \ \rho. \tag{4.3.9}$$

These coordinates are solutions to the *determining equations*, which, in turn, appear as necessary and sufficient conditions for the invariance of system (4.3.1)–(4.3.3), (4.3.6) with respect to the group with generator (4.3.8). Local (differential) determining equations can be stated and solved directly in terms of the generator (4.3.8). In this section we however use the canonical form

$$Y = \varkappa^1 \frac{\partial}{\partial f} + \varkappa^2 \frac{\partial}{\partial E} + \varkappa^3 \frac{\partial}{\partial j} + \varkappa^4 \frac{\partial}{\partial \rho}, \tag{4.3.10}$$

for the generator (4.3.3), which offers substantial advantages in the group analysis of the complete system of Vlasov–Maxwell equations because of non-locality of the system.

### 4.3.1.1 Non-relativistic Electron Gas: The Solution to the Local Determining Equations

The invariance conditions for Vlasov kinetic equation (4.3.1), the field equations (4.3.2), and (4.3.6) with respect to the group with canonical generator (4.3.10) are given by the six local determining equations

$$D_t(\varkappa^1) + v D_x(\varkappa^1) + \frac{e}{m} E D_v(\varkappa^1) + \frac{e}{m} \varkappa^2 f_v = 0,$$

$$D_x(\varkappa^2) = 4\pi \varkappa^4, \quad D_t(\varkappa^2) = -4\pi \varkappa^3, \tag{4.3.11}$$

$$D_v(\varkappa^2) = 0, \quad D_v(\varkappa^3) = 0, \quad D_v(\varkappa^4) = 0,$$

which should be solved with taking into account the fact that the group variables (4.3.9) and the corresponding derivatives are related by the manifold (4.3.1)–(4.3.3), (4.3.6) and (4.3.7). Here we use the standard notations (see, e.g. Chap. 1) for the operator of total differentiation $D_i$ with respect to the group variable indicated by the subscript. For example, the operator $D_v$ of total differentiation with respect to $v$ is given by

$$D_v \equiv \frac{\partial}{\partial v} + f_v \frac{\partial}{\partial f} + f_{vt} \frac{\partial}{\partial f_t} + f_{vx} \frac{\partial}{\partial f_x} + f_{vv} \frac{\partial}{\partial f_v} + \cdots. \qquad (4.3.12)$$

The solution of the system of local determining equations (4.3.11) is given by the following formulas for the coordinates $\varkappa$ of the generator (4.3.10):

$$
\begin{aligned}
\varkappa^1 &= \eta^1 - f_t \xi^1 - f_x \left[ x \left( A_4 + \frac{1}{2}\xi_t^1 \right) + \beta \right] - f_v \left[ v \left( A_4 - \frac{1}{2}\xi_t^1 \right) + \frac{1}{2} x \xi_{tt}^1 + \beta_t \right], \\
\varkappa^2 &= E \left[ A_4 - \frac{3}{2}\xi_t^1 \right] + \frac{m}{e} \left[ \frac{1}{2} x \xi_{ttt}^1 + \beta_{tt} \right] - E_t \xi^1 - E_x \left[ x \left( A_4 + \frac{1}{2}\xi_t^1 \right) + \beta \right], \\
\varkappa^3 &= j \left[ A_4 - \frac{5}{2}\xi_t^1 \right] + \rho \left[ \frac{1}{2} x \xi_{tt}^1 + \beta_t \right] + \frac{3}{8\pi} E \xi_{tt}^1 \\
&\quad - \frac{m}{4\pi e} \left[ \frac{1}{2} x \xi_{tttt}^1 + \beta_{ttt} \right] - j_t \xi^1 - j_x \left[ x \left( A_4 + \frac{1}{2}\xi_t^1 \right) + \beta \right], \\
\varkappa^4 &= -2\rho \xi_t^1 + \frac{m}{8\pi e}\xi_{ttt}^1 - \rho_t \xi^1 - \rho_x \left[ x \left( A_4 + \frac{1}{2}\xi_t^1 \right) + \beta \right].
\end{aligned}
\qquad (4.3.13)
$$

The coordinates (4.3.13) depend upon three arbitrary functions

$$\xi(t), \quad \beta(t), \quad \eta^1(f) \qquad (4.3.14)$$

and $A_4$ is an arbitrary constant. The group symmetry with the generator (4.3.10) and coordinates (4.3.13) admitted by the system of equations (4.3.1), (4.3.2), (4.3.6) will be referred to as the *intermediate* group symmetry of the complete system of the self-consistent field equations (4.3.1)–(4.3.3). The symmetry is generated only by the differential equations in the integro-differential Vlasov–Maxwell system and does not take into account integral terms in the material equations (4.3.3), which determine charge and current densities of electrons. The intermediate group symmetry (4.3.10), (4.3.13) plays an auxiliary role in obtaining the final equations for the coordinates $\xi$ and $\eta$ of the generator (4.3.8) of the desired Lie group. The charge density $\rho$ and the current density $j$ have a concrete physical meaning. By introducing them as independent group variables in the set (4.3.9) along with $t$, $x$, $v$, $f$, and $E$, we divide the group analysis of the local and the nonlocal part of the Vlasov–Maxwell system into two stages. The intermediate symmetry (4.3.10), (4.3.13) completes the local group analysis of the system. In what follows we shall see that the nonlocal determining equations appearing as invariance conditions for the material equations (4.3.3) with respect to the sought Lie group eliminate the arbitrary dependence of $\xi$, $\beta$ and $\eta^1$ on $t$ and $f$.

### 4.3.1.2  Non-relativistic Electron Gas: Nonlocal Determining Equations and Their Solutions

Since the material equations (4.3.3) are nonlocal (they involve integration of the distribution function $f$ and of the product $vf$ over the electron velocity $v$), the differentiation with respect to $f$ in the first term of the canonical generator (4.3.10) should be generalized so as to act not only on functions of $f$ but also on linear functionals (4.3.3) of $f$. Hence, we represent this term as the integral of variational derivative with respect to $f$ with weight $\varkappa^1$ over electron velocity $v$:

$$\varkappa^1 \frac{\partial}{\partial f} \equiv \int dv \varkappa^1(v) \frac{\delta}{\delta f(v)}. \tag{4.3.15}$$

For brevity, we indicate only the integration variable $v$ as an argument of $f$ and of $\varkappa^1$. Our shorthand notation implies that the coordinate $\varkappa^1(v)$ of the canonical generator in (4.3.15) stands for the following extended expression in (4.3.13), depending on integration variable $v$:

$$\varkappa^1(v) \equiv \eta^1(f(t, x, v)) - \xi^1 f_t(t, x, v) - \left[ x \left( A_4 + \frac{1}{2}\xi_t^1 \right) + \beta \right] f_x(t, x, v)$$

$$- \left[ v \left( A_4 - \frac{1}{2}\xi_t^1 \right) + \frac{1}{2}x\xi_{tt}^1 + \beta_t \right] f_v(t, x, v). \tag{4.3.16}$$

When applied to functions of $f$, the operator of the differentiation with respect to $f$ in (4.3.15) gives the usual result, i.e. it coincides with the ordinary differentiation with respect to $f$. When applied to linear functionals of $f$, i.e., to the charge and current densities (4.3.3), the derivative in (4.3.15) permits us to introduce the variational derivative on the right-hand side in (4.3.15) under the integral over $v$ together with the coordinate $\varkappa^1$ of the canonical generator (4.3.10).

Substituting (4.3.15) in (4.3.10) and using the a well-known identity

$$\frac{\delta f(v)}{\delta f(v')} = \delta(v - v'), \tag{4.3.17}$$

where $\delta$ is the Dirac delta-function we obtain the invariance conditions for the integral material equations (4.3.3) with respect to the Lie group with canonical generator (4.3.10), the nonlocal determining equations [7, 8]

$$\varkappa^4 - em \int dv \varkappa^1 = 0, \quad \varkappa^3 - em \int dv v \varkappa^1 = 0. \tag{4.3.18}$$

The integration in (4.3.18) is over all values of $v$, just as in (4.3.3) and (4.3.15). Let us consider the first of the two determining equations in (4.3.18) in more detail. Substituting the coordinates $\varkappa^1$ and $\varkappa^4$ from (4.3.13) into the determining equations in question and taking into account (4.3.3) for charge density $\rho$, we reduce the determining equations to the simple form

$$em \int dv [\eta^1(f(v)) + f(v)\mathcal{K}(t)] - \frac{m}{8\pi e}\xi_{ttt}^1(t) = 0. \tag{4.3.19}$$

The coefficient $\mathcal{K}$ in the product $\mathcal{K}f$ in the integrand on the left-hand side in (4.3.19) is independent of $v$ and $f$; specifically, we have

$$\mathcal{K}(t) = A_4 + \frac{3}{2}\xi_t^1. \tag{4.3.20}$$

The derivation of the determining equations (4.3.19) involves integrating by parts with respect to $v$, which removes the derivative $f_v$ from the integrand in the nonlocal term in the original determining equations. The resultant antiderivative $f$ is assumed to vanish at the ends of the infinite integration interval, that is,

$$f \to 0, \quad v \to \pm\infty. \tag{4.3.21}$$

The determining equations (4.3.19) is a linear nonhomogeneous integral equation for $\eta^1$, which can easily be solved. According to the general ideas of Lie technique, (4.3.19), as well as any determining equation, is an identity with respect to the group variable $f$. Therefore, it remains valid after differentiating with respect to $f$. Since the determining equations (4.3.19) is an integral equation, we should use variational differentiation with respect to $f$ rather than ordinary differentiation. Taking into account that nonhomogeneous term proportional to $\xi^1_{ttt}$ in (4.3.19) is independent of $f$, we obtain:

$$\frac{\delta}{\delta f(v')} \int dv \left[ \eta^1 \big( f(v) \big) + f(v) \mathcal{K}(t) \right] = 0. \tag{4.3.22}$$

The nonlocal equation (4.3.19), which is an identity with respect to $f$, should be combined with its differential corollary (4.3.22) in the sense that a solution to (4.3.22) is also a solution to (4.3.19). Introducing the variational derivative in (4.3.22) under the integral over $v$,

$$\int dv \left\{ \eta^1_f + \mathcal{K} \right\} \frac{\delta f(v)}{\delta f(v')} = 0. \tag{4.3.23}$$

and evaluating the integral over $v$ with the aid of the delta-function (4.3.17) that appears in the integrand, as a consequence of (4.3.19), we obtain a first-order ordinary differential equation for the dependence of the coordinate $\eta^1$ of the determining equations (4.3.8) on $f$:

$$\eta^1_f + \mathcal{K} = 0. \tag{4.3.24}$$

Its solution depends on one arbitrary constant

$$\eta^1 = -\mathcal{K} f + A. \tag{4.3.25}$$

Since the coordinate $\eta^1$ is independent of $t$, we immediately obtain the condition

$$\mathcal{K}_t = 0 \tag{4.3.26}$$

imposed on the coefficient $\mathcal{K}$ of the determining equations (4.3.19). It follows from (4.3.20) and (4.3.26) that

$$\xi^1_{tt} = 0. \tag{4.3.27}$$

As was mentioned above, it is necessary to consider (4.3.25) for $\eta^1$, appearing as a direct consequence of variational differentiation (4.3.9) of the determining equations (4.3.19) with respect to the distribution function $f$, together with (4.3.19):

$$em \int\limits_{-\infty}^{+\infty} dv A - \frac{m}{8\pi e} \xi^1_{ttt}(t) = 0. \tag{4.3.28}$$

In view of (4.3.27), the second term on the left-hand side in (4.3.28) is zero. Therefore, (4.3.28) is reduced to

$$Aem \int_{-\infty}^{+\infty} dv = 0,$$

(4.3.29)

whence follows that the integration constant $A$ in (4.3.25) is zero, that is, we have

$$\eta^1 = -\mathscr{K}f.$$

(4.3.30)

The integration of (4.3.30) yields

$$\xi^1(t) = A_1 + 2A_3 t.$$

(4.3.31)

We insert the explicit formula (4.3.31) for the dependence of $\xi^1$ on $t$ into expression (4.3.20) for the coefficient $\mathscr{K}$ and obtain

$$\mathscr{K} = 3A_3 + A_4,$$

(4.3.32)

whence follows the definite expression for the coordinate

$$\eta^1(f) = -(3A_3 + A_4)f.$$

(4.3.33)

Equations (4.3.31) and (4.3.33) are the basic result of solving the first nonlocal determining equations in (4.3.18) and define explicit dependence of $\xi$ and $\eta$ on $t$ and $f$ in the intermediate group symmetry (4.3.13). The second nonlocal determining equations in system (4.3.18) pertains to the invariance of electron current density with respect to the admitted Lie group. By substituting the extended expressions for the coordinates $\varkappa^3$ and $\varkappa^1$ from (4.3.13) into this determining equations, we easily reduce it to the following linear nonhomogeneous integral equation for $\eta^1$, which is similar to (4.3.19):

$$em \int dvv\left(\eta^1 + f\mathscr{K}\right) + \frac{3}{8\pi} E\xi_{tt}^1 - \frac{mx}{8\pi e}\xi_{tttt}^1 - \frac{m}{4\pi e}\beta_{ttt} = 0.$$

(4.3.34)

The passage from (4.3.18) to (4.3.34) involves integration by parts with respect to $v$. Here we take into account conditions (4.3.21), which state that the electron distribution function $f$ decays rapidly for large velocities. The coefficient $\mathscr{K}$ in the product $\mathscr{K}vf$ in the integrand on the left-hand side in (4.3.34) has the same form (4.3.20) as in (4.3.19). Hence, taking into account (4.3.27) and (4.3.30), we see that the determining equations (4.3.34) is reduced to $\beta_{ttt} = 0$, which implies

$$\beta(t) = A_2 + A_5 t + \frac{1}{2}A_6 t^2.$$

(4.3.35)

Substitution of (4.3.31), (4.3.33) and (4.3.35) into (4.3.13) yields canonical coordinates that satisfy determining equations (4.3.11), and (4.3.18), and therefore define the sought for group symmetry

$$\varkappa^1 = -A_1 f_t - A_2 f_x - A_3 (3f + 2t f_t + x f_x - v f_v) - A_4 (f + x f_x + v f_v)$$
$$- A_5 (t f_x + f_v) - A_6 \left( \frac{t^2}{2} f_x + t f_v \right),$$
$$\varkappa^2 = -A_1 E_t - A_2 E_x - A_3 (3E + 2t E_t + x E_x) - A_4 (-E + x E_x)$$
$$- A_5 t E_x - A_6 \left( \frac{t^2}{2} E_x - \frac{m}{e} \right),$$
$$\varkappa^3 = -A_1 j_t - A_2 j_x - A_3 (5j + 2t j_t + x j_x) - A_4 (-j + x j_x)$$
$$- A_5 (t j_x - \rho) - A_6 \left( \frac{t^2}{2} j_x - t \rho \right),$$
$$\varkappa^4 = -A_1 \rho_t - A_2 \rho_x - A_3 (4\rho + 2t \rho_t + x \rho_x) - A_4 x \rho_x$$
$$- A_5 t \rho_x - A_6 \left( \frac{t^2}{2} \right) \rho_x.$$

$$(4.3.36)$$

Formulas (4.3.36) refer to the following six basic generators, written in a non-canonical form [7]:

$$X_1 = \frac{\partial}{\partial t}, \quad X_2 = \frac{\partial}{\partial x},$$
$$X_3 = 2t \frac{\partial}{\partial t} + x \frac{\partial}{\partial x} - v \frac{\partial}{\partial v} - 3f \frac{\partial}{\partial f} - 3E \frac{\partial}{\partial E} - 5j \frac{\partial}{\partial j} - 4\rho \frac{\partial}{\partial \rho},$$
$$X_4 = x \frac{\partial}{\partial x} + v \frac{\partial}{\partial v} - f \frac{\partial}{\partial f} + E \frac{\partial}{\partial E} + j \frac{\partial}{\partial j}, \quad X_5 = t \frac{\partial}{\partial x} + \frac{\partial}{\partial v} + \rho \frac{\partial}{\partial j},$$
$$X_6 = \frac{t^2}{2} \frac{\partial}{\partial x} + t \frac{\partial}{\partial v} + \frac{m}{e} \frac{\partial}{\partial E} + t \rho \frac{\partial}{\partial j}.$$

$$(4.3.37)$$

The set of generators (4.3.37) span the six-dimensional Lie algebra

$$L_6 = \langle X_1, X_2, \dots, X_6 \rangle. \tag{4.3.38}$$

Generators (4.3.37) of the six-parametric continuous point Lie group admitted by the Vlasov–Maxwell equations (4.3.1)–(4.3.3), have clear physical meaning: the operators $X_1$ and $X_2$ describe translations along $t$ and $x$-axes, the generator $X_3$ and $X_4$ relate to dilations, which can be easily verified, and the generators $X_5$ define the Galilean transformations. The finite transformations corresponding to the generator $X_6$ have the following form for each of six variables (4.3.9):

$$\bar{t} = t; \quad \bar{x} = x + \frac{at^2}{2}; \quad \bar{v} = v + at; \quad \bar{f} = f;$$
$$\bar{E} = E + \frac{ma}{e}; \quad \bar{j} = j + at\rho; \quad \bar{\rho} = \rho.$$

$$(4.3.39)$$

In mechanics, the one-parameter transformation group with generator $X_6$ can be interpreted for the first three equations in (4.3.39) as the transformation of variables due to passing into a coordinate system moving linearly with constant acceleration $a = \text{const}$ with respect to the laboratory frame.

### 4.3.1.3 Including Electron Charge and Electron Mass into Group Transformations

The set of group variables (4.3.9) can be extended by involving the parameters $e$ and $m$ of the Vlasov–Maxwell equations (4.3.1)–(4.3.3) into the group transformations

$$X = \xi^1 \frac{\partial}{\partial t} + \xi^2 \frac{\partial}{\partial x} + \xi^3 \frac{\partial}{\partial v} + \xi^4 \frac{\partial}{\partial e} + \xi^5 \frac{\partial}{\partial m}$$
$$+ \eta^1 \frac{\partial}{\partial f} + \eta^2 \frac{\partial}{\partial E} + \eta^3 \frac{\partial}{\partial j} + \eta^4 \frac{\partial}{\partial \rho}. \tag{4.3.40}$$

The extension adds two more basis generator to the algebra (4.3.37), (4.3.38). They correspond to the dilations of electron charge and mass:

$$X_7 = e \frac{\partial}{\partial e} + m \frac{\partial}{\partial m} - 2f \frac{\partial}{\partial f}, \quad X_8 = m \frac{\partial}{\partial m} + E \frac{\partial}{\partial E} + j \frac{\partial}{\partial j} + \rho \frac{\partial}{\partial \rho}. \tag{4.3.41}$$

The operators (4.3.41) commute with each other and with all operator (4.3.37), so that the set (4.3.37), (4.3.41) is the eight-dimensional Lie algebra

$$L_8 = \langle X_1, X_2, \ldots, X_6, X_7, X_8 \rangle. \tag{4.3.42}$$

If the electron charge and mass are not invariant, then the general operator of the continuous point Lie group admitted by the Vlasov–Maxwell equations (4.3.1)–(4.3.3) is given by

$$X = \sum_{\alpha=1}^{8} A_\alpha(e, m) X_\alpha. \tag{4.3.43}$$

It corresponds to an infinite group with continual arbitrariness given by eight functions $A_\alpha$ depending on two of the nine variables

$$t, x, v, e, m, f, E, j, \rho \tag{4.3.44}$$

and can be obtained by solving local and nonlocal determining equations for the coordinates of the generator under the conditions

$$f = f(t, x, v, e, m); \quad E = E(t, x, e, m);$$
$$j = j(t, x, e, m); \quad \rho = \rho(t, x, e, m) \tag{4.3.45}$$

in a way similar to that given in the previous sections.

The infiniteness of the Lie group (4.3.44), (4.3.37), (4.3.41) is due to the fact that the parameters $e$ and $m$ that are arbitrary elements of the group classification are included in the set of the group variables (4.3.44). This procedure that looks trivial from the group analysis viewpoint is typical in "classical" renormalization group method in quantum field theory (for details see Sect. 4.7). In the similar manner to take into account the relativistic motion of electrons, we have to introduce a third parameter, namely, the light velocity in vacuum (denoted by $c$). We can pass to relativistic velocities also in the one-dimensional approximation with the same field equations (4.3.2). In doing so, the one-parameter Galilean group with the generator $X_5$ from (4.3.37) is transformed into the Lorentz group and is inherited (in the sense of [11, 12]) in an arbitrary order with respect to the parameter $v/c$, which takes into account the finiteness of the light velocity. This will be demonstrated in Sect. 4.3.2.

### 4.3.2 Relativistic Electron Gas

The one-dimensional system of self-consistent field equations (4.3.1)–(4.3.3) for charged relativistic electron gas is modified as follows:

$$f_t + vf_x + eEf_p = 0, \quad \rho = e\int dpf, \quad j = e\int dpfv. \qquad (4.3.46)$$

In contrast to (4.3.1) and (4.3.3), instead of electron velocity $v$ we use moment $p$, which can be expressed in terms of $v$ by the well-known equality

$$p = mv\gamma \equiv mv\big(1 - (v/c)^2\big)^{-1/2}, \qquad (4.3.47)$$

where $\gamma$ is the relativistic factor. Using (4.3.47) and passing from electron moment $p$ to electron velocity $v$ in (4.3.46), we obtain the equations

$$f_t + vf_x + \frac{e}{m}\gamma^{-3}Ef_v = 0, \qquad (4.3.48)$$

$$\rho = em\int\limits_{-c}^{+c} dv\gamma^3 f, \quad j = em\int\limits_{-c}^{+c} dv\gamma^3 fv, \qquad (4.3.49)$$

which differ from (4.3.1) and (4.3.3) in that the relativistic factor $\gamma > 1$ is taken into account. In finding symmetry of the integro-differential system of equations (4.3.48), (4.3.49), (4.3.2), and (4.3.6) we assume that not only time $t$, coordinate $x$, and electron velocity $v$ but also charge $e$, electron mass $m$, and light velocity $c$ are independent variables.

Omitting the calculations akin to that were done in the previous Sect. 4.3.1 we present the final expression for the group generator in the form of a linear combination of seven basic generators with the coefficients $A_\alpha$ that are arbitrary functions of three variables [7]:

$$X = \sum_{\alpha=1}^{7} A_\alpha(e, m, c)X_\alpha,$$

$$X_1 = \frac{\partial}{\partial t}, \quad X_2 = c\frac{\partial}{\partial x},$$

$$X_3 = t\frac{\partial}{\partial t} + x\frac{\partial}{\partial x} - 2f\frac{\partial}{\partial f} - E\frac{\partial}{\partial E} - 2j\frac{\partial}{\partial j} - 2\rho\frac{\partial}{\partial \rho},$$

$$X_4 = \frac{1}{c}\left(x\frac{\partial}{\partial t} + c^2 t\frac{\partial}{\partial x} + (c^2 - v^2)\frac{\partial}{\partial v} + \rho c^2\frac{\partial}{\partial j} + j\frac{\partial}{\partial \rho}\right), \qquad (4.3.50)$$

$$X_5 = x\frac{\partial}{\partial x} + v\frac{\partial}{\partial v} + c\frac{\partial}{\partial c} - f\frac{\partial}{\partial f} + E\frac{\partial}{\partial E} + j\frac{\partial}{\partial j},$$

$$X_6 = m\frac{\partial}{\partial m} + E\frac{\partial}{\partial E} + j\frac{\partial}{\partial j} + \rho\frac{\partial}{\partial \rho},$$

$$X_7 = e\frac{\partial}{\partial e} + m\frac{\partial}{\partial m} - 2f\frac{\partial}{\partial f}.$$

The generators (4.3.50) span the seven-dimensional Lie algebra

$$L_7 = \langle X_1, X_2, \ldots, X_7 \rangle \tag{4.3.51}$$

with numerical structural constants. The last three generators in (4.3.50), which determine the dilations of electron charge and mass and of light velocity, commute with all remaining generator in (4.3.50) and with one another. The first four generators in (4.3.50) form the four-dimensional subalgebra

$$L_4 = \langle X_1, X_2, X_3, X_4 \rangle. \tag{4.3.52}$$

The finite transformations given by solutions to the Lie equations for the generator $X_4$ (the Lorentz transformations) correspond to hyperbolic rotations in the planes $(ct, x)$ and $(cp, j)$, and to the linear-fractional transformation of electron velocity $v$ with group parameter $a$:

$$\bar{t} = t \cosh(ac) + (x/c) \sinh(ac), \quad \bar{x} = x \cosh(ac) + ct \sinh(ac),$$

$$\bar{v} = (v + c \tanh(ac))(1 + (v/c) \tanh(ac))^{-1},$$

$$\bar{\rho} = \rho \cosh(ac) + (j/c) \sinh(ac), \quad \bar{j} = j \cosh(ac) + c\rho \sinh(ac), \tag{4.3.53}$$

$$\bar{e} = e, \quad \bar{m} = m, \quad \bar{c} = c, \quad \bar{f} = f, \quad \bar{E} = E.$$

The generator $X_4$ from (4.3.50) and its finite transformations in the form (4.3.53) extends the Galilean generator $X_5$ from (4.3.37) to the relativistic domain of electron velocities. Comparing the algebras (4.3.37) and (4.3.50) of the point symmetry groups we see that transition from non-relativistic to relativistic electron gas deletes the generator $X_6$ from (4.3.37).

The algebra (4.3.50), (4.3.51) is fairly consistent with the physical ideas on the symmetry of system (4.3.48), (4.3.2) and (4.3.49), developed in the theory of plasma. The characteristic feature of the system is in that the relativistic effects are taken into account for electron motion but the finite value of light velocity $c$ is ignored in the field equations (4.3.2) in one dimensional approximation. However, we can extend the scope of the method by taking into account the three-dimensional relativistic motion of electrons in self-coordinated electric field $E$ and magnetic field $B$ obeying the Maxwell equations. This is done in Sect. 4.4.1.

### 4.3.3 Collisionless Non-relativistic Electron–Ion Plasma

In this section we turn to a model that contains two plasma particle species, namely electrons and ions. It means that the basic system of equations should be supplemented by the kinetic equation for the ion distribution function $\bar{f}$ and the corresponding items in the field equations,

$$f_t + vf_x + \frac{e}{m}Ef_v = 0, \quad \bar{f}_t + v\bar{f}_x + \frac{\bar{e}}{\bar{m}}E\bar{f}_v = 0,$$

$$\rho = \int dv(emf + \bar{e}\bar{m}\bar{f}), \quad j = \int dvv(emf + \bar{e}\bar{m}\bar{f}), \tag{4.3.54}$$

$$E_x = 4\pi\rho, \quad E_t + 4\pi j = 0.$$

The solution of the local and nonlocal determining equations for non-relativistic Vlasov–Maxwell equations for electron–ion plasma is fulfilled in the same root as for the electron gas model. The final result is given as the general element of the Lie algebra of point symmetry operators of the Vlasov–Maxwell equations (4.3.54) is determined by the linear combination [8]

$$X = \sum_{\alpha=1}^{9} A_\alpha(e, m, \bar{e}, \bar{m}) X_\alpha, \tag{4.3.55}$$

$$X_1 = \frac{\partial}{\partial t}, \quad X_2 = \frac{\partial}{\partial x},$$

$$X_3 = 2t\frac{\partial}{\partial t} + x\frac{\partial}{\partial x} - v\frac{\partial}{\partial v} - 3f\frac{\partial}{\partial f} - 3\bar{f}\frac{\partial}{\partial \bar{f}} - 3E\frac{\partial}{\partial E} - 5j\frac{\partial}{\partial j} - 4\rho\frac{\partial}{\partial \rho},$$

$$X_4 = t\frac{\partial}{\partial x} + \frac{\partial}{\partial v} + \rho\frac{\partial}{\partial j},$$

$$X_5 = x\frac{\partial}{\partial x} + v\frac{\partial}{\partial v} - f\frac{\partial}{\partial f} - \bar{f}\frac{\partial}{\partial \bar{f}} + E\frac{\partial}{\partial E} + j\frac{\partial}{\partial j}, \tag{4.3.56}$$

$$X_6 = \frac{1}{em}\frac{\partial}{\partial f} - \frac{1}{\bar{e}\bar{m}}\frac{\partial}{\partial \bar{f}}, \quad X_7 = e\frac{\partial}{\partial e} + m\frac{\partial}{\partial m} - 2f\frac{\partial}{\partial f},$$

$$X_8 = m\frac{\partial}{\partial m} + \bar{m}\frac{\partial}{\partial \bar{m}} + E\frac{\partial}{\partial E} + j\frac{\partial}{\partial j} + \rho\frac{\partial}{\partial \rho},$$

$$X_9 = \bar{e}\frac{\partial}{\partial \bar{e}} + \bar{m}\frac{\partial}{\partial \bar{m}} - 2\bar{f}\frac{\partial}{\partial \bar{f}}.$$

## 4.3.4 Collisionless Relativistic Electron–Ion Plasma

This section presents the result of the symmetry group calculation for the relativistic analogue of equations discussed in the previous section:

$$f_t + vf_x + \frac{e}{m\gamma^3}Ef_v = 0, \quad \bar{f}_t + v\bar{f}_x + \frac{\bar{e}}{\bar{m}\gamma^3}E\bar{f}_v = 0,$$
$$E_x = 4\pi\rho, \quad E_t + 4\pi j = 0, \tag{4.3.57}$$
$$\rho = \int dv\gamma^3(emf + \bar{e}\bar{m}\,\bar{f}), \quad j = \int dv\gamma^3 v(emf + \bar{e}\bar{m}\,\bar{f}).$$

The Lie group admitted by the Vlasov–Maxwell equations (4.3.57) is a one-dimensional analog of the group with algebra (4.3.50) provided that the parameters $e, m, \bar{e}, \bar{m}$ and $c$ are invariant [8]:

$$L_5 = \langle X_1, X_2, X_3, X_4, X_5 \rangle, \tag{4.3.58}$$

$$X_1 = \frac{\partial}{\partial t}, \quad X_2 = c\frac{\partial}{\partial x},$$

$$X_3 = t\frac{\partial}{\partial t} + x\frac{\partial}{\partial x} - 2f\frac{\partial}{\partial f} - 2\bar{f}\frac{\partial}{\partial \bar{f}} - E\frac{\partial}{\partial E} - 2j\frac{\partial}{\partial j} - 2\rho\frac{\partial}{\partial \rho},$$

$$X_4 = \frac{1}{c}\left( x\frac{\partial}{\partial t} + c^2 t\frac{\partial}{\partial x} + (c^2 - v^2)\frac{\partial}{\partial v} + c^2\rho\frac{\partial}{\partial j} + j\frac{\partial}{\partial \rho} \right), \quad (4.3.59)$$

$$X_5 = \frac{1}{em}\frac{\partial}{\partial f} - \frac{1}{\bar{e}\bar{m}}\frac{\partial}{\partial \bar{f}}.$$

Here the first, second, and fourth generators coincide with those of algebra $L_4$ for the relativistic electron gas. The dilation generator $X_3$ in (4.3.59) differs from the corresponding generator in (4.3.50) by the term $(-2\bar{f}\partial_{\bar{f}})$ containing the ion partition function $\bar{f}$. The quasi-neutrality operator $X_5$ in (4.3.59) is new as compared with the four-dimensional "electron" algebra (4.3.52) in Sect. 4.3.2. Taking into consideration transformations of parameters, we obtain the four generators

$$X_6 = c\frac{\partial}{\partial c} + x\frac{\partial}{\partial x} + v\frac{\partial}{\partial v} - f\frac{\partial}{\partial f} - \bar{f}\frac{\partial}{\partial \bar{f}} + E\frac{\partial}{\partial E} + j\frac{\partial}{\partial j},$$

$$X_7 = m\frac{\partial}{\partial m} + \bar{m}\frac{\partial}{\partial \bar{m}} + E\frac{\partial}{\partial E} + j\frac{\partial}{\partial j} + \rho\frac{\partial}{\partial \rho},$$

$$X_8 = \bar{e}\frac{\partial}{\partial \bar{e}} + \bar{m}\frac{\partial}{\partial \bar{m}} - 2\bar{f}\frac{\partial}{\partial \bar{f}},$$

$$X_9 = e\frac{\partial}{\partial e} + m\frac{\partial}{\partial m} - 2f\frac{\partial}{\partial f}$$

$$(4.3.60)$$

in addition to the basis (4.3.59).

The general element of the Lie point algebra is a linear combination of nine generators with coefficients that are arbitrary scalar functions of five variables,

$$X = \sum_{\alpha=0}^{9} A_\alpha(e, m, \bar{e}, \bar{m}, c)X_\alpha. \quad (4.3.61)$$

We omit the calculations that lead to (4.3.59)–(4.3.60), since they just reproduce the calculations made above.

Sections 4.3.3 and 4.3.4 demonstrate the point symmetry of kinetic equations of collisionless electron–ion plasma. Two additional Lie groups admitted by the Vlasov–Maxwell equations of quasi-neutral plasma are presented in the next sections. In contrast to present section, these equations correspond to a simplified model of electron plasma, i.e., we consider ions as a positively charged background neutralizing the negative charge of the electron plasma. Thus we omit the kinetic Vlasov equations for the ion distribution function and describe ions by means of "hydrodynamic" parameters.

## 4.3.5 Non-relativistic Electron Plasma Kinetics with a Moving and Stationary Ion Background

The non-relativistic one-dimensional equations of self-consistent fields for electron plasma with moving positive homogeneous ion background neutralizing the charge of electrons read

$$f_t + vf_x + \frac{e}{m}Ef_v = 0; \quad E_x = 4\pi\rho, \quad E_t + 4\pi j = 0,$$

$$\rho = em \int dv f + \bar{e}n, \quad j = em \int dv vf + \bar{e}nu. \tag{4.3.62}$$

Here $f$ is the partial function of non-relativistic electrons with charge $e < 0$ and mass $m$. The parameters $\bar{e}$, $n$, and $u$ correspond to the ion charge ($\bar{e} > 0$), ion density $n$, and ion velocity $u$, respectively. Unlike the case of the electron–ion plasma (Sect. 4.3.3), the ion mass $\bar{m}$ is not involved in (4.3.62) and the ion motion is described by the term $\bar{e}nu$ in the plasma current density $j$. Group analysis of (4.3.62) give rise to a ten-dimensional Lie algebra $L_{10}$ with numerical structural constants [8]:

$$L_{10} = \langle X_1, X_2, \ldots, X_{10}\rangle, \tag{4.3.63}$$

$$X_1 = \frac{1}{\omega}\frac{\partial}{\partial t}, \quad X_2 = \frac{u}{\omega}\frac{\partial}{\partial x},$$

$$X_3 = (x - ut)\frac{\partial}{\partial x} + (v - u)\frac{\partial}{\partial v} - f\frac{\partial}{\partial f} + E\frac{\partial}{\partial E} + (j - u\rho)\frac{\partial}{\partial j},$$

$$X_4 = \sin(\omega t)\frac{\partial}{\partial x} + \omega\cos(\omega t)\frac{\partial}{\partial v} + 4\pi\bar{e}n\sin(\omega t)\frac{\partial}{\partial E} + \omega\cos(\omega t)(\rho - \bar{e}n)\frac{\partial}{\partial j},$$

$$X_5 = \cos(\omega t)\frac{\partial}{\partial x} - \omega\sin(\omega t)\frac{\partial}{\partial v} + 4\pi\bar{e}n\cos(\omega t)\frac{\partial}{\partial E} - \omega\sin(\omega t)(\rho - \bar{e}n)\frac{\partial}{\partial j},$$

$$X_6 = ut\frac{\partial}{\partial x} + u\frac{\partial}{\partial v} + u\frac{\partial}{\partial u} + u\rho\frac{\partial}{\partial j},$$

$$X_7 = 2t\frac{\partial}{\partial t} + x\frac{\partial}{\partial x} - v\frac{\partial}{\partial v} - 4n\frac{\partial}{\partial n} - u\frac{\partial}{\partial u} - 3f\frac{\partial}{\partial f} - 3E\frac{\partial}{\partial E} - 5j\frac{\partial}{\partial j} - 4\rho\frac{\partial}{\partial\rho},$$

$$X_8 = e\frac{\partial}{\partial e} + m\frac{\partial}{\partial m} - 2f\frac{\partial}{\partial f},$$

$$X_9 = m\frac{\partial}{\partial m} + n\frac{\partial}{\partial n} + E\frac{\partial}{\partial E} + j\frac{\partial}{\partial j} + \rho\frac{\partial}{\partial\rho}, \quad X_{10} = \bar{e}\frac{\partial}{\partial\bar{e}} - n\frac{\partial}{\partial n}.$$

$$\tag{4.3.64}$$

Here $\omega$ is the well-known Langmuir electron frequency

$$\omega = \left(-\frac{4\pi e\bar{e}n}{m}\right)^{1/2}.$$

The general element $X$ of the Lie algebra is a linear combination of all generators

$$X = \sum_{\alpha=1}^{10} A_\alpha(e, m, \bar{e}, n, u) X_\alpha, \tag{4.3.65}$$

with coefficients $A_\alpha$, which are arbitrary functions of the five variables

$$e, m, \bar{e}, n, u. \tag{4.3.66}$$

Parameters (4.3.66) are invariants of the first five generators (4.3.64) and when the ion velocity is zero, $u = 0$, these generators correspond to the result obtained in [5].

The additional terms, which are missing in generators obtained in [5], take into account the transformation of the plasma current density $j$ (which is equal to the electron current in the limit (4.3.62)), while the plasma charge density $\rho$ is invariant. These terms are the prolongation of the group in [5] to the nonlocal variables

$$\rho = em \int dv f + \bar{e} n, \quad j = em \int dv v f, \tag{4.3.67}$$

and can be omitted in case we consider the group of transformations in the space of group variables $\{t, x, v, f, E\}$.

The generator $X_6$ in (4.3.64) is due to the nonzero ion velocity $u$ included in the set of variables (4.3.66) together with all variables involved in group transformations. By doing this we preserve an analog of the Galilean subgroup in the admitted Lie group, which is absent in the five-parameter group [5] (here $a$ is a group parameter):

$$t' = t, \quad x' = x + (e^a - 1)ut, \quad v' = v + (e^a - 1)u, \quad e' = e, \quad m' = m,$$

$$\bar{e}' = \bar{e}, \quad n' = n, \quad u' = ue^a, \quad f' = f, \tag{4.3.68}$$

$$E' = E, \quad \rho' = \rho, \quad j' = j + (e^a - 1)u\rho.$$

This example shows the importance of including parameters into group transformations.

## 4.3.6 Relativistic Electron Plasma Kinetics with a Moving Ion Background

In this section we present the result of the symmetry group calculation for relativistic equations generalizing (4.3.62):

$$f_t + v f_x + \frac{e}{m\gamma^3} E f_v = 0; \quad E_x = 4\pi\rho, \quad E_t + 4\pi j = 0,$$

$$\rho = em \int dv \gamma^3 f + \bar{e} n, \quad j = em \int dv \gamma^3 v f + \bar{e} n u, \tag{4.3.69}$$

$$\gamma \equiv \left[1 - (v/c)^2\right]^{-1/2}.$$

An infinite symmetry group admitted by (4.3.69) is given by a linear combination of the eight generators [8]

$$X = \sum_{\alpha=1}^{8} A_\alpha(e, m, \bar{e}, n, u, c) X_\alpha, \tag{4.3.70}$$

$$X_1 = \frac{\partial}{\partial t}, \quad X_2 = c \frac{\partial}{\partial x},$$

$$X_3 = t \frac{\partial}{\partial t} + x \frac{\partial}{\partial x} - 2n \frac{\partial}{\partial n} - 2f \frac{\partial}{\partial f} - E \frac{\partial}{\partial E} - 2j \frac{\partial}{\partial j} - 2\rho \frac{\partial}{\partial \rho},$$

$$X_4 = \frac{1}{c} \left[ x \frac{\partial}{\partial t} + c^2 t \frac{\partial}{\partial x} + (c^2 - v^2) \frac{\partial}{\partial v} + (c^2 - u^2) \frac{\partial}{\partial u} \right.$$

$$\left. + un \frac{\partial}{\partial n} + c^2 \rho \frac{\partial}{\partial j} - j \frac{\partial}{\partial \rho} \right], \tag{4.3.71}$$

$$X_5 = e \frac{\partial}{\partial e} + m \frac{\partial}{\partial m} - 2f \frac{\partial}{\partial f},$$

$$X_6 = m \frac{\partial}{\partial m} + n \frac{\partial}{\partial n} + E \frac{\partial}{\partial E} + j \frac{\partial}{\partial j} + \rho \frac{\partial}{\partial \rho}, \quad X_7 = \bar{e} \frac{\partial}{\partial \bar{e}} - n \frac{\partial}{\partial n},$$

$$X_8 = c \frac{\partial}{\partial c} + x \frac{\partial}{\partial x} + v \frac{\partial}{\partial v} + u \frac{\partial}{\partial u} - f \frac{\partial}{\partial f} + E \frac{\partial}{\partial E} + j \frac{\partial}{\partial j},$$

This example again shows the importance of the inclusion of the parameters into the group transformations: the six generators in (4.3.71) are due to the noninvariance of the parameters.

### 4.3.7 Non-relativistic Electron–Ion Plasma in Quasi-neutral Approximation

Essential progress in studying dynamics of plasma expansion and acceleration of ions was achieved by use of quasi-neutral approximation [13, 14], suitable for descriptions of plasma flows with characteristic scale of density variation which is large in comparison with Debye length for plasma particles. In this approximation charge and current densities in plasma are set equal to zero, that essentially simplifies the initial model with non-local terms. Thus, instead of the complete system of Vlasov–Maxwell equations (4.1.1)–(4.1.3) with the corresponding material relations here we will only use the kinetic equations for particle distribution functions for various species

$$f_t^\alpha + v f_x^\alpha + (e_\alpha/m_\alpha) E(t, x) f_v^\alpha = 0 \tag{4.3.72}$$

with additional non-local restrictions imposed on them, which arise from vanishing conditions for the current and the charge densities

$$\int dv \sum_\alpha e_\alpha f^\alpha = 0, \quad \int dv v \sum_\alpha e_\alpha f^\alpha = 0. \tag{4.3.73}$$

At that the electric field $E$ is expressed through the moments of distribution functions:

$$E(t, x) = \left( \int dv v^2 \partial_x \sum_\alpha e_\alpha f^\alpha \right) \left( \int dv \sum_\alpha \frac{e_\alpha^2}{m_\alpha} f^\alpha \right)^{-1}. \tag{4.3.74}$$

Equations (4.3.72), (4.3.73) describe one-dimensional dynamics of a plasma, which is inhomogeneous upon the coordinate $x$; thus the distribution functions of particles $f^\alpha$ depend upon $t$, $x$ and the velocity component $v$ in the directions of plasma inhomogeneity.

The group of point Lie transformations admitted by system (4.3.72) and (4.3.73) is specified by the following set of infinitesimal operators [14]:

$$X_1 = \frac{\partial}{\partial t}, \quad X_2 = \frac{\partial}{\partial x}, \quad X_3 = t \frac{\partial}{\partial t} - v \frac{\partial}{\partial v}, \quad X_4 = x \frac{\partial}{\partial x} + v \frac{\partial}{\partial v},$$

$$X_5 = \sum_\alpha f^\alpha \frac{\partial}{\partial f^\alpha}, \quad X_6 = t \frac{\partial}{\partial x} + \frac{\partial}{\partial v},$$

$$X_7 = t^2 \frac{\partial}{\partial t} + tx \frac{\partial}{\partial x} + (x - vt) \frac{\partial}{\partial v},$$

$$X_\alpha = \frac{1}{Z_{\alpha+1}} \frac{\partial}{\partial f^{\alpha+1}} - \frac{1}{Z_\alpha} \frac{\partial}{\partial f^\alpha}$$

$$\tag{4.3.75}$$

with the general element of the algebra represented by their linear combination

$$X = \sum_{j=1}^{7} c_j X_j + \sum_\alpha b_\alpha X_\alpha. \tag{4.3.76}$$

In the operators $X_\alpha$ in system (4.3.75), $Z_\alpha = e_\alpha / |e|$ is the charge number of the particle species $\alpha$, and the index $\alpha + 1$ denotes the particle species that follows $\alpha$. The operators $X_\alpha$ exist only in plasma with the number of particle types larger than or equal to two and their number is less than the number of plasma components by one. Transformation of charge and mass of particles are not included in (4.3.75).

The method for calculating the admitted symmetry group used here qualitatively differs from the method used earlier in Sect. 4.3 in that the electric field $E$ is treated not as one of the dependent variables but as an unknown function of the variables $t$ and $x$, $E(t, x)$. This case of finding the symmetry logically follows from the simpler, quasineutral model of plasma (4.3.72), (4.3.73) in contrast to the complete system of Vlasov–Maxwell equations (4.3.1)–(4.3.3). It is easy to verify that the translation operators $X_1$ and $X_2$, the Galilean transformation operator $X_6$, and the quasineutrality operators $X_\alpha$ are contained in the symmetry group obtained in Sect. 4.3.3 by a different method without assuming that $E$ is an arbitrary function of two variables to be determined. The two dilation generators specified in (4.3.56) are obtained by combining the three expansion operators $X_3$, $X_4$, and $X_5$ from (4.3.75) and by adding the contributions responsible for the dilation transformations of the electric field $E$, charge density $\rho$, and electric current density $j$. The projective group operator $X_7$ is new among the generators (4.3.75). Since here, in contrast to (4.3.54), we chose a different normalization of the particle distribution functions, the quasineutrality generators, $X_\alpha$, contrary to (4.3.56) contain factors that do not depend on particle mass.

## 4.4 Group Analysis of Three Dimensional Collisionless Plasma Kinetic Equations

In this section we calculate the point symmetry of the self-consistent field equations for three dimensional kinetic models of collisionless plasma. In the first subsection the group analysis is fulfilled for the three-dimensional kinetic equations of relativistic electron gas. In the second one the same is done for the model of quasi-neutral multi-species plasma.

### 4.4.1 Relativistic Electron Gas Kinetics

We start with the equations of kinetic theory for collisionless relativistic electron gas, described by the system of equations (4.1.1)–(4.1.3) where only one particle species, electrons, are taken into account. As in one-dimensional case, (4.1.1)–(4.1.3) should be supplemented by additional differential constraints

$$E_v = 0, \quad B_v = 0, \quad j_v = 0, \quad \rho_v = 0, \qquad (4.4.1)$$

which explicitly show that electromagnetic fields and momenta of the distribution function do not depend on the electron velocity $v$.

The canonical group generator $Y$ of the continuous point Lie group admitted by system (4.1.1)–(4.1.3), (4.4.1) has the form

$$Y = \varkappa^1 \frac{\partial}{\partial f} + \vec{\varkappa}^2 \frac{\partial}{\partial E} + \vec{\varkappa}^3 \frac{\partial}{\partial B} + \vec{\varkappa}^4 \frac{\partial}{\partial j} + \varkappa^5 \frac{\partial}{\partial \rho}, \qquad (4.4.2)$$

where the first term is given by the following three-dimensional relativistic analog of representations (4.3.15) in Sect. 4.3.1:

$$\varkappa^1 \frac{\partial}{\partial f} = \int d\boldsymbol{v} \varkappa^1(\boldsymbol{v}) \frac{\delta}{\delta f(\boldsymbol{v})}. \qquad (4.4.3)$$

As in (4.1.2), the integration domain in this formula is the sphere $|\boldsymbol{v}| < c$ of radius $c$. The procedure of symmetry group construction is similar to that in the one-dimensional case though calculus are a little bit more tedious here. As a result we get the group that is represented by the following basic generators [6, 8] (for convenience, they are written in a non-canonical form):

$$X_0 = \frac{\partial}{\partial t}, \quad X_i = c\frac{\partial}{\partial x_i},$$

$$Y_i = \frac{1}{c}\left[ x_i \frac{\partial}{\partial t} + c^2 t \frac{\partial}{\partial x_i} + (c^2 \delta_{is} - v_i v_s)\frac{\partial}{\partial v_s} \right.$$

$$\left. - c e_{isk} B_s \frac{\partial}{\partial E_k} + c e_{isk} E_s \frac{\partial}{\partial B_k} + c^2 \rho \frac{\partial}{\partial j_i} + j_i \frac{\partial}{\partial \rho} \right],$$

$$Z_i = e_{isk}\left( x_s \frac{\partial}{\partial x_k} + v_s \frac{\partial}{\partial v_k} + E_s \frac{\partial}{\partial E_k} + B_s \frac{\partial}{\partial B_k} + j_s \frac{\partial}{\partial j_k} \right), \qquad (4.4.4)$$

$$X_4 = t\frac{\partial}{\partial t} + x_s\frac{\partial}{\partial x_s} - 2f\frac{\partial}{\partial f} - E_s\frac{\partial}{\partial E_s}$$
$$- B_s\frac{\partial}{\partial B_s} - 2j_s\frac{\partial}{\partial j_s} - 2\rho\frac{\partial}{\partial \rho}.$$

Here summation is performed over repeated indices, $\delta_{is}$ and $e_{isk}$ are the Kronecker symbols of the second and the third order, $1 \leq i, s, k \leq 3$.

Generators (4.4.4) form the 11-dimensional Lie algebra

$$L_{11} = \langle X_0, X, Y, Z, X_4 \rangle \tag{4.4.5}$$

and any infinitesimal operator in (4.4.4) has a simple physical meaning. The generators $X_0$ and $X$ correspond to time and space translations, respectively. The operator $Y$ generates Lorentz transformations, which do not involve the distribution function $f$, e.g., hyperbolic rotations in the planes $(ct, x)$ and $(c\rho, j)$ and linear-fractional transformations of the electron velocity $v$. Lorentz transformations of vectors $E$ and $B$ correspond to the transformation of the 4-tensor of electromagnetic field (see [15, §22, 23]). The operator $Z$ generates rotations. The operator $X_4$ generates dilations, and it is the only group transformations in (4.4.4) which involve $f$.

The 10-dimensional algebra of the Poincaré group,

$$L_{10} = \langle X_0, X, Y, Z \rangle \tag{4.4.6}$$

is included in (4.4.5), $L_{10} \subset L_{11}$, and it also appears in the independent (local) group analysis of the Maxwell equations (4.1.2),

$$X = \xi^1\frac{\partial}{\partial t} + \xi^2\frac{\partial}{\partial x} + \eta^2\frac{\partial}{\partial E} + \eta^3\frac{\partial}{\partial B} + \eta^4\frac{\partial}{\partial j} + \eta^5\frac{\partial}{\partial \rho}, \tag{4.4.7}$$

as a subalgebra of the 16-dimensional algebra

$$L_{16} = \langle X_0, X, Y, Z, U_0, U, \bar{X}_4, \bar{X}_5 \rangle \tag{4.4.8}$$

of the conformal group admitted by (4.1.2). Here the scalar $U_0$ and the vector $U$ operators are given by

$$U_0 = \frac{1}{c^2}\left[\frac{1}{2}(c^2t^2 + x^2)\frac{\partial}{\partial t} + c^2tx_i\frac{\partial}{\partial x_i} - c(2ctE_i - e_{isk}B_sx_k)\frac{\partial}{\partial E_i}\right.$$
$$\left. - c(2ctB_i + e_{isk}E_sx_k)\frac{\partial}{\partial B_i} + c^2(-3tj_i + \rho x_i)\frac{\partial}{\partial j_i} + (-3t\rho c^2 + j_ix_i)\frac{\partial}{\partial \rho}\right],$$

$$U_i = \frac{1}{c}\left[tx_i\frac{\partial}{\partial t} + \left(x_ix_s + \frac{1}{2}(c^2t^2 - x^2)\delta_{is}\right)\frac{\partial}{\partial x_s} + \left(x_kE_i - (E \cdot x)\delta_{ik}\right.\right. \tag{4.4.9}$$
$$\left. - 2x_iE_k - cte_{isk}B_s\right)\frac{\partial}{\partial E_k} + \left(x_kB_i - (B \cdot x)\delta_{ik} - 2x_iB_k + cte_{isk}E_s\right)\frac{\partial}{\partial B_k}$$
$$\left. + \left(x_kj_i - (j \cdot x)\delta_{ik} - 3x_ij_k + c^2t\rho\delta_{ik}\right)\frac{\partial}{\partial j_k} + (tj_i - 3\rho x_i)\frac{\partial}{\partial \rho}\right].$$

The two last (scalar) generators in (4.4.8) generate the dilations

$$\bar{X}_4 = t\frac{\partial}{\partial t} + x_i\frac{\partial}{\partial x_i} - 2E_i\frac{\partial}{\partial E_i} - 2B_i\frac{\partial}{\partial B_i} - 3j_i\frac{\partial}{\partial j_i} - 3\rho\frac{\partial}{\partial \rho},$$

$$\bar{X}_5 = E_i\frac{\partial}{\partial E_i} + B_i\frac{\partial}{\partial B_i} + j_i\frac{\partial}{\partial j_i} + \rho\frac{\partial}{\partial \rho}.$$

(4.4.10)

The invariance conditions for the kinetic Vlasov equation (4.1.1) violates the conformal part (4.4.9) of group (4.4.8) when the intermediate group symmetry is taken into account. Adding the dilation operators (4.4.10) and "correcting" the sum by taking into account the dilation of $f$ we obtain a "prototype" of the generator $X_4$ in the algebra (4.4.5). Thus, using the relation between the algebras $L_{10}$, $L_{11}$ and $L_{16}$ we can interpret the result (4.4.5) in terms of the group symmetry of the Maxwell equations (4.1.2). The nonlocal determining equations yield the contribution $(-2f\partial_f)$ into the generator $X_4$ in (4.4.6); this term cannot be obtained from the standard group analysis, but it is easily reproduced from physical considerations.

Including parameters $e$, $m$, and $c$ in the set of group variables of the system under consideration we add three scalar generators to (4.4.6) and thereby take into account dilations of the electron charge, mass, and the light velocity $c$ in vacuum:

$$X_5 = m\frac{\partial}{\partial m} - 2f\frac{\partial}{\partial f} + E_s\frac{\partial}{\partial E_s} + B_s\frac{\partial}{\partial B_s} + j_s\frac{\partial}{\partial j_s} + \rho\frac{\partial}{\partial \rho},$$

$$X_6 = e\frac{\partial}{\partial e} + m\frac{\partial}{\partial m} - 4f\frac{\partial}{\partial f},$$

(4.4.11)

$$X_7 = c\frac{\partial}{\partial c} + x_s\frac{\partial}{\partial x_s} + v_s\frac{\partial}{\partial v_s} - 3f\frac{\partial}{\partial f} + E_s\frac{\partial}{\partial E_s} + B_s\frac{\partial}{\partial B_s} + j_s\frac{\partial}{\partial j_s}.$$

Then the Lie group of the Vlasov–Maxwell equations (4.1.1)–(4.1.3) becomes infinite and the common element $X$ of the operator algebra depending on 14 scalar functions of three variables $e$, $m$, and $c$ is given by [8]

$$X = \sum_{\alpha=0}^{7} A_\alpha(e, m, c)X_\alpha + b(e, m, c)Y + g(e, m, c)Z.$$

(4.4.12)

The group analysis of the equations of collisionless electron gas (single-component charged plasma) performed in the present section is supplemented in the next section by the group analysis of kinetic equations of a quasi-neutral multi-component (electron–ion) plasma.

## 4.4.2 Relativistic Electron–Ion Plasma Kinetic Equations

In this Section we point to the distinctive features that arise for the symmetry group of a multi-species electron–ion plasma (with $k > 1$ particle species), as compared to algebra (4.4.5). Starting with Vlasov–Maxwell equations (4.1.1)–(4.1.3) in the most general form with $1 \leq \alpha \leq k$ and adding (4.4.1) we come after fulfilling the procedure used above to the following $11 + (k-1)$-parameter Lie group [6, 8]

$$L_{11+(k-1)} = \langle X_0, X, Y, Z, X_4, X^\mu \rangle, \quad 2 \leq \mu \leq k, \qquad (4.4.13)$$

including the Poincaré group as a subgroup. The first ten (scalar) generators of the algebra (4.4.13) are listed in (4.4.5) and span the algebra (4.4.8). The infinitesimal operator $X_4$ in (4.4.13), in contrast to $X_4$ in (4.4.9), includes dilations of all distribution functions $f^\alpha$:

$$X_4 = t\frac{\partial}{\partial t} + x_s\frac{\partial}{\partial x_s} - 2\sum_{\mu=1}^{k} f^\mu \frac{\partial}{\partial f^\mu} - E_s\frac{\partial}{\partial E_s}$$

$$- B_s\frac{\partial}{\partial B_s} - 2j_s\frac{\partial}{\partial j_s} - 2\rho\frac{\partial}{\partial \rho}. \qquad (4.4.14)$$

The algebra $L_{11+(k-1)}$ contains $k-1$ new operators not included in $L_{11}$ in (4.4.9); these are the "quasi-neutrality operators"

$$X^\mu = \frac{1}{e^1(m^1)^3}\frac{\partial}{\partial f^1} - \frac{1}{e^\mu(m^\mu)^3}\frac{\partial}{\partial f^\mu}, \quad 2 \leq \mu \leq k, \qquad (4.4.15)$$

typical for the multi-component plasma. The quasi-neutrality generator (4.4.15) determines consistent translation transformations of the distribution functions $f^\mu$.

Including $2k+1$ parameters (masses and charges of particles and light velocity in vacuum) of multi-component plasma equations in the set of group variables yields $2k+1$ additional generators of dilations ($1 \leq \lambda, \nu \leq k$)

$$X_5 = c\frac{\partial}{\partial c} + x_s\frac{\partial}{\partial x_s} + v_s\frac{\partial}{\partial v_s} - 3\sum_{q=1}^{k} f^q \frac{\partial}{\partial f^q} + E_s\frac{\partial}{\partial E_s} + B_s\frac{\partial}{\partial B_s} + j_s\frac{\partial}{\partial j_s},$$

$$X_6 = \sum_{q=1}^{k} m^q \frac{\partial}{\partial m^q} - 2\sum_{q=1}^{k} f^q \frac{\partial}{\partial f^q} + E_s\frac{\partial}{\partial E_s} + B_s\frac{\partial}{\partial B_s} + j_s\frac{\partial}{\partial j_s} + \rho\frac{\partial}{\partial \rho},$$

$$X^\lambda = e^\lambda \frac{\partial}{\partial e^\lambda} + m^\lambda \frac{\partial}{\partial m^\lambda} - 4f^\lambda \frac{\partial}{\partial f^\lambda},$$

$$X^\nu = e^\nu \frac{\partial}{\partial e^\nu} + m^\nu \frac{\partial}{\partial m^\nu} - 4f^\nu \frac{\partial}{\partial f^\nu}.$$

$$(4.4.16)$$

Then the Lie group admitted by the Vlasov–Maxwell equations (4.1.1)–(4.1.3) becomes infinite and its general element $X$ depends on $3k+9$ arbitrary scalar functions of the $2k+1$ group variables $e^\alpha$, $m^\alpha$, and $c$:

$$X = \sum_{\alpha=0}^{6} A_\alpha(e^\alpha, m^\alpha, c)X_\alpha + b(e^\alpha, m^\alpha, c)Y + g(e^\alpha, m^\alpha, c)Z$$

$$+ \sum_{\mu=2}^{k} A_\mu(e^\alpha, m^\alpha, c)X^\mu + \sum_{\lambda=1}^{k} A_\lambda(e^\alpha, m^\alpha, c)X^\lambda + \sum_{\nu=1}^{k} A_\nu(e^\alpha, m^\alpha, c)X^\nu.$$

$$(4.4.17)$$

## 4.5  Symmetry of Vlasov–Maxwell Equations in Lagrangian Variables

This section is devoted to calculation of the symmetry group for the system of equations (4.1.5)–(4.1.7) that presents the Lagrangian formulation of the known Vlasov–Maxwell equations (4.1.1)–(4.1.3). The infinitesimal operator of the admitted local group of point one-parameter transformations in a standard form

$$X = \xi^1 \frac{\partial}{\partial t} + \xi^2 \frac{\partial}{\partial r} + \xi^3 \frac{\partial}{\partial w} + \sum_\alpha \eta^{1\alpha} \frac{\partial}{\partial N^\alpha} + \sum_\alpha \eta^{2\alpha} \frac{\partial}{\partial V^\alpha}$$

$$+ \eta^3 \frac{\partial}{\partial E} + \eta^4 \frac{\partial}{\partial B} + \eta^5 \frac{\partial}{\partial j} + \eta^6 \frac{\partial}{\partial \rho}, \qquad (4.5.1)$$

where coordinates $\xi^i$ and $\eta^k$ depend upon $t$, $r$, $w$, $N^\alpha$, $V^\alpha$, $E$, $B$, $j$ and $\rho$. In the canonical form this operator is given as:

$$Y = \sum_\alpha \varkappa^{1\alpha} \frac{\partial}{\partial N^\alpha} + \vec{\varkappa}^{2\alpha} \frac{\partial}{\partial V^\alpha} + \vec{\varkappa}^3 \frac{\partial}{\partial E} + \vec{\varkappa}^4 \frac{\partial}{\partial B} + \vec{\varkappa}^5 \frac{\partial}{\partial j} + \varkappa^6 \frac{\partial}{\partial \rho},$$

$$\varkappa^{1\alpha} = \eta^{1\alpha} - \mathcal{D} N^\alpha, \quad \vec{\varkappa}^{2\alpha} = \eta^{2\alpha} - \mathcal{D} V^\alpha,$$

$$\vec{\varkappa}^3 = \eta^3 - \mathcal{D} E, \quad \vec{\varkappa}^4 = \eta^4 - \mathcal{D} B,$$

$$\vec{\varkappa}^5 = \eta^5 - \mathcal{D} j, \quad \varkappa^6 = \eta^6 - \mathcal{D} \rho,$$

$$\mathcal{D} \equiv \xi^1 \partial_t - (\xi^2 \cdot \nabla_r) - (\xi^3 \cdot \nabla_w).$$

$$(4.5.2)$$

The current and charge densities in (4.1.6) are moments of functions $N^\alpha$ and $V^\alpha$ and, similar to electric and magnetic fields in Maxwell equations (4.1.3), do not depend upon the plasma particles velocity. This lead to additional differential constraints

$$E_w = 0; \quad B_w = 0; \quad j_w = 0; \quad \rho_w = 0, \qquad (4.5.3)$$

that are obvious from the physical point of view, however essential for calculating symmetries of Vlasov–Maxwell equations.

Following the procedure, fulfilled in the preceding section, we obtain the continuous Lie point transformation group for Vlasov–Maxwell equations (with Lagrangian velocity), which we present in a non-canonical form (compare to (4.4.5), (4.4.13) in Sect. 4.4)

$$P_0 = i \frac{\partial}{\partial t}, \quad P = i \frac{\partial}{\partial r},$$

$$\mathbf{B} = r\frac{\partial}{\partial t} + c^2 t\frac{\partial}{\partial r} - c\left[\mathbf{B} \times \frac{\partial}{\partial \mathbf{E}}\right] + c\left[\mathbf{E} \times \frac{\partial}{\partial \mathbf{B}}\right] + c^2\rho\frac{\partial}{\partial j} + j\frac{\partial}{\partial \rho}$$

$$+ \sum_\alpha \left(N^\alpha V^\alpha \partial_{N^\alpha} + c^2 \partial_{V^\alpha} - V^\alpha\left(V^\alpha \cdot \frac{\partial}{\partial V^\alpha}\right)\right),$$

$$\mathbf{R} = \left[r \times \frac{\partial}{\partial r}\right] + \left[V^\alpha \times \frac{\partial}{\partial V^\alpha}\right] + \left[\mathbf{E} \times \frac{\partial}{\partial \mathbf{E}}\right] + \left[\mathbf{B} \times \frac{\partial}{\partial \mathbf{B}}\right] + \left[j \times \frac{\partial}{\partial j}\right],$$

$$\mathrm{D} = t\frac{\partial}{\partial t} + r\frac{\partial}{\partial r} - 2\sum_\alpha N^\alpha\frac{\partial}{\partial N^\alpha} - \mathbf{E}\frac{\partial}{\partial \mathbf{E}} - \mathbf{B}\frac{\partial}{\partial \mathbf{B}} - 2j\frac{\partial}{\partial j} - 2\rho\frac{\partial}{\partial \rho},$$

$$X_\infty = \xi\frac{\partial}{\partial w} - \left(5\frac{(w \cdot \xi)}{c^2}\gamma^2 + (\nabla_w \cdot \xi)\right)\sum_\alpha N^\alpha\frac{\partial}{\partial N^\alpha}.$$

$$(4.5.4)$$

The operators in (4.5.4) has a simple physical interpretation: $P_\mu = (P_0, \mathbf{P})$, where $\mu = 0, 1, 2, 3$, specify translation in time and translation along the three components of radius-vector $r$, $\mathbf{B}$ defines Lorentz transformations, consisting of hyperbolic rotations (boosts) in the $\{ct, r\}$ and $\{c\rho, j\}$ planes, linear-fractional transformations of the velocity $V^\alpha$, transformations of the density $N^\alpha$ and transformations of components of the 4-tensor of the electromagnetic field (see, e.g., §24, 25 in [15]), while $\mathbf{R}$ specifies circular rotations. These ten (scalar) operators define the Poincaré group:[3]

$$L_{10} = \langle P_0, \mathbf{P}, \mathbf{B}, \mathbf{R}\rangle.$$

In (4.5.4) this is supplemented by the operator D, specifying dilations, and the operator of the infinite subgroup $X_\infty$ (see also [16] and [17] (p. 419, vol. 2)), specifying the consistent transformations of Lagrangian velocity and the density of the plasma particles. Thus, provided parameters $e_\alpha, m_\alpha$ and $c$ are not involved in transformations the continuous Lie point group, admitted by Vlasov–Maxwell equations with Lagrangian velocity, is defined by the 11-dimensional subalgebra, specified by the algebra $L_{10}$ of the Poincaré group and the one-dimensional algebra with the dilation operator D, and the infinite-dimensional subalgebra with the operator $X_\infty$.

To end of this section we prolong the generators (4.5.4) to the space of Fourier variables for functions, independent of Lagrangian velocity $w$. From a point of initial representation, specifying of the Fourier transformation, say, of a charge density

$$\tilde\rho(\omega, k) = \int dt\, dr\, \rho(t, r)\exp(i\omega t - ikr), \qquad (4.5.5)$$

---

[3]Frequently the six operators specifying hyperbolic and circular rotations in $(c^2 t, x^k)$ and $(x^j, x^k)$ planes, respectively $(j, k = 1, 2, 3; r = (x^1, x^2, x^3))$, are written in a universal form using the operators $M_{\mu\nu}$, where $M_{0k} = i\mathrm{B}_{0k}$ and $M_{jk} = i\mathrm{R}_{jk}$. The three operators $(M_{23}, M_{31}, M_{12})$ are components of the vector-operator $\mathbf{M} = [r \times \mathbf{P}]$.

is equivalent to introduction of a new non-local variable. Similar to Sect. 4.2.4 to fulfill the procedure of prolongation of Lie point group operator (4.5.1) on a non-local variable, we rewrite down this operator in the canonical form (4.5.2) and formally prolong it on the non-local variable $\tilde{\rho}(\omega, \mathbf{k})$

$$\tilde{Y} \equiv Y + \tilde{\varkappa}^6 \frac{\partial}{\partial \tilde{\rho}}. \tag{4.5.6}$$

The integral relation between $\varkappa^6$ and $\tilde{\varkappa}^6$ results while applying the operator (4.5.2) to (4.5.5). Here we consider it as the definition of the variable $\tilde{\rho}$

$$\tilde{\varkappa}^6 = \int \mathrm{d}t \, \mathrm{d}\mathbf{r} \, \varkappa^6 \exp(i\omega t - i\mathbf{k}\mathbf{r}). \tag{4.5.7}$$

Substituting $\varkappa^6$ from (4.5.5), (4.5.6) into (4.5.7) and calculating the integrals obtained (integrating by parts), we get the desired coordinate $\tilde{\varkappa}^6$. For example, for the operator of time translations $P_0$ the coordinate $\varkappa^6 = -i\rho_t$ after substitution into (4.5.5) yields the following expression for the coordinate $\tilde{\varkappa}^{1e} = -\omega\tilde{\rho}$ in Fourier variables. Other coordinates of a canonical operator are calculated in a similar way. Inserting these results into (4.5.6), restricting the group to Fourier variables not containing dependencies upon Lagrangian velocity $\mathbf{w}$ (i.e. leaving in (4.5.7) only the contributions responsible for transformation of these variables in Fourier representation) and returning back to non-canonical representation, we obtain the following set of operators for 11-parametric Lie point group in $\{\omega, \mathbf{k}\}$ representation (see also [16])

$$\tilde{P}_0 = -\omega \left( \tilde{E}\frac{\partial}{\partial \tilde{E}} + \tilde{B}\frac{\partial}{\partial \tilde{B}} + \tilde{j}\frac{\partial}{\partial \tilde{j}} + \tilde{\rho}\frac{\partial}{\partial \tilde{\rho}} \right);$$

$$\tilde{P} = \mathbf{k} \left( \tilde{E}\frac{\partial}{\partial \tilde{E}} + \tilde{B}\frac{\partial}{\partial \tilde{B}} + \tilde{j}\frac{\partial}{\partial \tilde{j}} + \tilde{\rho}\frac{\partial}{\partial \tilde{\rho}} \right);$$

$$\tilde{B} = c^2 \mathbf{k}\frac{\partial}{\partial \omega} + \omega\frac{\partial}{\partial \mathbf{k}} - c \left[ \tilde{B} \times \frac{\partial}{\partial \tilde{E}} \right] + c \left[ \tilde{E} \times \frac{\partial}{\partial \tilde{B}} \right] + c^2 \tilde{\rho}\frac{\partial}{\partial \tilde{j}} + \tilde{j}\frac{\partial}{\partial \tilde{\rho}};$$

$$\tilde{R} = \left[ \mathbf{k}, \frac{\partial}{\partial \mathbf{k}} \right] + \left[ \tilde{E} \times \frac{\partial}{\partial \tilde{E}} \right] + \left[ \tilde{B} \times \frac{\partial}{\partial \tilde{B}} \right] + \left[ \tilde{j} \times \frac{\partial}{\partial \tilde{j}} \right];$$

$$\tilde{D} = -\omega\frac{\partial}{\partial \omega} - \mathbf{k}\frac{\partial}{\partial \mathbf{k}} + 3\tilde{E}\frac{\partial}{\partial \tilde{E}} + 3\tilde{B}\frac{\partial}{\partial \tilde{B}} + 2\tilde{j}\frac{\partial}{\partial \tilde{j}} + 2\tilde{\rho}\frac{\partial}{\partial \tilde{\rho}}.$$

$$\tag{4.5.8}$$

Formulas (4.5.8) supplement the group (4.5.4) by the appropriate transformations of variables in Fourier-space. For example, Lorentz transformations with the operator $\mathbf{B}$ are supplemented with hyperbolic rotations in $\{\omega, c\mathbf{k}\}$ and $\{c\tilde{\rho}, \tilde{j}\}$ planes and transformations of the 4-tensor of the Fourier-components of the electromagnetic field.

## 4.6 Vlasov-Type Equations: Symmetries of the Benney Equations

### 4.6.1 Different Forms of the Benney Equations

The Benney equations referred to by the name of the author of a pioneering work [18] appear in long wavelength hydrodynamics of an ideal incompressible fluid of a finite depth in a gravitational field. From the group theoretical point of view they are of particular interest due to the existence of an infinite set of conservation laws obtained in [18]. The latter property of the Benney equations emphasizes their significance that goes far beyond an interesting example of an integrable system of hydrodynamic equations.

In practice, the Benney equations are used in various representation. One of them is the kinetic Benney equation (a kinetic equation with a self-consistent field):

$$f_t + vf_x - A^0_x f_v = 0, \qquad A^0(t, x) = \int\limits_{-\infty}^{+\infty} f(t, x, v)\,dv. \qquad (4.6.1)$$

This equation appears as a unique representative of a set of hierarchy of kinetic equations of Vlasov-type [19]. A detailed study of its group properties will lead to better understanding of the symmetry properties of kinetic equations of collisionless plasma, namely the Vlasov–Maxwell equations.

Another form of the Benney equations is an infinite set of coupled equations

$$A^i_t + A^{i+1}_x + i A^0_x A^{i-1} = 0, \qquad i \geq 0 \qquad (4.6.2)$$

for a countable set of functions $A^i$ of two independent variables, time $t$ and the spatial coordinate $x$. In terms of hydrodynamics these functions appear as averaged values (with respect to the depth) of integer powers $i \geq 0$ of the horizontal component of the liquid flow velocity. The corresponding integrals that describe this averaging are taken over the vertical coordinate in the limits from the flat bottom up to the free liquid surface. Solutions, Hamiltonian structure and conservation laws for (4.6.2) were discussed in details in [20, 21].

From the kinetic point of view the system (4.6.2) can be treated as a system of equations for moments of the distribution function $f$ that obeys the kinetic Benney equation (4.6.1)

$$A^i(t, x) = \int\limits_{-\infty}^{+\infty} v^i f\,dv, \qquad i \geq 0. \qquad (4.6.3)$$

This fact with the explicit formulation of the Benney equation (4.6.1) was first stated independently in [22, 23]. The Lagrangian change of the Euler velocity $v$,

$$v = V(t, x, u) \qquad (4.6.4)$$

yields one more representation for Benney equations (4.6.1):

$$f_t + Vf_x = 0, \qquad V_t + VV_x = -A^0_x, \qquad A^0(t, x) = \int V_u f(t, x, u)\,du. \qquad (4.6.5)$$

Equations (4.6.3) are readily converted into the hydrodynamic-type form

$$n_t + (nV)_x = 0, \quad V_t + VV_x = -A^0_x, \quad A^0 = \int n(t, x, u) du, \quad (4.6.6)$$

if one employs the "density" $n$ depending on the Lagrangian velocity $u$:

$$n = f(t, x, u) V_u. \quad (4.6.7)$$

Using the form (4.6.6) of the Benney equations an infinite set of conservation laws were constructed in [22] with the densities regarded as functions of the Lagrangian velocity $u$.

The knowledge of the complete Lie–Bäcklund symmetry for the Benney equations in different representations (4.6.1)–(4.6.6) can clarify the question of structure of solutions and conservation laws for these equations. This statement is partially confirmed by the fact that one of the main results of the works [20, 21], namely the higher order Benney equations, can be re-formulated in terms of the first order Lie–Bäcklund group, admitted by the system (4.6.2). Unfortunately, the complete description of the Lie–Bäcklund symmetry for (4.6.2) is not available in the literature. This section is devoted to calculating an infinite (countable) part of the Lie point symmetries of the moment equations (4.6.2).

## 4.6.2  Lie Subgroup and Lie–Bäcklund Group: Statement of the Problem

A Lie subgroup, admitted by the kinetic Benney equation (4.6.1) in the space of four variables

$$t, \ x, \ v, \ f \quad (4.6.8)$$

is defined by five basic infinitesimal operators

$$X_1 = \frac{\partial}{\partial t}, \quad X_2 = \frac{\partial}{\partial x}, \quad X_3 = t\frac{\partial}{\partial x} + \frac{\partial}{\partial v},$$

$$X_4 = t\frac{\partial}{\partial t} - v\frac{\partial}{\partial v} - f\frac{\partial}{\partial f}, \quad X_5 = x\frac{\partial}{\partial x} + v\frac{\partial}{\partial v} + f\frac{\partial}{\partial f}. \quad (4.6.9)$$

With the less computation difficulties this group can be obtained using the approach developed in Sect. 4.3.1.

Prolongation of infinitesimal operators (4.6.9) on nonlocal variables (4.6.3) extends the set of variables (4.6.8) up to a countable set

$$t, \ x, \ v, \ f, \ A^0, \ \ldots, \ A^i, \ \ldots. \quad (4.6.10)$$

In the latter case infinitesimal operators (4.6.9) rewritten in the canonical form and restricted on the sub-manifold

$$t, \ x, \ A^0, \ \ldots, \ A^i, \ \ldots \quad (4.6.11)$$

are given by the following expressions

$$X_1 = \sum_{i=0}^{\infty} (A_x^{i+1} + i A^{i-1} A_x^0) \frac{\partial}{\partial A^i}; \quad X_2 = \sum_{i=0}^{\infty} A_x^i \frac{\partial}{\partial A^i};$$

$$X_3 = \sum_{i=0}^{\infty} (i A^{i-1} - t A_x^i) \frac{\partial}{\partial A^i};$$

$$X_4 = \sum_{i=0}^{\infty} [(i+2) A^i - t(A_x^{i+1} + i A^{i-1} A_x^0)] \frac{\partial}{\partial A^i}; \tag{4.6.12}$$

$$X_5 = \sum_{i=0}^{\infty} [(i+2) A^i - x A_x^i] \frac{\partial}{\partial A^i}.$$

It can be easily checked that infinitesimal operators (4.6.12) are admitted by Benney equations (4.6.2) and it goes without saying that they directly result from the group analysis of Benney equations (4.6.2). Just in this way (i.e., using the method of moments) infinitesimal operators (4.6.9) were first obtained in [19] by using non-canonical form of infinitesimal operators (4.6.12) with the subsequent passage to the representation (4.6.9) in the space of variables (4.6.10).

### 4.6.3 Incompleteness of the Point Group: Statement of the Problem

It is evident, however, that the subgroup (4.6.12) does not exhaust the complete group symmetry of Benney equations (4.6.2). The incompleteness of the result (4.6.12) is obvious form many points of view. Here we shall only point on the non-conformity of finite dimension of the algebra (4.6.12) to the infinite set of conservation laws for Benney equations, and on the infinite extension of the point symmetry group for Benney equations in the form of (4.6.5), (4.6.6) with Lagrangian velocity. Here of principle significance for us is the following statement [24]: *the group (4.6.12) is incomplete not only from the standpoint of Lie–Bäcklund symmetry for Benney equations but also from the standpoint of the Lie point symmetry.* The validity of the statement can be proved by direct solving of determining equations for the first order Lie–Bäcklund group (contact group, that is not reduced to point one)

$$D_t(\varkappa^i) + D_x(\varkappa^{i+1}) + i A^{i-1} D_x(\varkappa^0) + i A_x^0 \varkappa^{i-1} = 0, \quad i \geq 0, \tag{4.6.13}$$

where coordinates $\varkappa^i$ of canonical operator

$$X = \sum_{i=0}^{\infty} \varkappa^i \frac{\partial}{\partial A^i}, \tag{4.6.14}$$

depend upon the countered set of group variables

$$t, x; \ A^0, \ldots, A^j, \ldots; \ A_x^0, \ldots, A_x^j, \ldots; \quad j \geq 0. \tag{4.6.15}$$

To prove the above statement one can consider only partial solutions of determining equations (4.6.12)

$$\varkappa^i = \eta^i(A^0, \ldots, A^j, \ldots); \quad i, j \geq 0, \tag{4.6.16}$$

that depend upon moments $A^j$, $j \geq 0$, and does not depend upon $t$, $x$. It appears that thanks to these infinitesimal operators (4.6.14), (4.6.16) an infinite extension of the group (4.6.12) takes place. Now the problem is to find these operators.

### 4.6.4  Determining Equations and Their Solution

Before proceeding further we write determining equations of first-order Lie–Bäcklund group, admitted by a more infinite system of coupling equations for functions $A^i(t, x)$ with the arbitrary element $\varphi(A^0)$

$$A^i_t + A^{i+1}_x + i A^{i-1}\big[\varphi(A^0)\big]_x = 0, \qquad i \geq 0. \tag{4.6.17}$$

For the coordinates $\varkappa^i$ of canonical infinitesimal operator (4.6.14) the following chains of determining equations are valid which result from splitting (4.6.17) with respect to second derivatives:

$$
\begin{aligned}
&\varkappa^{i+1}_{A^0_x} + i\varphi_1 A^{i-1}\varkappa^0_{A^0_x} = \sum_{j=0}^{\infty} j\varphi_1 A^{j-1}\varkappa^i_{A^j_x}, \quad i \geq 0; \\
&\varkappa^{i+1}_{A^j_x} + i\varphi_1 A^{i-1}\varkappa^0_{A^j_x} = \varkappa^i_{A^{j-1}_x}; \quad i \geq 0,\ j \geq 1, \\
&\varkappa^i_t + \varkappa^{i+1}_x + i\varphi_1 A^{i-1}\varkappa^0_x + A^0_x\big(i\varphi_1\varkappa^{i-1} + i\varphi_2 A^{i-1}\varkappa^0\big) \\
&\quad + \sum_{j=0}^{\infty}\big[i\varphi_1 A^{i-1}A^j_x\varkappa^0_{A^j} - \big(A^{j+1}_x + j\varphi_1 A^0_x A^{j-1}\big)\varkappa^i_{A^j} + A^j_x\varkappa^{i+1}_{A^j}\big] \\
&\quad - \sum_{j=0}^{\infty} j A^0_x\big(\varphi_1 A^{j-1}_x + \varphi_2 A^0_x A^{j-1}\big)\varkappa^i_{A^j_x} = 0, \quad i \geq 0.
\end{aligned}
\tag{4.6.18}
$$

Here $\varphi_1$ and $\varphi_2$ are the first and the second derivatives of the function $\varphi$ with respect to its argument. From the various standpoints at list three distinct values of the function $\varphi$ are specified. In case $\varphi(A^0) = A^0$ we come to kinetic Benney equations (4.6.2), whereas for $\varphi = a(A^0)^2$ extension of the admitted point group takes place thanks to projective transformations in $t$, $x$-plane (see [19]). For $\varphi = a \ln A^0$ the corresponding kinetic equation

$$f_t + vf_x - a\frac{A^0_x}{A^0}f_v = 0, \qquad A^0 = \int\limits_{-\infty}^{+\infty} dv f, \tag{4.6.19}$$

that gives rise to the discussed system of equations for moments, is of special interest in plasma theory. It appears as the equation for the distribution function of plasma ions, while electrons obey the Boltzmann distribution. More complicated dependencies of $\varphi(A^0)$ upon $A^0$ can also be of interest in plasma physics for non-Boltzmann distribution functions for hot electrons. Equation (4.6.19) was studied in details in [25].

For the Benney equations (4.6.2) the determining equations (4.6.18) are rewritten in the following form

$$\varkappa^{i+1}_{A^0_x} + i A^{i-1} \varkappa^0_{A^0_x} - \sum_{j=0}^{\infty} j A^{j-1} \varkappa^i_{A^j_x} = 0, \quad i \geq 0,$$

$$\varkappa^{i+1}_{A^{j+1}_x} - \varkappa^i_{A^j_x} + i A^{i-1} \varkappa^0_{A^{j+1}_x} = 0, \quad i \geq 0, \ j \geq 0.$$

$$\varkappa^i_t + \varkappa^{i+1}_x + i A^{i-1} \varkappa^0_x + A^0_x \left( i \varkappa^{i-1} - \sum_{j=0}^{\infty} j A^{j-1} \varkappa^i_{A^j} - \sum_{j=0}^{\infty} (j+1) A^j_x \varkappa^i_{A^{j+1}_x} \right)$$

$$+ i A^{i-1} \sum_{j=0}^{\infty} A^j_x \varkappa^0_{A^j} + \sum_{j=0}^{\infty} A^j_x \varkappa^{i+1}_{A^j} - \sum_{j=0}^{\infty} A^{j+1}_x \varkappa^i_{A^j} = 0, \quad i \geq 0.$$

$$(4.6.20)$$

Under conditions (4.6.16) the determining equations (4.6.20) are split and reduced to two infinite chains of equalities, namely one-dimensional (vector) and two-dimensional (tensor):

$$\eta^{i+1}_{A^0} - \sum_{j=0}^{\infty} j A^{j-1} \eta^i_{A^j} + i A^{i-1} \eta^0_{A^0} + i \eta^{i-1} = 0, \quad i \geq 0;$$

$$\eta^{i+1}_{A^{k+1}} - \eta^i_{A^k} + i A^{i-1} \eta^0_{A^{k+1}} = 0, \quad i \geq 0, \ k \geq 0.$$

$$(4.6.21)$$

The apparent difficulty in analytical solving of the given system of determining equations (4.6.21) is due to a "nonlocal" nature of the second term in the vector chain in the form of an infinite sum with respect to index $j \geq 0$. The measure of this non-locality is characterized by a number of nonzero components of tensor $\eta^i_j$. But in fact in case of an overdetermined system (4.6.21) we obtain a finite upper value of the summation index $j < \infty$, which depends upon the other index $i$ of this tensor.[4] Then we come to a much more simplified (but equivalent) formulation of the system (4.6.21)

$$\eta^{i+1}_{A^0} - \sum_{j=0}^{i-2} j A^{j-1} \eta^i_{A^j} + i \eta^{i-1} = 0, \quad \eta^{i+1}_{A^i} = 0, \quad i \geq 0;$$

$$\eta^{i+1}_{A^{k+1}} = \eta^i_{A^k}, \quad \eta^i_{A^{i+k}} = 0, \quad i \geq 0, \ k \geq 0.$$

$$(4.6.22)$$

Before proceeding to enumerating all solutions of the system of determining equations (4.6.22), we present here yet another form of the chain in (4.6.22)

$$\eta^{i+1}_{A^0} - \sum_{j=0}^{i-2} j A^{j-1} \eta^{i-j}_{A^0} + i \eta^{i-1} = 0, \quad i \geq 0. \tag{4.6.23}$$

This form can be employed to clarify the general structure of these solutions on basis of the corresponding generating functions.

---

[4] For more details we refer the reader to [24].

### 4.6.5 Discussion of the Solution of the Determining Equations

The integrability procedure in itself for determining equations (4.6.22) is of no difficulties. For example the first six coordinates $\eta^i$ $(0 \leq i \leq 5)$ of the desired infinitesimal operator (4.6.14), (4.6.16) are given by the following formulas for the general solutions of determining equations (4.6.22) that depend upon six arbitrary constants $C^j$ $(0 \leq j \leq 5)$ and are described by polynomials in moments $A^l$

$$\eta^0 = C^0, \quad \eta^1 = C^1, \quad \eta^2 = C^2 - C^0 A^0, \quad \eta^3 = C^3 - 2C^1 A^0 - C^0 A^1,$$
$$\eta^4 = C^4 - 3C^2 A^0 - 2C^1 A^1 + C^0 \left[ -A^2 + (A^0)^2 \right], \tag{4.6.24}$$
$$\eta^5 = C^5 - 4C^3 A^0 - 3C^2 A^1 + C^1 \left[ -2A^2 + 3(A^0)^2 \right] + C^0 \left( -A^3 + 2A^0 A^1 \right).$$

It appears that the polynomial dependence of any solution $\eta^i$ of determining equations (4.6.22) upon moments $A^j$ is a general property of components of the vector $\eta^i$ for any $i \geq 0$. The example (4.6.24) demonstrates that the procedure of obtaining solutions of determining equations (4.6.22) is reduced to their enumeration. To be concrete, we assume the following scheme of indicating of the $k$-th basic solution $\eta_k^i$ of determining equations (4.6.22) for the coordinate $\eta^i$:

$$\eta_k^i = \begin{cases} 0, & i < k; \\ 1, & i = k; \\ 0, & i = k+1; \end{cases} \qquad \left[ \eta_k^i \right] = i - k, \ i \geq k + 2; \quad i, k \geq 0. \tag{4.6.25}$$

In the solutions (4.6.25) this scheme demands quit definite choice of values of integration constants $C^j$ in the form of Kronecker symbols

$$C^j = \delta_{jk}; \quad j, k \geq 0. \tag{4.6.26}$$

The last of the four equalities for $\eta_k^i$ in (4.6.25) (in square brackets) indicates the homogeneity degree $(i - k)$ of the polynomial "tail" of the solution $\eta^i$ for $i \geq k + 2$ in accordance with the attributed to any of the moments $A^i$ of the order $i$ the homogeneity degree, which is equal to positive number $(i + 2)$ (see e.g. [20])

$$\left[ A^i \right] = i + 2, \quad i \geq 0. \tag{4.6.27}$$

For instance, the component $\eta_1^5$ of the basis solution $\eta_1^i$ of determining equations (4.6.22) in accordance with (4.6.24), (4.6.25) and (4.6.26) has the homogeneity degree equal to four

$$\eta_1^5 = -2A^2 + 3(A^0)^2; \quad \left[ \eta_1^5 \right] = 4. \tag{4.6.28}$$

The indexing of the presented infinite (countable) vectors $\eta^i$ by one more integral number $k \geq 0$ yields the desired representation of all linear independent solutions of determining equations (4.6.22) in the form of tensor of the second rank (matrix) $\eta_k^i$, in which the lower index $k \geq 0$ indicates the index of the basis infinitesimal operator in the general element of an infinite Lie algebra under consideration

$$X = \sum_{i,k=0}^{\infty} C^k \eta_k^i \frac{\partial}{\partial A^i}. \tag{4.6.29}$$

Under the conditions (4.6.25) the integration of determining equations (4.6.22) for the given basis vector $\eta_k^i$ for a fixed value $k \geq 0$ is carried out with boundary conditions, that are imposed by requirements (4.6.25) in a single way.

The representation of matrix $\eta_k^i$ for different lines are as follows ($i$ is the column number, $k$ is the line number)

$$\eta_k^i = \{0, \ldots, 0, 1, 0, -(k+1)A^0, -(k+1)A^1, \ldots\}. \qquad (4.6.30)$$

Here zeroes preceding unity describe matrix elements, which exist only for $i < k$, i.e. which are located below the principle diagonal $i = k$, that contains only units. The first nearest upper off-diagonal $i = k+1$ also contains only zeroes. Expressions for elements from the second $i = k+2$ and the third $i = k+3$ upper off-diagonals are given in (4.6.30) explicitly: they contain monomials, the homogeneity degree of which is equal to 2 and 3 respectively, while the numerical coefficient $(k+1)$ is defined by the line number.

In general, any one of the nonzero off-diagonals $i = k+s$ with the number $s \geq 2$ is presented by polynomials with the homogeneity degree equal to $s$. This "line scheme" (4.6.30) is readily illustrated by a pictorial rendition of elements of the high left block of the discussed matrix ($0 \leq i \leq 5, 0 \leq k \leq 3$)

$$\eta_k^i = \begin{pmatrix} 1 & 0 & -A^0 & -A^1 & -A^2 + (A^0)^2 & -A^3 + 2A^0 A^1 & \cdots \\ 0 & 1 & 0 & -2A^0 & -2A^1 & -2A^2 + 3(A^0)^2 & \cdots \\ 0 & 0 & 1 & 0 & -3A^0 & -3A^1 & \cdots \\ 0 & 0 & 0 & 1 & 0 & -4A^0 & \cdots \\ \cdots & \cdots & \cdots & \cdots & \cdots & \cdots & \cdots \end{pmatrix}.$$

$$(4.6.31)$$

As a more illustrative example we present here the element $\eta_1^i$ of the matrix (4.6.30) with sufficiently high column number $i = 10$ and the homogeneity degree 9, that is located in the line with $k = 1$ (the second from above)

$$\eta_1^{10} = -2A^7 + 6A^5 A^0 + 6A^4 A^1 + 6A^3 A^2 - 12A^3 (A^0)^2$$
$$- 24A^2 A^1 A^0 - 4(A^1)^3 + 20A^1 (A^0)^3. \qquad (4.6.32)$$

### 4.6.6 Illustrative Example for Matrix Elements

A much more comprehensive idea of definite expressions of matrix elements $\eta_k^i$ is given by the following list of elements (with the previous result included) of the first 11 columns ($0 \leq i \leq 10$) and 4 lines ($0 \leq k \leq 3$) of matrix $\eta_k^i$, which define the $k$-th basic solution of determining equations (4.6.22) for vectors $\eta_k^i$ of the canonical infinitesimal operator (4.6.14), (4.6.16). The lower index "$k$" is omitted for simplicity.

(0) $k = 0$; $\eta^0 = 1$, $\eta^1 = 0$, $[\eta^i] = i$, $i \geq 2$.

$\eta^2 = -A^0$,

$\eta^3 = -A^1$,

$$\eta^4 = -A^2 + (A^0)^2,$$
$$\eta^5 = -A^3 + 2A^0 A^1,$$
$$\eta^6 = -A^4 + 2A^0 A^2 + (A^1)^2 - (A^0)^3,$$
$$\eta^7 = -A^5 + 2A^0 A^3 + 2A^2 A^1 - 3A^1 (A^0)^2,$$
$$\eta^8 = -A^6 + 2A^0 A^4 + 2A^3 A^1 + (A^2)^2 - 3A^2 (A^0)^2$$
$$\qquad - 3A^0 (A^1)^2 + (A^0)^4,$$
$$\eta^9 = -A^7 + 2A^0 A^5 + 2A^4 A^1 + 2A^3 A^2 - 3A^3 (A^0)^2$$
$$\qquad - 6A^0 A^1 A^2 - (A^1)^3 + 4A^1 (A^0)^3,$$
$$\eta^{10} = -A^8 + 2A^0 A^6 + 2A^5 A^1 + A^4 [2A^2 - 3(A^0)^2] + A^3 [A^3 - 6A^0 A^1]$$
$$\qquad + A^2 [-3(A^1)^2 - 3A^0 A^2 + 4(A^0)^3] + 6(A^1)^2 (A^0)^2 - (A^0)^5.$$

(1) $k = 1$; $\eta^0 = 0$, $\eta^1 = 1$, $\eta^2 = 0$, $[\eta^i] = i - 1$, $i \geq 3$.

$$\eta^3 = -2A^0,$$
$$\eta^4 = -2A^1,$$
$$\eta^5 = -2A^2 + 3(A^0)^2,$$
$$\eta^6 = -2A^3 + 6A^0 A^1,$$
$$\eta^7 = -2A^4 + 6A^0 A^2 + 3(A^1)^2 - 4(A^0)^3,$$
$$\eta^8 = -2A^5 + 6A^0 A^3 + 6A^2 A^1 - 12A^1 (A^0)^2,$$
$$\eta^9 = -2A^6 + 6A^0 A^4 + 6A^3 A^1 + A^2 [3A^2 - 12(A^0)^2]$$
$$\qquad - 12A^0 (A^1)^2 + 5(A^0)^4,$$
$$\eta^{10} = -2A^7 + 6A^0 A^5 + 6A^4 A^1 + 6A^3 [A^2 - 2(A^0)^2]$$
$$\qquad - 24A^0 A^1 A^2 + A^1 [-4(A^1)^2 + 20(A^0)^3].$$

(2) $k = 2$; $\eta^0 = 0$, $\eta^1 = 0$, $\eta^2 = 1$, $\eta^3 = 0$, $[\eta^i] = i - 2$, $i \geq 4$.

$$\eta^4 = -3A^0,$$
$$\eta^5 = -3A^1,$$
$$\eta^6 = -3A^2 + 6(A^0)^2,$$
$$\eta^7 = -3A^3 + 12A^0 A^1,$$
$$\eta^8 = -3A^4 + 12A^0 A^2 + 6(A^1)^2 - 10(A^0)^3,$$
$$\eta^9 = -3A^5 + 12A^0 A^3 + 12A^2 A^1 - 30A^1 (A^0)^2,$$
$$\eta^{10} = -3A^6 + 12A^0 A^4 + 12A^3 A^1 + 6(A^2)^2$$
$$\qquad - 30A^0 (A^1)^2 + 15(A^0)^4 - 30A^2 (A^0)^2.$$

(3) $k = 3$; $\eta^0 = 0$, $\eta^1 = 0$, $\eta^2 = 0$, $\eta^3 = 1$, $\eta^4 = 0$, $[\eta^i] = i - 3$, $i \geq 5$.

$$\eta^5 = -4A^0,$$
$$\eta^6 = -4A^1,$$
$$\eta^7 = -4A^2 + 10(A^0)^2,$$

$$\eta^8 = -4A^3 + 20A^0 A^1,$$
$$\eta^9 = -4A^4 + 20A^0 A^2 + 10(A^1)^2 - 20(A^0)^3,$$
$$\eta^{10} = -4A^5 + 20A^0 A^3 + 20A^2 A^1 - 60A^1 (A^0)^2.$$

To conclude, we present a result of calculation of the infinite (countable) part of Lie point group admitted by the system of Benney equations — moment equations (4.6.2). In standard (non-canonical representation) the point Lie group of Benney equations (4.6.2) is described by the infinitesimal operator

$$X = \xi^1 \frac{\partial}{\partial t} + \xi^2 \frac{\partial}{\partial x} + \sum_{i=0}^{\infty} \eta^i \frac{\partial}{\partial A^i}, \qquad (4.6.33)$$

where coordinates $\xi$ and $\eta$ obey the system of determining equations

$$\eta_{A^0}^{i+1} - \sum_{j=0}^{\infty} j A^{j-1} \eta_{A_x^j}^i + i \eta^{i-1} + i A^{i-1}\left(\eta_{A^0}^0 + \xi_t^1 - \xi_x^2\right)$$

$$+ (i+1)A^i \xi_x^1 - \xi_t^2 \delta_{i,0} = 0,$$

$$\eta_{A^{k+1}}^{i+1} - \eta_{A^k}^i + i A^{i-1}\left(\eta_{A^{k+1}}^0 + \xi_x^1 \delta_{0,k}\right) \qquad (4.6.34)$$

$$+ (\xi_t^1 - \xi_x^2)\delta_{i,k} + \xi_x^1 \delta_{i+1,k} - \xi_t^2 \delta_{i,k+1} = 0,$$

$$\eta_t^i + \eta_x^{i+1} + i A^{i-1} \eta_x^0 = 0, \qquad i, k \geq 0.$$

Determining equations (4.6.34) result from (4.6.20) in account of relationships between coordinates of infinitesimal operators (4.6.33) and (4.6.14)

$$\varkappa^i = \eta^i + \xi^1 (A_x^{i+1} + i A^{i-1} A_x^0) - \xi^2 A_x^i. \qquad (4.6.35)$$

Infinitesimal operators (4.6.12), that were presented above, gives rise to the following coordinates

$$\xi^1 = K^4 + K^5 t, \quad \xi^2 = K^1 + K^2 t + K^3 x,$$
$$\eta^i = i A^{i-1} K^2 + (i+2)A^i (K^3 - K^5). \qquad (4.6.36)$$

The problem of finding coordinates of the operator (4.6.33) was first treated in [19], where only these solutions, namely (4.6.9), (4.6.12) and (4.6.36), were described. The main result described in Sect. 4.6 is that point symmetries of Benney equations (4.6.2) are exhausted by formulas (4.6.12) and solutions of determining equations (4.6.22), i.e. determining equations (4.6.34) do not have any other solutions. Solutions of determining equations (4.6.22) which are responsible for the infinite part of the point group probably have not been known so far [24].

As a next step it seems intriguing to generalize the result (4.6.35), i.e. to find the first order Lie–Bäcklund group admitted by Benney equations (4.6.2) with coordinates $\varkappa^i$ of the canonical infinitesimal operator (4.6.14), that has the linear form

$$\varkappa^i = \eta^i + \sum_{j=0}^{\infty} \eta^{i,j} A_x^j, \qquad i \geq 0. \qquad (4.6.37)$$

Though the unique existence of the linear form (4.6.37) as well as the complete solution of determining equations[5] for the tensor $\eta^{i,j}$ has not yet been obtained, all known facts are in agreement with this linear form. In particular, results of [20, 21] mentioned above are consistent with the following expression for the tensor $\eta^{i,j}$ of the linear form

$$\eta_s^{i,j} = \sum_{k=0}^{\infty} k H_{A^k}^s \delta_{i+k,j+1} + s \sum_{k=0}^{s-j-2} (i+k) A^{i+k-1} H_{A^{j+k+1}}^{s-1}; \quad i, \ j, \ s \geq 0.$$

(4.6.38)

Here $s$ is the number of the basis solution (similar to that used for $\eta^i$ in (4.6.28)), $H^s$ is a polynomial of the homogeneity degree $(s+2)$ in moments $A^i$. Compatibility conditions for determining equations for the tensor $\eta^{i,j}$ give rise to many relationships for $H^s$, for example

$$\sum_{j=0}^{\infty} j A^{j-1} H_{A^j}^s = s H^{s-1}, \quad s \geq 0.$$

(4.6.39)

An explicit form for the polynomial $H^7$ is presented below just to illustrate the aforesaid

$$H^7 = A^7 + 7A^5 A^0 + 7A^4 A^1 + 7A^3 A^2 + 21 A^3 (A^0)^2 + 42 A^2 A^1 A^0$$
$$+ 7(A^1)^3 + 35 A^1 (A^0)^3.$$

(4.6.40)

Comparison between formulas (4.6.32) and (4.6.40) shows that they differ only in numerical values (and signs) of coefficients. The generating function for polynomials $H^s$ is given in [20, 21]. So constructing of a recursion operator, which relates solutions of determining equations (4.6.22) for the point group to the solutions of the determining equations for the first order Lie–Bäcklund symmetry defined by the linear form (4.6.37) with coefficients given by (4.6.38) in particular is of principal interest.

## 4.7 Symmetries in Application to Plasma Kinetic Theory. Renormalization Group Symmetries for Boundary Value Problems and Solution Functionals

The above Sects. 4.3–4.6 deal with calculating symmetries for systems of integro-differential (nonlocal) equations while this section gives illustrations of symmetry applications to problems of mathematical physics with nonlocal equations.

---

[5]For simplicity these equations are omitted here.

### 4.7.1 Introduction to Renormgroup Symmetries

In mathematical physics a solution of a physical problem usually appears as a solution of some boundary value problem. Note that the symmetry of boundary value problem solutions is closely related to RenormGroup (RG) symmetry, introduced in mathematical physics in the beginning of the 1990s [26, 27] (see also reviews [28, 29, 32]). As for the notion of Renormalization Group, or briefly RenormGroup, this was imported to mathematical physics from the most complicated part of theoretical physics, quantum field theory. Recall that the (Lie transformation) group structure discovered by Stueckelberg and Petermann in the early 1950s in calculation results in renormalized quantum field theory and the exact symmetry of solutions related to this structure were used in 1955 by Bogoliubov and Shirkov to develop a regular method for improving approximate solutions of quantum field problems, the RG method. This method is based on the use of the infinitesimal form of the exact group property of a solution to improve a perturbative (that is, obtained by means of the perturbation theory) representation of this solution. The improvement of the approximation properties of a solution turns out to be most efficient in the presence of a singularity, because the correct structure of the singularity is then recovered.

In extending the RG conceptions in quantum field theory to boundary value problems of classical mathematical physics the main achievement was the development of a regular algorithm (see the reviews [28–32]) for finding symmetries of the RG type by means of the modern group analysis. The existence of such an algorithm eliminates the usual deficiency of the RG approach beyond the scope of quantum field theory problems: finding the group property of solutions requires using special-purpose methods of analysis, usually nonstandard, in each particular case. The new algorithm has the same aim of finding an improved solution (in comparison with the initial approximate solution) as the algorithm of Bogoliubov's RG method, but in finding symmetries of a solution of a boundary value problem it uses a scheme of calculations similar to that of the modern group analysis. The attribute 'renormalization group' thus points to similarities existing between these symmetries and the symmetries in quantum field theory related to the operation of renormalization of masses and charges (coupling constants).

Initially [26, 28, 29], applying the RG algorithm was mainly limited to problems based on differential equations, although this algorithm can be used formally in any problem for which a regular way of calculating symmetries for the basic equations can be specified. Hence, transition to such objects, which until recently were not a subject of group analysis, in particular, to integral and integro-differential equations, essentially expands the area of the RG symmetry applications [30–32].

In problems with involved equations, e.g., in transfer theory with integro-differential Boltzmann equation or in quantum field theory with an infinite chain of coupled integro-differential Dyson–Schwinger equations, only some solution components or their integrated characteristics satisfy a sufficiently simple symmetry. Thus, in the one-velocity plane transfer problem, the RG property is related [33] to the asymptotics of the "density of particles, moving deep into the medium" $n_+(x)$,

$x \to \infty$, not entering the Boltzmann equation.[6] In such problems, integral relations form the problem skeleton. But they can appear as some independent objects for applying the RG symmetry constructed for solutions of differential equations. Frequently, not the solution itself in its entire range of the variables and parameters but rather some integral characteristic, a solution functional, is of physical interest. This characteristic can appear, for example, as a result of averaging (integrating) over one of the independent variables or of transition to a new integral representation, for example, a Fourier representation.

This section is structured as follows. In Sect. 4.7.2, one finds an introductory example of the RG algorithm in mathematical physics, illustrated by a solution of a simple boundary value problem. In Sect. 4.7.3, a general scheme for constructing the RG algorithm, valid for models with both local (differential) and nonlocal terms, including integral and integro-differential equations, is described. Section 4.7.4 gives several examples of application of the RG algorithm.

## 4.7.2  RG Symmetry: An Idea of Construction and Its Simple Realization

We preface the description of the RG algorithm with the following simple argument.

Let the Lie group $G$ with generator

$$X = \xi^t \frac{\partial}{\partial t} + \xi^x \frac{\partial}{\partial x} + \eta \frac{\partial}{\partial y} \qquad (4.7.1)$$

be defined for the system of the first-order partial differential equations

$$y_t = F(t, x, y, y_x). \qquad (4.7.2)$$

The typical boundary value problem for (4.7.2) is the Cauchy problem with boundary manifold defined by

$$t = 0, \qquad y = \psi(x). \qquad (4.7.3)$$

Solution of this Cauchy problem is the $G$-invariant solution iff for any generator (4.7.1), function $\psi$ satisfies the equation [34, §29]

$$\eta(0, x, \psi) - \xi^x(0, x, \psi)\psi_x - \xi^t(0, x, \psi)F(0, x, \psi, \psi_x) = 0. \qquad (4.7.4)$$

The solution of Cauchy problem (4.7.2), (4.7.3) coincides with orbit of the group $G$, and the boundary manifold is *not* the invariant manifold of the group.

This example gives an instructive idea for constructing generators of RG symmetries. The milestones here are (a) considering the boundary value problem in the extended space of group variables that involve parameters of boundary conditions in group transformations, (b) calculating the admitted group using the infinitesimal

---

[6]This is representable as the integral $\int_0^1 n(x, \vartheta) \, d\cos\vartheta$ of the kinetic equation solution $n(x, \vartheta)$.

approach, (c) checking the invariance condition akin to (4.7.4) to find the symmetry group with the orbit that coincides with the boundary value problem solution, and (d) using the RG symmetry to find the improved (renormalized) solution of the boundary value problem.

The complete algorithm [28, 29, 31, 32] will be described in detail in the next section; here we only give a general grasp of the problem using a trivial example, the boundary value problem for the Hopf equation

$$v_t + v v_x = 0, \quad v(0, x) = \varepsilon U(x), \tag{4.7.5}$$

where $U$ is an invertible function of $x$ and the parameter $\varepsilon$ defines the initial amplitude at the boundary $t = 0$. For small values of $t \ll 1/\varepsilon$, i.e., near the boundary, $t \to 0$, a perturbation theory (PT) solution of (4.7.5) has the form of a truncated power series in $\varepsilon t$,

$$v = \varepsilon U - \varepsilon^2 t U U_x + O(t^2). \tag{4.7.6}$$

It is obvious that this solution is invalid for large distances from the boundary, when $\varepsilon t U_x \simeq 1$. The RG symmetry gives a way to improve the perturbation theory result and restore the correct structure of the boundary value problem solution in the vicinity of a singularity (in the event that such singularity appears for some finite value of $t$).

It is convenient to introduce the new function $u = v/\varepsilon$ and rewrite (4.7.5) in the form

$$u_t + \varepsilon u u_x = 0, \quad u(0, x) = U(x). \tag{4.7.7}$$

In order to calculate the renormgroup symmetries, we add the parameter $\varepsilon$ to the list of the independent variables and consider the manifold (termed in general the *basic manifold*) given by (4.7.7) in the space of variables $\{t, x, \varepsilon, u, u_t, u_x\}$. Then we calculate the generator

$$X = \xi^t \frac{\partial}{\partial t} + \xi^x \frac{\partial}{\partial x} + \xi^\varepsilon \frac{\partial}{\partial \varepsilon} + \eta \frac{\partial}{\partial u} \tag{4.7.8}$$

of the group admitted by the first equation in (4.7.7) and obtain the following coordinates of the generator (4.7.8):

$$\xi^t = \psi^1, \quad \xi^x = \varepsilon u \psi^1 + \psi^2 + x(\psi^3 + \psi^4), \quad \xi^\varepsilon = \varepsilon \psi^4, \quad \eta = u \psi^3,$$
$$\tag{4.7.9}$$

where $\psi^i$, $i = 2, 3, 4$, are arbitrary functions of $\varepsilon$, $u$, and $x - \varepsilon u t$ and $\psi^1$ being an arbitrary function of all the group variables. These formulas define an infinite-dimensional Lie algebra with four generators

$$X_1 = \psi^1 \left( \frac{\partial}{\partial t} + \varepsilon u \frac{\partial}{\partial x} \right), \quad X_2 = \psi^2 \frac{\partial}{\partial x},$$
$$X_3 = \psi^3 \left( x \frac{\partial}{\partial x} + u \frac{\partial}{\partial u} \right), \quad X_4 = \psi^4 \left( \varepsilon \frac{\partial}{\partial \varepsilon} + x \frac{\partial}{\partial x} \right). \tag{4.7.10}$$

Suppose that a particular solution of boundary value problem (4.7.7),

$$S \equiv u - W(t, x, \varepsilon) = 0,$$

which defines an invariant manifold of group (4.7.8), (4.7.9) is known. The corresponding invariance condition evaluated on frame $S$ is similar to (4.7.4):

$$XS_{|[S]} \equiv (W - xW_x)\psi^3 - W_x\psi^2 - (\varepsilon W_\varepsilon + xW_x)\psi^4 = 0. \quad (4.7.11)$$

The term with $\psi^1$ does not give any input in (4.7.11) since it is proportional to $W_t + \varepsilon WW_x$ and vanishes on the solutions of (4.7.7). Equation (4.7.11) is valid for all $t$. Hence, it remains valid for $t \to 0$, when $W$ is replaced with approximate solution, which follows from (4.7.6),

$$W = U - \varepsilon t U U_x + O(t^2). \quad (4.7.12)$$

In this limit, $t \to 0$, condition (4.7.11) gives a relation between the $\psi^i$, $i = 2, 3, 4$ (no restrictions are imposed on $\psi^1$), that can be easily prolonged on $t \neq 0$,

$$\psi^2 = -\chi(\psi^3 + \psi^4) + (u/U_\chi)\psi^3, \quad \chi = x - \varepsilon ut, \quad (4.7.13)$$

where the derivative $U_\chi$ should be expressed, due to the boundary condition, either in terms of $\chi$ or $u$. By substituting (4.7.13) in (4.7.9), we obtain a group of a smaller dimension with generators

$$R_1 = \psi^1 \left( \frac{\partial}{\partial t} + \varepsilon u \frac{\partial}{\partial x} \right),$$

$$R_2 = u\psi^3 \left[ (\varepsilon t + 1/U_\chi) \frac{\partial}{\partial x} + \frac{\partial}{\partial u} \right], \quad (4.7.14)$$

$$R_3 = \varepsilon \psi^4 \left( tu \frac{\partial}{\partial x} + \frac{\partial}{\partial \varepsilon} \right).$$

The above procedure, which transforms (4.7.10) to (4.7.14), is the *restriction of the group* (4.7.8) *on a particular solution.*

The boundary value problem solution defines a manifold, that, by construction, turns to be invariant for any generator $R_i$. Hence, (4.7.14) defines the desired RG symmetries. This means that the boundary value problem solution can be constructed by use any of generators in (4.7.14), the generator $R_3$ for example. Without loss of generality, we choose $\varepsilon \psi^4 = 1$ and obtain the finite RG transformations ($a$ is a group parameter)

$$x' = x + atu, \quad \varepsilon' = \varepsilon + a, \quad t' = t, \quad u' = u, \quad (4.7.15)$$

where $t$ and $u$ are invariants of the RG transformations while the transformations of $\varepsilon$ and $x$ are translations, which also depend on $t$ and $u$ in the case of $x$. For $\varepsilon = 0$, in view of (4.7.6), we have $x = H(u)$, where $H(u)$ is a function inverse to $U(x)$. Eliminating $a, t, u$ from (4.7.15) and omitting the primes on variables, we obtain the desired solution of boundary value problem (4.7.7) in the implicit form

$$x - \varepsilon tu = H(u). \quad (4.7.16)$$

This in fact is the improved perturbation theory solution (4.7.6), which is valid not only for small $\varepsilon t \ll 1$, provided dependence (4.7.16) can be resolved uniquely. Depending upon $H(u)$ it gives either proper singular behavior at some finite $t \to t_{sing}$ or correct asymptotic behavior at $t \to \infty$.

**Example 4.7.1** One example of the first option is the solution of the boundary value problem for the linear function $U(x) = x$. This yields the solution $v = \varepsilon x (1 + \varepsilon t)^{-1}$, which remains finite as $t \to \infty$.

**Example 4.7.2** For the second option, we can select, for instance, a sine wave $U(x) = -\sin x$ at the boundary. Then solution (4.7.16) describes the well-known distortion of the initial profile of a sine wave, transforming it into a saw-tooth shape [35, Chap. 6, §1], with a singularity forming at a finite distance $t_{sing} = 1/\varepsilon$ from the boundary.

We note that for finding solution (4.7.16) of the boundary value problem we use *only* the known symmetry of the solution and the corresponding perturbation theory (PT).

The peculiarity of the procedure for constructing RG symmetries is the multi-choice first step, which depends on how the boundary conditions are formulated and the form in which the admitted group is calculated. For example, instead of calculating the Lie point symmetry group, we can consider the Lie–Bäcklund symmetries (see Sect. 1.5 in Chap. 1) with the canonical generator $R = \varkappa \partial_u$, where $\varkappa$ depends not only on $t$, $x$, $\varepsilon$, and $u$ but also on higher-order derivatives of $u$. We can seek $\varkappa$ in the form of a power series in $\varepsilon$, and invariance condition (4.7.11) is formulated as vanishing of $\varkappa$ at $t = 0$. Depending on the choice of the zeroth-order term representation, we obtain either an infinite or a truncated power series for $\varkappa$, for example, a form linear in $\varepsilon$,

$$R = \varkappa \frac{\partial}{\partial u}, \quad \varkappa = 1 - \frac{u_x}{U_x(u)} - \varepsilon t u_x. \tag{4.7.17}$$

This RG generator (4.7.17) is equivalent to the Lie point generator $R_2$ in (4.7.14) and therefore gives the same result.

Another possibility for calculating RG symmetries for boundary value problem (4.7.7) is offered by taking some additional differential constraints consistent with boundary conditions and input equations into account. For example, when the boundary condition in (4.7.7) is linear in its argument, $U(x) = x$, the differential constraint can be chosen as $u_{xx} = 0$; this equality reflects the invariance of the original equation with respect to the second-order Lie–Bäcklund symmetry group. Calculating the Lie point symmetry group for the joint system of this constraint and the Hopf equation gives another way to find RG symmetries for boundary value problem (4.7.7).

The above example demonstrates the key features of the RG algorithm in mathematical physics. The details of the general approach are discussed in the next section.

**Fig. 4.1** Scheme of RG
algorithm

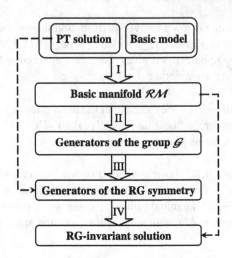

### 4.7.3 Renormgroup Algorithm

The general construction scheme of the RG algorithm (shown in Fig. 4.1) is given
as four consecutive steps [28–32]:

   I. constructing the basic *manifold* $\mathcal{RM}$,
  II. calculating the admitted (*symmetry*) group $\mathcal{G}$,
 III. *restricting* it on the particular boundary value problem solution and constructing
     $\mathcal{RG}$, and
 IV. seeking an *analytic solution*.

#### 4.7.3.1 Basic Manifold $\mathcal{RM}$

The initial issue is to construct the RG symmetry and appropriate transformations
that involve the parameters of partial solution. Therefore, the purpose of step **I** is
to include all the parameters, both from the equations and from the boundary conditions on which a particular solution depends, in group transformations in one or
another way. This purpose is achieved by constructing a special manifold $\mathcal{RM}$
given by a system that consists of $s$ $k$th-order differential equations and $q$ nonlocal
relations

$$F_\sigma(z, u, u_{(1)}, \ldots, u_{(k)}) = 0, \quad \sigma = 1, \ldots, s, \quad (4.7.18)$$

$$F_\sigma(z, u, u_{(1)}, \ldots, u_{(r)}, J(u)) = 0, \quad \sigma = 1 + s, \ldots, q + s. \quad (4.7.19)$$

The nonlocal variables $J(u)$ here are introduced by integral objects,

$$J(u) = \int \mathscr{F}(u(z)) \, dz. \quad (4.7.20)$$

The presence of relations (4.7.19) in the system determining $\mathcal{RM}$ characterizes the
basic difference between the case of a nonlocal problem and the case of a boundary
value problem for differential equations, for which $\mathcal{RM}$ is a differential manifold.

### 4.7.3.2 Admitted Group $\mathscr{G}$

Step **II** is to calculate the widest admitted group $\mathscr{G}$ for system (4.7.18), (4.7.19). In application to an $\mathscr{RM}$ defined only by system of differential equations (4.7.18), the question is about a local group of transformations in a space of differential functions $\mathscr{A}$, for which system (4.7.18) remains unchanged. This group is defined by the generator of form (4.7.8) prolonged on all higher-order derivatives,

$$X = \xi^i \frac{\partial}{\partial z^i} + \eta^\alpha \frac{\partial}{\partial u^\alpha} + \zeta_i^\alpha \frac{\partial}{\partial u_i^\alpha} + \zeta_{i_1 i_2}^\alpha \frac{\partial}{\partial u_{i_1 i_2}^\alpha} + \cdots, \qquad (4.7.21)$$

where $\xi^i([z, u])$, $\eta^\alpha([z, u]) \in \mathscr{A}$ and

$$\zeta_i^\alpha = D_i(\varkappa^\alpha) + \xi^j u_{ij}^\alpha, \quad \zeta_{i_1 i_2}^\alpha = D_{i_1} D_{i_2}(\eta^\alpha - \xi^i u_i^\alpha) + \xi^j u_{ji_1 i_2}^\alpha.$$

Meanwhile, the classical Lie algorithm using the infinitesimal approach seems to be inapplicable to a manifold $\mathscr{RM}$ set by system (4.7.18), (4.7.19). The issue is that the $\mathscr{RM}$ in this case is not determined *locally* in the space of differential functions. Therefore, the main advantage of the Lie computational algorithm, namely, representation of the determining equations as an over-determined system of equations is not realized here. Furthermore, the procedure for prolongation the group operator of point transformations on nonlocal variables is not defined in the framework of classical group analysis.

In modifying the RG algorithm, we rely on the direct method for calculating symmetries described in Chaps. 2 and 4. Therefore, constructing the symmetries for the nonlocal equations also appears as an algorithmic procedure. This is the generalization of the second step of the algorithm to the case where $\mathscr{RM}$ is an integral or integro-differential manifold.

### 4.7.3.3 Restriction of the Admitted Group on Solutions

The group $\mathscr{G}$ found in step **II** and determined by operators (4.7.21) is generally wider than the RG of interest, which is related to a particular solution of a boundary value problem. Hence, to obtain the RG symmetry, we need step **III**, *restricting* the group $\mathscr{G}$ on a manifold determined by this particular solution. From the mathematical standpoint, this procedure consists in checking the vanishing conditions for a linear combination of coordinates $\varkappa_j^\alpha$ of a canonical operator equivalent to (4.7.21) on some particular boundary value problem solution $U^\alpha(z)$,

$$\left\{ \sum_j A^j \varkappa_j^\alpha \equiv \sum_j A^j \left( \eta_j^\alpha - \xi_j^i u_i^\alpha \right) \right\}_{|u^\alpha = U^\alpha(z)} = 0. \qquad (4.7.22)$$

The form of the condition set by relation (4.7.22) is common for any solution of the boundary value problem, but how the restriction procedure of a group is realized may differ in each partial case. In the general scheme (given at the beginning of the section), it is related to the dashed arrow connecting the "initial object" (a perturbation theory solution of a particular boundary value problem) to the object arising as a result of step **III**.

In calculating combination (4.7.22) on a particular solution $U^\alpha(z)$, the latter is transformed from a system of differential equations for group invariants to algebraic relations. Note two consequences of step **III**. First, the restriction procedure results in a set of relations between $A^j$ and thus "links" the coordinates of various group operators $X_j$ admitted by $\mathcal{RM}$ (4.7.18), (4.7.19). Second, it (partially or completely) eliminates an arbitrariness that can arise in the values of the coordinates $\xi^i$ and $\eta^\alpha$ in the case of an infinite group $\mathcal{G}$.

As a rule, the procedure of restricting the group $\mathcal{G}$ reduces its dimension. After performing this procedure a general element (4.7.21) of a new group $\mathcal{RG}$ is represented by a linear combination of new generators $R_i$ with coordinates $\hat{\xi}^i$ and $\hat{\eta}^\alpha$ and arbitrary constants $B^j$:

$$X \Rightarrow R = \sum_j B^j R_j, \quad R_j = \hat{\xi}^i_j \frac{\partial}{\partial x^i} + \hat{\eta}^\alpha_j \frac{\partial}{\partial u^\alpha}. \tag{4.7.23}$$

The set of operators $R_j$, each containing the required solution of a problem in the invariant manifold, defines a group of transformations $\mathcal{RG}$, which we also call RenormGroup.

### 4.7.3.4 Renormgroup Invariant Solutions

The three steps described above completely form the regular algorithm for constructing the RG symmetry, but to finish a final step is needed. This step **IV** uses the RG symmetry operators to find analytic expressions for new, improved boundary value problem solutions (compared with the input perturbative solution).

From the mathematical standpoint, realizing this step involves use of *RG-invariance* conditions set by a *joint* system of equations (4.7.18) and (4.7.19) and the vanishing conditions for a linear combination of the coordinates $\hat{\varkappa}^\alpha_j$ of the canonical operator equivalent to (4.7.23),

$$\sum_j R^j \hat{\varkappa}^\alpha_j \equiv \sum_j B^j \left( \hat{\eta}^\alpha_j - \hat{\xi}^i_j u^\alpha_i \right) = 0. \tag{4.7.24}$$

The need to use $\mathcal{RM}$ in constructing the boundary value problem solution is shown in the scheme by the dashed arrow connecting these objects.

Specification of step **IV** concludes the description of the regular algorithm of RG symmetries construction for models with integro-differential equations. We note that last the two steps are basically the same as for models with differential equations. The next sections contains a set of examples showing the ability of the upgraded RG algorithm.

### 4.7.4 Examples of RG Symmetries in Plasma Theory

#### 4.7.4.1 Nonlinear Dielectric Permittivity of Plasma

Nonlinearity of electrodynamics of real medium is due to nonlinear relation between the induced current and charge density inside the medium and the electromagnetic field. This relation, named the material equation, originates from a dependence of an electric induction vector upon the electromagnetic field (see [36], p. 48). The induction vector $D(t, r)$ is related to the electric field $E(t, r)$ and the current density $j(t, r)$ via an equality, which in Fourier representation has the following form (here variables "with tildes" are used to distinguish the Fourier representation from the usual space-time representation):

$$\tilde{D}(\omega, k) = \tilde{E}(\omega, k) + i \frac{4\pi}{\omega} \tilde{j}(\omega, k). \tag{4.7.25}$$

In an effort to describe weak-turbulent plasma, processes of particle-wave scattering, parametric instabilities, generation of harmonics, and etc., the material equation is represented as a series in positive powers of electromagnetic fields. Hence, the current density $\tilde{j}(\omega, k)$ is expressed as a sum

$$\tilde{j}(\omega, k) = \sum_l \tilde{j}^{(l)}(\omega, k), \quad \tilde{j}^{(l)}(\omega, k) \backsim O(\tilde{E}^l). \tag{4.7.26}$$

In view of time and spatial dispersion the relation between the induced current and the field appears as integral, nonlocal, that results in the material equation which in Fourier representation has the following form [36]:[7]

$$\tilde{D}_i(\omega, k) = \varepsilon_{ij}(\omega, k)\tilde{E}_j(\omega, k) + \sum_{n=2}^{\infty} \int \delta(\omega - \omega_1 - \cdots - \omega_n)$$
$$\times \delta(k - k_1 - \cdots - k_n)\varepsilon_{ij_1\ldots j_n}(\omega_1, k_1; \ldots; \omega_n, k_n)$$
$$\times \tilde{E}_{j_1}(\omega_1, k_1) \ldots \tilde{E}_{j_n}(\omega_n, k_n) \, d\omega_1 \, dk_1 \ldots d\omega_n \, dk_n. \tag{4.7.27}$$

We compare (4.7.26) and (4.7.25) with (4.7.27) to establish a relation between the current density $\tilde{j}^{(l)}$ of the appropriate order $l \geq 2$ and multi-index tensors of nonlinear dielectric permittivity of plasma $\varepsilon_{ij_1\ldots j_n}$, which are kernels of nonlinear (with respect to electromagnetic field) integral terms in series (4.7.27).

Usually, without use of the RG algorithm, the nonlinear dielectric permittivity for hot plasma is obtained by iterating the Vlasov kinetic equation for the distribution function of particles $f(t, r, v)$ (4.1.1) with a stationary and homogeneous in coordi-

---

[7]Here the bottom index specifies on a corresponding tensor component, instead of designating a derivative.

nate $r$ background distributions $f_0(v)$ in powers of a self-consistent electromagnetic field (here we omit an index of particles):

$$f(t, r, v) = f_0(v) + \sum_{l \geq 1} f^{(l)}(t, r, v), \quad f^{(l)} \smile O(E^l),$$

$$j^{(l)}(t, r) = em^3 \int f^{(l)} \gamma^5 v \, dv.$$
(4.7.28)

As for the nonlinear dielectric permittivity for cold plasma it is usually obtained by iterations of more simple equations of collisionless hydrodynamics for density $N(t, r)$ and velocity $V(t, r)$ of particles (written down here for one sort of particles in non-relativistic approach)

$$N_t + \operatorname{div}(NV) = 0, \quad V_t + (V \cdot \nabla)V = \frac{e}{m}\left\{ E + \frac{1}{c}[V \times B] \right\}, \quad (4.7.29)$$

in which the electric $E$ and the magnetic field $B$ obey Maxwell equations (4.1.3), and charge $\rho$ and current $j$ densities have the form

$$\rho = eN, \quad j = eNV. \tag{4.7.30}$$

In the right-hand part of (4.7.30) summation upon various species of plasma particles is implied, however for simplification of notations the index of species is omitted and only one sort of particles, for example electrons is underlined further.

It is commonly accepted, that formulas for the nonlinear dielectric permittivity in hot plasma are more general, than in cold (see, for example, [36], Chap. 2) and they are reduced to the last in that specific case, when the distribution function of plasma particles upon momentum in the initial equilibrium state is represented by the Dirac delta-function, $f_0(v) = \delta(v)$. With growth of the order of nonlinearity ($l \geq 4$) an algebraic procedure of symmetrization for nonlinear dielectric permittivity tensors becomes more cumbersome in hot plasma, than in cold. The use of RG algorithm allows to establish a one-to-one correspondence between tensors of the nonlinear dielectric permittivity in cold and hot plasma in any order of nonlinearity $l$ and also specifies a way of obtaining expressions for tensors of the nonlinear dielectric permittivity in hot plasma from appropriate "cold" expressions.

For this purpose we present a current density of the given order $\tilde{j}^{(l)}(\omega, k)$ in hot plasma as a convolution of two functions, the partial current density $\hat{j}^{(l)}(\omega, k, w)$, which depends on the Lagrangian velocity of particles $w$, and an equilibrium velocity distribution function of particles in absence of electromagnetic fields $f_0(w)$,

$$\tilde{j}^{(l)}(\omega, k) = \int f_0(w) \hat{j}^{(l)}(\omega, k, w) \, dw. \tag{4.7.31}$$

An expression for the partial current density for $f_0(w) = \delta(w)$, i.e. in cold plasma ($w = 0$), is obtained by iterating (4.7.29), (4.7.30) with respect to the self-consistent field, while a transition from $\hat{j}^{(l)}(\omega, k, 0)$ to $\hat{j}^{(l)}(\omega, k, w)$ with arbitrary $w \neq 0$ is carried out with the help of group of transformations, defined by the appropriate RG symmetry operator.

Since the procedure of construction of the multi-index nonlinear dielectric permittivity tensor in hot plasma from the appropriate expressions in cold plasma is identical for the permittivity tensor of any order we illustrate it by using linear with respect to a self-consistent electric field $E$ material relations in non-relativistic plasma. In cold plasma Fourier-components of the partial current $\hat{j}^{(1)}(\omega, k, 0)$ and charge $\hat{\varrho}^{(1)}(\omega, k, 0)$ densities, which are linear in the field $\tilde{E}(\omega, k)$, are obtained by linearizing (4.7.29), (4.7.30) on a background of the homogeneous and equilibrium electron density $n_{e0}$ and are determined by well-known relations

$$\hat{j}^{(1)}(\omega, k, 0) = i\frac{e^2 n_{e0}}{m\omega}\tilde{E}; \quad \hat{\varrho}^{(1)}(\omega, k, 0) = i\frac{e^2 n_{e0}}{m\omega^2}(k \cdot \tilde{E}). \quad (4.7.32)$$

The use of the latter in (4.7.25) gives a scalar dielectric permittivity for cold homogeneous non-relativistic plasma,

$$\varepsilon(\omega, k) = 1 - \frac{4\pi e^2 n_e}{m\omega^2}. \quad (4.7.33)$$

Expressions (4.7.32) define zero-order terms in expansion of the partial current density $\hat{j}^{(l)}(\omega, k, w)$ in powers of plasma particles velocity $w$. For obtaining the next terms of this series one should use the kinetic description of plasma. Here it appears more convenient to use instead of Vlasov equations (4.1.1) with the Euler velocity $v$ the non-relativistic hydrodynamic analogue (4.1.5) of Vlasov equations with Lagrangian velocity $w$ and the equilibrium distribution function $f_0(w)$. Such (Lagrangian) formulation of the kinetic description of plasma results from a non-relativistic limit of (4.1.5), and coincides in the form with (4.7.29), with that, however, an essential difference, that as against (4.7.29) the density $N(t, r, w)$ and the velocity $V(t, r, w)$ now depend upon Lagrangian velocity as well and in the homogeneous non-perturbed plasma state obey the "initial" conditions at $t = t_0 = -\infty$

$$N(t_0, r, w) = n_{e0}f_0(w), \quad V(t_0, r, w) = w;$$

$$E(t_0, r) = B(t_0, r) = 0, \quad \int f_0 \, dw = 1. \quad (4.7.34)$$

In a non-relativistic limit material relations (4.1.6) also become simpler (we use different normalization for the distribution function here, hence material relations do not contain mass multipliers)

$$\rho(t, r) = e\int N \, dw, \quad j(t, r) = e\int NV \, dw. \quad (4.7.35)$$

Linearizing the equations of plasma kinetics in Lagrangian variables on the background of the basic state (4.7.34) results to the following formulas for corrections to the partial current density for small values of $w$:

$$\hat{j}^{(1)}(\omega, k, w)$$
$$= i\frac{e^2 n_{e0}}{m\omega}\left\{\tilde{E} + \frac{1}{\omega}\left(w(k \cdot \tilde{E}) + k(w \cdot \tilde{E})\right)\right\} + O(w^2). \quad (4.7.36)$$

To prolong this formula on any nonzero values of $w$ we employ the RG symmetry operator which is constructed from the Lie group of point transformations (4.5.4), admitted by plasma kinetic equations. Two operators of the admitted group are of interest for us, namely, the operator of translations in Lagrangian velocity, which results from the operator $X_\infty$, and the operator of Galilean transformations, which is a non-relativistic analogue of the operator of Lorentz transformations $\mathbf{B}$ in the set (4.5.4),

$$Z_1 = \frac{\partial}{\partial \boldsymbol{w}}, \quad Z_2 = t\frac{\partial}{\partial \boldsymbol{r}} + \frac{\partial}{\partial \boldsymbol{V}} - \frac{1}{c}\left[\boldsymbol{B} \times \frac{\partial}{\partial \boldsymbol{E}}\right] + \rho\frac{\partial}{\partial \boldsymbol{j}}. \qquad (4.7.37)$$

Let us proceed in the operator $Z_2$ from the velocity $\boldsymbol{V}$ and the density $N$ to the partial current and charge densities, $\hat{\boldsymbol{j}}$ and $\hat{\varrho}$, prolong the operator obtained on Fourier variables and combine it with the operator of translations $Z_1$. As a result we get the operator that leaves the partial current density (4.7.36) invariant at $\boldsymbol{w} \to 0$, i.e. the required RG symmetry operator

$$R = k\frac{\partial}{\partial \omega} + \frac{\partial}{\partial \boldsymbol{w}} - \frac{1}{c}\left[\tilde{\boldsymbol{B}} \times \frac{\partial}{\partial \tilde{\boldsymbol{E}}}\right] + \hat{\varrho}\frac{\partial}{\partial \hat{\boldsymbol{j}}}. \qquad (4.7.38)$$

The operator (4.7.38) is related to a three-parameter group with the vector parameter $\boldsymbol{w}$, and its final transformations (the variables with primes here correspond to transformed variables)

$$\omega' = \omega + \boldsymbol{kw}; \quad (\beta'_{is}/\omega')\tilde{E}'_s = (1/\omega)\tilde{E}_i; \quad \hat{\varrho}' = \hat{\varrho}; \quad \hat{j}'_i = \beta'_{si}\hat{j}_s;$$

$$k' = k; \quad \tilde{\boldsymbol{B}}' = \tilde{\boldsymbol{B}} = (c/\omega)\,[k \times \tilde{\boldsymbol{E}}]; \quad \beta_{is} = \delta_{is} + k_i w_s/(\omega - \boldsymbol{kw}), \qquad (4.7.39)$$

give the required relationship between the value of the partial current density $\hat{j}(\omega, k, 0)$ at $\boldsymbol{w} = 0$ (in cold plasma) and the analogous value of the partial current density $\hat{j}(\omega, k, \boldsymbol{w})$ with any $\boldsymbol{w} \neq 0$. When integrating over velocity $\boldsymbol{w}$ with the "weight" $f_0(\boldsymbol{w})$, following (4.7.31), we get an expression for a current density of the given order in hot plasma which defines the appropriate multi-index nonlinear dielectric permittivity tensor of plasma.

**Example 4.7.3** In particular, in the linear in the electric field approximation the use of (4.7.32) leads to the relationship

$$\hat{j}_i^{(1)}(\omega, k, \boldsymbol{w}) = \frac{ie^2 n_{e0}}{m\omega}\beta_{si}\beta_{sa}\tilde{E}_a(\omega, k). \qquad (4.7.40)$$

Substitution of (4.7.40) into (4.7.31) and the further use of $\tilde{j}_i^{(1)}(\omega, k)$ in (4.7.25) gives the required expression for the tensor of the linear dielectric permittivity for hot homogeneous non-relativistic plasma in the absence of external fields with the equilibrium distribution function $f_0(\boldsymbol{w})$

$$\varepsilon_{ab}(\omega, k) = \delta_{ab} - \frac{4\pi e^2 n_{e0}}{m\omega^2}\int f_0(\boldsymbol{w})\beta_{sa}\beta_{sb}\,d\boldsymbol{w}. \qquad (4.7.41)$$

Formula (4.7.41), which arises from the scalar equality (4.7.33) as a result of application of RG transformations to partial current density in cold plasma with the subsequent integration over the group parameter, illustrates an opportunity of obtaining a tensor of dielectric permittivity of hot plasma from the appropriate "cold" expression [27].

**Example 4.7.4** RG symmetry generator (4.7.38) results from symmetry operators admitted by the plasma kinetic equations after their subsequent prolongation on solution functionals, partial current and a charge densities in Fourier representation. Thus a linear in the electromagnetic field approximation used above is not an essential restriction, as relations between transformed (primed) and non-transformed partial current and a charge density remains linear under group transformations (4.7.39). It means, that it is also possible to apply transformations (4.7.39) to partial current and a charge densities of any order $l$, i.e. the offered RG scheme allows to build nonlinear dielectric permittivity tensors of any order in hot plasma proceeding from the appropriate "cold" expressions for the nonlinear dielectric permittivity. Omitting intermediate calculations, we present a result of such construction

$$\varepsilon_{ij_1...j_n}(\omega_1, \mathbf{k}_1; \ldots; \omega_n, \mathbf{k}_n) = \int f_0(\mathbf{w})\bar{\varepsilon}_{ab_1...b_n}(\Omega_1, \mathbf{k}_1; \ldots; \Omega_n, \mathbf{k}_n)$$

$$\times \frac{\Omega \Omega_1 \ldots \Omega_n}{\omega \omega_1 \ldots \omega_n} \beta_{ai}(\omega, \mathbf{k})\beta_{b_1 j_1}(\omega_1, \mathbf{k}_1) \ldots \beta_{b_n j_n}(\omega_n, \mathbf{k}_n)\,\mathrm{d}\mathbf{w};$$

$$n \geq 2; \tag{4.7.42}$$

$$\omega = \omega_1 + \cdots + \omega_n; \quad \mathbf{k} = \mathbf{k}_1 + \cdots + \mathbf{k}_n;$$

$$\Omega \equiv (\omega - \mathbf{k}\mathbf{w}), \quad \Omega_i \equiv (\omega_i - \mathbf{k}_i\mathbf{w}), \quad i = 1, \ldots, n.$$

Here $\bar{\varepsilon}$ corresponds to the nonlinear dielectric permittivity tensor in cold collisionless plasma without external fields. For example, for the nonlinearity of the second order it is determined by the formula

$$\bar{\varepsilon}_{isj}(\Omega_1, \mathbf{k}_1; \Omega_2, \mathbf{k}_2)$$

$$= -\frac{4\pi i e^3 n_{e0}}{2!m^2 \Omega \Omega_1 \Omega_2}\left(\frac{k_i}{\Omega}\delta_{js} + \frac{k_{1s}}{\Omega_1}\delta_{ij} + \frac{k_{2j}}{\Omega_2}\delta_{is}\right). \tag{4.7.43}$$

The similar result can be obtained and for relativistic plasma, however thus it is necessary to use not the three-parameter group of Galilean transformations, but the six-parameter group including Lorentz transformations and rotations.

### 4.7.4.2 Adiabatic Expansion of Plasma Bunches

Here RG algorithm is applied to the problem of expansion of plasma bunches and related generation of the accelerated particles. The mechanisms and characteristics of ions triggered by the interaction of a short-laser-pulse with plasma are of current interest because of their possible applications to the novel-neutron-source development and isotope production. In the near future ultra-intense laser pulses will be used for ion beam generation with energies useful for proton therapy, fast ignition inertial confinement fusion, radiography, neutron-sources.

The commonly recognized effect responsible for ion acceleration is charge separation in the plasma due to high-energy electrons, driven by the laser inside the target. During the plasma expansion, the kinetic energy of the fast electrons transforms into the energy of electrostatic field, which accelerates ions and their energy is expected to be at the level of the hot-electron energy. The mathematical model describing this phenomenon is based on plasma kinetic equations with a self-consistent field (4.1.1)–(4.1.3), which is rather complicated for analytical treatment. However, to describe plasma flows with characteristic scale of density variation large compared to Debye length for plasma particles, the quasi-neutral approximation is used. In this approximation charge and current densities in plasma are set equal to zero, that essentially simplifies the initial model with nonlocal terms. Instead of the system of Vlasov–Maxwell equations (4.1.1), (4.1.3) with the corresponding material equations here we use only the kinetic equations for particle distribution functions for various species (4.3.72) with additional nonlocal restrictions imposed on them, which arise from vanishing conditions for the current and the charge densities (4.3.73). Initial conditions for solutions of (4.3.72) and (4.3.73) correspond to distribution functions for electrons and ions, specified at $t = 0$

$$f^\alpha\big|_{t=0} = f_0^\alpha(x, v). \qquad (4.7.44)$$

Equations (4.3.72), (4.3.73) describe one-dimensional dynamics of a plasma bunch, which is inhomogeneous upon the coordinate $x$; thus the distribution functions of particles $f^\alpha$ depend upon $t$, $x$ and the velocity component $v$ in the directions of plasma inhomogeneity. Analytical study of such yet simplified model represents the essential difficulties, but due to application of RG algorithm it is possible not only to construct solution at various initial particle distribution functions but also to find the law of variation of particles density without calculations of distribution functions for particles in an explicit form [14, 32].

For construction of RG symmetries we consider (step **I**) a set of local (4.3.72) and nonlocal (4.3.73) equations as $\mathcal{RM}$, in which the electric field $E(t, x)$ appears as some arbitrary function to be found of its variables. Calculating the Lie group of point transformations admitted by this manifold (step **II**) is given by (4.3.75), and in particular contains the generator of time translations and the projective group generator. Precisely these operators enables to construct a class of exact solutions to the initial problem that are of interest, as a linear combination of the operator of time translations and the operator of the projective group leaves the approximate perturbation theory solution of the initial value problem $f^\alpha = f_0^\alpha(x, v) + O(t)$ invariant at $t \to 0$, i.e. it is the RG symmetry operator,

$$R = (1 + \Omega^2 t^2)\frac{\partial}{\partial t} + \Omega^2 t x \frac{\partial}{\partial x} + \Omega^2(x - vt)\frac{\partial}{\partial v}, \qquad (4.7.45)$$

which results from the group restriction procedure (step **III**), for spatially symmetric initial distribution functions with the zero average velocity. It is possible to treat the constant $\Omega$ as the ratio of a characteristic sound velocity $c_s$ to initial inhomogeneity scale of the density of electrons, $L_0$.

Invariants of the RG generator (4.7.45) are two combinations, $x/\sqrt{1 + \Omega^2 t^2}$ and $v^2 + \Omega^2(x - vt)^2$, and particle distribution functions $f^\alpha$. Hence, solutions of initial value problem at any time $t \neq 0$ (step **IV**) are expressed via these invariants in terms of initial values (4.7.44),

$$f^\alpha = f_0^\alpha(I^{(\alpha)}), \quad I^{(\alpha)} = \frac{1}{2}\left(v^2 + \Omega^2(x - vt)^2\right) + \frac{e_\alpha}{m_\alpha}\Phi_0(x'). \quad (4.7.46)$$

Here the dependence of $\Phi_0$ upon the variable $x' = x/\sqrt{1 + \Omega^2 t^2}$ is defined by quasi-neutral conditions (4.3.73), and the electric field $E = -\Phi_x$ is found with the help of the potential

$$\Phi(t, x) = \Phi_0(x')\left(1 + \Omega^2 t^2\right)^{-1}. \quad (4.7.47)$$

Formulas (4.7.46) give the solution to the initial value problem (4.3.72), (4.3.73). However, for practical applications we need frequently more rough characteristic of plasma dynamics, for example, a density of particles (ions) of the given species $n^q(t, x)$ which can be calculated using the appropriate distribution function:

$$n^q(t, x) = \int\limits_{-\infty}^{\infty} f^q(t, x, v)\, dv. \quad (4.7.48)$$

In view of the complex dependence upon the invariant $I^{(\alpha)}$ it is not always possible to carry out direct integration of a distribution function over velocity in the analytical form, therefore here the procedure of prolongation of the operator on solution functionals described in Sect. 4.2.1.4 comes to the aid. As the density $n^q(t, x)$ is a linear functional of $f^q$, the prolongation of the operator (4.7.45) on the functional of the solution (4.7.48) in the narrowed space of variables $\{t, x, n^q\}$ gives the following RG operator

$$R = (1 + \Omega^2 t^2)\frac{\partial}{\partial t} + \Omega^2 tx\frac{\partial}{\partial x} - \Omega^2 tn^q\frac{\partial}{\partial n^q}. \quad (4.7.49)$$

The solution of Lie equations for the operator $R$ in view of initial conditions (4.7.44) gives relations between invariants of this operator, namely one of the combinations $J = x/\sqrt{1 + \Omega^2 t^2}$ already given for the operator (4.7.45) and the product $J^q = n^q\sqrt{1 + \Omega^2 t^2}$ for arbitrary $t \neq 0$ with their values at $t = 0$: $J_{|t=0} = x'$, $J^q_{|t=0} = \mathcal{N}_q(x')$. This relationship immediately leads to the formulas that characterize spatial-temporal distribution of the density of ions of a given species in terms of the initial density distribution

$$n^q = \frac{1}{\sqrt{1 + \Omega^2 t^2}}\mathcal{N}_q\left(\frac{x}{\sqrt{1 + \Omega^2 t^2}}\right),$$

$$\mathcal{N}_q(x') = \int\limits_{-\infty}^{\infty} f_0^q(I^{(q)})\, dv. \quad (4.7.50)$$

**Example 4.7.5** We illustrate general results with reference to expansion of a plasma slab, consisting of cold ($\alpha = c$) and hot ($\alpha = h$) electrons and of two ion species ($q = 1, 2$). Let initially (at $t = 0$) ions are characterized by Maxwellian distribution functions with densities $n_{10}, n_{20} \ll n_{10}$ and temperatures $T_1, T_2$, and the distribution function of electrons looks like two-temperature Maxwellian distribution with the appropriate densities $n_{c0}$ and $n_{h0} \ll n_{c0}$ ($n_{c0} + n_{h0} = Z_1 n_{10} + Z_2 n_{20}$) and temperatures $T_c$ and $T_h \gg T_c$ of hot and cold components. From the physical point of view such choice of initial conditions refer to an expansion of the target consisting of heavy ions with a small impurity of light ions adsorbed on a surface (for example, protons) which preliminary was heated quickly by a short pulse of laser radiation with formation of a group of hot electrons. Then the solution of the initial problem (4.7.46) is represented as:

$$
f^e = \frac{n_{c0}}{\sqrt{2\pi}\, v_{Tc}} \exp\left(-\frac{I^{(c)}}{v_{Tc}^2}\right) + \frac{n_{h0}}{\sqrt{2\pi}\, v_{Th}} \exp\left(-\frac{I^{(h)}}{v_{Th}^2}\right),
$$

$$
f^q = \frac{n_{q0}}{\sqrt{2\pi}\, v_{Tq}} \exp\left(-\frac{I^{(q)}}{v_{Tq}^2}\right), \quad v_{T\alpha}^2 = \frac{T_\alpha}{m_\alpha}, \quad q = 1,2,
$$

(4.7.51)

where invariants $I^{(\alpha)}$ are given by relations:

$$
\frac{I^{(c)}}{v_{Tc}^2} = \mathscr{E} + \frac{(1 + \Omega^2 t^2)}{2v_{Tc}^2}(v - u)^2, \quad \frac{I^{(h)}}{v_{Th}^2} = \mathscr{E}\frac{T_c}{T_h} + \frac{(1 + \Omega^2 t^2)}{2v_{Th}^2}(v - u)^2,
$$

$$
\frac{I^{(q)}}{v_{Tq}^2} = -\mathscr{E}\left(\frac{Z_q T_{c0}}{T_{q0}}\right) + \frac{U^2}{2v_{Tq}^2}\left(1 + \frac{Z_q m_e}{m_q}\right) + \frac{(1 + \Omega^2 t^2)}{2v_{Tq}^2}(v - u)^2.
$$

(4.7.52)

Here $u = xt\Omega^2/(1 + \Omega^2 t^2)$ is a local velocity of plasma particles, $U = x\Omega/\sqrt{1 + \Omega^2 t^2}$, and a potential $\Phi$ is expressed via the function $\mathscr{E}$,

$$
\mathscr{E} = \frac{e\Phi}{T_c}(1 + \Omega^2 t^2) + \frac{U^2}{2v_{Tc}^2},
$$

(4.7.53)

that is obtained from the transcendental equation,

$$
n_{c0} = \sum_{q=1,2} Z_q n_{q0} \exp\left[\left(1 + \frac{Z_q T_c}{T_q}\right)\mathscr{E} - \frac{U^2}{2v_{Tq}^2}\left(1 + \frac{Z_q m_e}{m_q}\right)\right]
$$

$$
- n_{h0} \exp\left[\left(1 - \frac{T_c}{T_h}\right)\mathscr{E}\right].
$$

(4.7.54)

Formulas (4.7.51)–(4.7.54) completely define the behavior of distribution functions of all particle species considered in the given example when studying the expansion of a plasma slab. At that the space-temporal distribution of the ion density of the given species is determined by formulas (4.7.50), in which the ion density $\mathcal{N}_q$ for the initial distribution functions specified above has the form

$$
\mathcal{N}_q = n_{q0} \exp\left[\mathscr{E}\left(\frac{Z_q T_{c0}}{T_{q0}}\right) - \frac{U^2}{2v_{Tq}^2}\left(1 + \frac{Z_q m_e}{m_q}\right)\right], \quad q = 1,2, \quad (4.7.55)
$$

where the relation between the function $\mathscr{E}$ with the variable $U$ still is from (4.7.54).

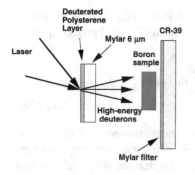
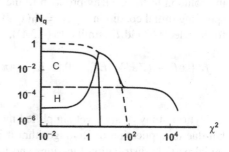

**Fig. 4.2** *Left panel*: typical experimental setup for registration of fast ions from the foil under ultra short laser pulses (from Ref. [37]). *Right panel*: the "universal" density $N_q$ of plasma ions — carbon ions (curves ($C$)) and protons (curves ($H$)) — versus the dimensionless "coordinate" $\chi^2 = (x/L_0)^2/(1 + \Omega^2 t^2)$. *Dotted curves* with *short* and *long strokes* show the dependencies of a dimensionless density for hot and cold electrons

On Fig. 4.2 we illustrate the typical "density" distribution (4.7.55) for a plasma slab, consisting of cold and hot electrons and two ions species: carbon ions $C^{+4}$ ($q = 1$) and protons $H^{+1}$ ($q = 2$). Block curves show dependence of a dimensionless "universal" density of plasma ions $N_q = (n_{q0}/n_{c0})\mathcal{N}_q$, referred to the maximal density of cold electrons, upon the dimensionless "coordinate" $\chi^2 = (x/L_0)^2/(1 + \Omega^2 t^2)$, referred to the characteristic initial density scale of ions $L_0$. "Universality" of this dependencies is the direct consequence of a relation which exists between invariants of the RG operator (4.7.49). Dotted curves give the distribution of the dimensionless density of cold electrons (short strokes), $(n_c/n_{c0})\sqrt{1 + \Omega^2 t^2}$ and hot electrons (long strokes), $(n_h/n_{c0})\sqrt{1 + \Omega^2 t^2}$, respectively.

Similar results are obtained for more complex distribution functions [14] and beyond the scope of the model used for the one-dimensional expansion, for example for spherically-symmetric expansion of a plasma bunch [38].

### 4.7.4.3 Coulomb Explosion of a Cluster in Ultra-short Laser Pulses

In this section we apply RG symmetry to the model that is used in a plasma kinetic theory for describing the Coulomb explosion of sub-micron plasmas in the field of multi-terrawatt femto-second laser pulses. Recent developments in this field have enabled examination of the fundamental physics of Coulomb explosion of nanoscale targets and ion acceleration at multi-MeV energies in different geometries of laser-plasma interaction experiments [39–41]. The mechanisms and characteristics of ions triggered by the interaction of a short-laser-pulse with plasma are of current interest because of their possible applications to the novel-neutron-source development, x-ray source, proton radiography, and isotope production.

The macroscopic state of cluster particles is governed by distribution functions $f$ (for cluster ions with mass $M$ and charge $Ze$), that dependents on time $t$, a co-ordinate $x$ of a particle, and its velocity $v$ (for simplicity we consider the one-dimensional plane geometry). Evolution of distribution functions is described by the solution to the Cauchy problem to the Vlasov kinetic equation with the corresponding initial condition $f|_{t=0} = f_0(x, v)$, supplemented by the Poisson equation for the electric field $E$ (similar to (4.3.1)),

$$f_t + vf_x + (Ze/M)Ef_v = 0, \quad E_x - 4\pi Ze \int dvf = 0, \quad f|_{t=0} = f_0(x, v).$$
(4.7.56)

Analytical study of such yet simplified model represents the essential difficulties, but due to application of RG algorithm it is possible to obtain solution at various initial particle distribution functions and find particles density, mean velocity and energy spectra. To construct RG symmetries we consider a set of local and non-local equations in (4.7.56) and the evident constraint $E_v = 0$ as $\mathscr{RM}$. The Lie group of point transformations admitted by this manifold consists of six generators

$$X_0 = \frac{\partial}{\partial t}; \quad X_1 = \frac{\partial}{\partial x}; \quad X_2 = t\frac{\partial}{\partial x} + \frac{\partial}{\partial v};$$

$$X_3 = x\frac{\partial}{\partial x} + v\frac{\partial}{\partial v} - f\frac{\partial}{\partial f} + E\frac{\partial}{\partial E};$$

$$X_4 = 2t\frac{\partial}{\partial t} + x\frac{\partial}{\partial x} - v\frac{\partial}{\partial v} - 3f\frac{\partial}{\partial f} - 2E\frac{\partial}{\partial E};$$

$$X_5 = (t^2/2)\frac{\partial}{\partial x} + t\frac{\partial}{\partial v} + (M/Ze)\frac{\partial}{\partial E},$$
(4.7.57)

describing time and space translations, $X_0$ and $X_1$, Galilean boosts, $X_2$, dilations, $X_3$ and $X_4$, and the generator $X_5$. Finite transformations defined by $X_5$ correspond to passing into a coordinate system moving linearly with constant acceleration with respect to the laboratory coordinate system. Two commutating generators in the above list (4.7.57), namely generator of Galilean boosts and generator of the transition to a uniformly accelerated frame, appear as the required RG symmetry generators [31],

$$R_1 = (t^2/2)\frac{\partial}{\partial x} + t\frac{\partial}{\partial v} + (M/Ze)\frac{\partial}{\partial E}, \quad R_2 = t\frac{\partial}{\partial x} + \frac{\partial}{\partial v}. \quad (4.7.58)$$

Successive application of finite transformations defined by theses generators shifts initial coordinates $h$ and velocities $v$ for any particle in the phase space to new values,

$$R(t, h, v) = h + vt + (Ze/2M)E(h)t^2, \quad U(t, h, v) = v + (Ze/M)E(h)t,$$
(4.7.59)

and the function $E(h)$ is defined by initial conditions (we assume the electric field to vanish at $x = 0$)

$$E(h) = 4\pi Ze \int_0^h dy \int_{-\infty}^\infty dvf_0(y, v). \quad (4.7.60)$$

The "partial" distribution function, specified by values $h$ and $v$, is the invariant of RG symmetry generators (4.7.58). Hence, the distribution function which is the solution to (4.7.56) is obtained by integrating this "partial" distribution function over all initial parameters, i.e. initial velocities and coordinates of plasma particles,

$$f(t, x, v) = \int\limits_{-\infty}^{\infty} dv \int\limits_{-\infty}^{\infty} dh f_0(v, h)\delta(x - R(t, h, v))\delta(v - U(t, h, v)).$$

(4.7.61)

For "cold" cluster particles, $f_0 \propto \delta(v)$, we need only one RG generator, $R_1$, to construct the solution of a boundary value problem. The zero and the first moments of the distribution function yield the density and the mean velocity distributions of the cluster ions, which enable to estimate the maximum energy of the accelerated ions, the ion energy spectrum and the relation between this spectrum and the initial ion density distribution [39, 40]. The similar approach to the spherical geometry [41] shows that the inhomogeneity of the initial cluster density distribution leads to the solution singularity at finite time interval even for initially immovable ions.

### 4.7.4.4 Renormgroup Algorithm Using Functionals

We consider some boundary value problem for local equations and assume that we are interested in an integral characteristic of the solution, given by a linear functional of this solution $J(u)$, say by (4.7.20). We also assume that for a particular solution $u$ of this boundary value problem, the RG algorithm has been used to find an RG symmetry with a generator $R$. To find RG symmetry generator for the functional $J(u)$, we prolong the RG symmetry operator $R$ on nonlocal variable (4.7.20) in much the same way as in Sect. 4.2.4. Considering the prolonged operator $R$ in the narrowed space of the variables defining the solution functional, we obtain the required infinitesimal RG symmetry operator for integral characteristic $J(u)$.

To demonstrate how formulas (4.7.20) and (4.2.22) actually work for functionals of solutions we consider a boundary value problems for a system of two nonlinear first-order partial differential equations for functions $v$ and $n > 0$:

$$v_t + vv_x = \alpha\varphi(n)n_x, \quad n_t + vn_x + nv_x = 0,$$
$$v(0, x) = \alpha W(x), \quad n(0, x) = N(x),$$

(4.7.62)

with constant $\alpha$ and a nonlinearity function $\varphi$ of the variable $n$. Depending on the sign of $\alpha\varphi(n)$, these equations are of either the hyperbolic ($\alpha\varphi(n) < 0$) or the elliptic ($\alpha\varphi(n) > 0$) type. In the first case, (4.7.62) corresponds to the standard equations of gas dynamics for one-dimensional planar isentropic motion of gas with the density $n$ and velocity $v$. The second case relates to equations of quasi-Chaplygin media.[8]

---

[8]The term 'quasi-Chaplygin media' is used in the discussion of nonlinear phenomena developing in accordance with the mathematical scenario for the Chaplygin gas, i.e., the gas with a negative

To calculate the RG symmetries for (4.7.62) it appears convenient to rewrite these equations in the hodograph variables $\tau = nt$ and $\chi = x - vt$,

$$\tau_v - \psi(n)\chi_n = 0, \quad \chi_v + \tau_n = 0, \quad \psi = n/\alpha\varphi. \tag{4.7.63}$$

Then the RG symmetry is given by the canonical Lie–Bäcklund operator [42]

$$R = f\frac{\partial}{\partial\tau} + g\frac{\partial}{\partial\chi}, \tag{4.7.64}$$

with coordinates $f$ and $g$ that are linear functions of variables $\tau$ and $\chi$ and their derivatives with respect to $n$ up to a fixed order $s$. Following the RG algorithm one should add the invariance conditions $f = 0$, and $g = 0$, to the basic manifold (4.7.63) and solve the resulting system of equations to get the solution to the boundary value problem (4.7.62). This procedure may appear complicated in the case of cumbersome formulas for coordinates $f$ and $g$ of RG symmetry generator (4.7.64).

In analyzing (4.7.62) for various physical problems such as a light beam behavior in a nonlinear medium the appearance of a solution singularity on the axis $x = 0$ represents the most attracting physical effect. This effect can be understood without knowledge of a complete solution by applying the RG algorithm to a functional of the solution, $n^0(t) \equiv n(t, 0)$, the value of the variable $n$ on the axis $x = 0$. As the RG symmetry generator (4.7.64) is defined in the space of hodograph variables it is convenient to use another functional of the solution introduced by a formal relationship

$$\tau^0 = \int \delta(v)\tau(v, n)\,dv. \tag{4.7.65}$$

Using (4.7.65) in (4.2.22) gives the coordinate $f^0$ of the canonical RG generator for the functional $\tau^0$. Here we present two simple illustrations.

**Example 4.7.6** Consider a solution of the boundary value problem for (4.7.62) with $\alpha = 1$, $\varphi(n) = 1$ for $W(x) = 0$ and $N(x) = \cosh^{-2}(x)$. The RG symmetry generator for this boundary value problem is defined by (4.7.64) in which coordinates $f$ and $g$ are given as

$$f = 2n(1 - n)\tau_{nn} - n\tau_n - 2nv(\chi_n + n\chi_{nn}) + nv^2\tau_{nn}/2, \tag{4.7.66}$$
$$g = 2n(1 - n)\chi_{nn} + (2 - 3n)\chi_n + v(2n\tau_{nn} + \tau_n) + (v^2/2)(n\chi_{nn} + \chi_n).$$

For RG symmetry (4.7.66), a solution exists on a finite interval $0 \le t \le t_{sing}$, until a singularity occurs on the axis $x = 0$ at $t = t_{sing} = 1/2$, when $v_x(t_{sing}, 0) \to \infty$ and the value of $n$ remains finite, $n(t_{sing}, 0) = 2$:

$$v = -2nt\tanh(x - vt), \quad n^2t^2 = n\cosh^2(x - vt) - 1. \tag{4.7.67}$$

---

adiabatic exponent. At first glance, such a model looks like the standard model of gas dynamics, but it corresponds to the negative first derivative of the 'pressure' with respect to the 'density.' A characteristic feature of quasi-Chaplygin media is a universal mathematical form of various nonlinear effects accompanying the development of an instability.

From the physical standpoint, solution (4.7.67), which was previously obtained in [43], describes the evolution of a planar light beam in a medium with a cubic nonlinearity (a quasi-Chaplygin medium) for the boundary condition $N(x) = \cosh^{-2}(x)$. The quantities $n$ and $v$ define the intensity and the eikonal derivative of the beam.

Prolongation of the RG symmetry generator (4.7.64), (4.7.66) on functional (4.7.65) gives the generator in the space $\{n, \tau^0\}$

$$R = f^0 \frac{\partial}{\partial \tau^0}, \tag{4.7.68}$$

with the coordinate

$$f^0 = 2n(1-n)\tau_{nn} - n\tau_n. \tag{4.7.69}$$

The RG invariance condition $f^0 = 0$ for operator (4.7.68) leads to an ordinary second-order differential equation for the function $\tau^0(n)$, which must be solved with initial conditions $\tau^0(1) = 0$, and $\tau_n^0 \sqrt{n-1}|_{n\to 1} = 1/2$ that follows from the original equations (4.7.63) for $v = 0$. This solution,

$$\tau^0 = \sqrt{n-1}, \tag{4.7.70}$$

results from (4.7.67) as well, but the method is simpler and solution (4.7.67) is not explicitly required.

**Example 4.7.7** Turn now to a solution of the boundary value problem for (4.7.62) with $\alpha = -1$, $\varphi(n) = 1/n$ for $W(x) = 0$ and $N(x) = \exp(-x^2)$. The RG symmetry generator for this boundary value problem is defined by (4.7.64) with the following coordinates $f$ and $g$

$$f = -n^2 \ln n \tau_{nn} - (n/2)\tau_n + \tau/2 + v(n^3 \chi_{nn} + (3/2)n^2 \chi_n),$$
$$g = -n^2 \ln n \chi_{nn} + (n/2)(1 + 4\ln n)\chi_n + \chi/2 + v(n\tau_{nn} + \tau_n/2). \tag{4.7.71}$$

For RG symmetry (4.7.71), the solution describes a monotonic evolution (decrease) with time $t$ of the density $n \geq 0$, while the particle velocity continues to be linearly dependent on the coordinate:

$$v = x\sqrt{2}q e^{-q^2/2}, \quad n = e^{-q^2/2}\exp(-x^2 e^{-q^2}),$$
$$t = (\sqrt{\pi}/2)\operatorname{erfi}(q/\sqrt{2}). \tag{4.7.72}$$

Solution (4.7.72), which was discussed in [44], describes an expanding plasma layer with the initial density distribution $N(x) = \exp(-x^2)$.

Prolongation of the RG symmetry generator (4.7.64), (4.7.71) on functional (4.7.65) gives the generator (4.7.68) in the space $\{n, \tau^0\}$ though with a different coordinate

$$f^0 = -n^2 \ln n \tau_{nn} - (n/2)\tau_n + \tau/2. \tag{4.7.73}$$

On account of (4.7.73) the RG invariance condition $f^0 = 0$ for operator (4.7.68) leads to an ordinary second-order differential equation for the function $\tau^0(n)$, which must be solved with initial conditions $\tau^0(1) = 0$, and $\tau_n^0 \sqrt{1-n}\big|_{n\to 1} = -1/2$ that follows from the original equations (4.7.63) for $v = 0$. This solution,

$$\tau^0 = \frac{\sqrt{\pi}}{2} n \mathrm{erfi}\left(\sqrt{\ln \frac{1}{n}}\right), \tag{4.7.74}$$

correlates with (4.7.72) for $v = 0$.

In conclusion we notice that expressions (4.7.70) and (4.7.74) result from the complete solutions as well. However, the RG algorithm for functionals presents here an elegant way of obtaining these formulas without calculating the complete solutions to boundary value problems.

# References

1. Vlasov, A.A.: The vibrational properties of an electron gas. J. Exp. Theor. Phys. **8**(3) (1938) 291–317 (in Russian); see also Sov. Phys. Usp. **10**, 721–733 (1968)
2. Lewak, G.J.: More-uniform perturbation theory of the Vlasov equation. J. Plasma Phys. **3**, 243–253 (1969)
3. Pustovalov, V.V., Chernikov, A.A.: Functional averaging and kinetics of plasma in Lagrangean variables. Preprint No. 171, P.N. Lebedev Physical Institute, AN USSR (1980) (in Russian)
4. Pustovalov, V.V., Romanov, A.B., Savchenko, M.A., Silin, V.P., Chernikov, A.A.: One method for solving the Vlasov kinetic equation. Sov. Phys., Lebedev Inst. Rep. **12**, 28–32 (1976)
5. Taranov, V.B.: On the symmetry of one-dimensional high frequency motions of a collisionless plasma. Sov. J. Tech. Phys. **21**, 720–726 (1976)
6. Kovalev, V.F., Krivenko, S.V., Pustovalov, V.V.: Group symmetry of the kinetic equations of a collisionless plasma. JETP Lett. **55**(4), 256–259 (1992)
7. Kovalev, V.F., Krivenko, S.V., Pustovalov, V.V.: Group analysis of the Vlasov kinetic equation, I. Differ. Equ. **29**(10), 1568–1578 (1993)
8. Kovalev, V.F., Krivenko, S.V., Pustovalov, V.V.: Group analysis of the Vlasov kinetic equation, II. Differ. Equ. **29**(11), 1712–1721 (1993)
9. Grigor'ev, Yu.N., Meleshko, S.V.: Group analysis of integro-differential Boltzmann equation. Sov. Phys. Dokl. **32**, 874–876 (1987)
10. Volterra, V.: Theory of Functional and of Integral and Integro-Differential Equations. Blackie, London (1929). Edited by Fantappie, L. Translated by Long, M. Also available as: Volterra, V.: Theory of Functionals and of Integral and Integro-Differential Equations. Dover, New York (1959). Russian translation: Nauka, Moscow (1982)
11. Baikov, V.A., Gazizov, R.K., Ibragimov, N.H.: Perturbation methods in group analysis. J. Sov. Math. **55**(1), 1450 (1991)
12. Baikov, V.A., Gazizov, R.K., Ibragimov, N.Kh.: Perturbation methods in group analysis. In: Itogi Nauki i Tekhniki. Ser. Sovrem. Probl. Mat. Nov. Dostizh., vol. 34, pp. 85–147. VINITI, Moscow (1989) (in Russian). J. Sov. Math. **55**(1), 1450–1490 (1991)
13. Dorozhkina, D.S., Semenov, V.E.: Exact solution of Vlasov equations for quasineutral expansion of plasma bunch into vacuum. Phys. Rev. Lett. **81**, 2691–2694 (1998)
14. Kovalev, V.F., Bychenkov, V.Yu., Tikhonchuk, V.T.: Particle dynamics during adiabatic expansion of a plasma bunch. JETP **95**(2), 226–241 (2002)
15. Landau, L.D., Livshitz, E.M.: Course of Theoretical Physics, vol. 2, The Classical Theory of Fields. Nauka, Moscow (1973)
16. Kovalev, V.F., Krivenko, S.V., Pustovalov, V.V.: Symmetry group of Vlasov–Maxwell equations in plasma theory. In: Proceedings of the International Conference "Symmetry in Nonlinear Mathematical Physics", July 3–8, 1995, Kiev, Ukraina, V. 2. J. Nonlinear Math. Phys. **3**(1–2), 175–180 (1996).

17. Ibragimov, N.H. (ed.): CRC Handbook of Lie Group Analysis of Differential Equations, vol. 1: Symmetries, Exact Solutions and Conservation Laws (1994); vol. 2: Applications in Engineering and Physical Sciences (1995); vol. 3: New Trends in Theoretical Developments and Computational Methods (1996). CRC Press, Boca Raton

18. Benney, D.J.: Some properties of long nonlinear waves. Stud. Appl. Math. **L11**(1), 45–50 (1973)

19. Krasnoslobodtzev, A.V.: Gas dynamic and kinetic analogies in the theory of vertically inhomogeneous shallow water. Trans. Inst. Gen. Phys. USSR Acad. Sci. **18**, 33–71 (1989) (in Russian)

20. Kupershmidt, B.A., Manin, Yu.I.: Long-wave equation with free boundary. I. Conservation laws and solutions. Funct. Anal. Appl. **11**(3), 188–197 (1977)

21. Kupershmidt, B.A., Manin, Yu.I.: Long-wave equation with free boundary. II. Hamiltonian structure and higher equations. Funct. Anal. Appl. **12**(1), 20–29 (1978)

22. Zakharov, V.E.: Benney equation and quasi-classical approximation in the method of the inverse problem. Funct. Anal. Appl. **14**(2), 89–98 (1980)

23. Gibbons, J.: Collisionless Boltzmann equations and integrable moment equations. Physica D3 **3**(3), 503–511 (1981)

24. Ibragimov, N.H., Kovalev, V.F., Pustovalov, V.V.: Symmetries of integro-differential equations: a survey of methods illustrated by the Benney equations. Nonlinear Dyn. **28**, 135–153 (2002). Preprint math-ph/0109012

25. Gurevich, A.V., Pitaevski, L.P.: Nonlinear dynamics of a rarefied plasmas and ionospheric aerodynamics. In: Problems of Plasma Theory, vol. 10, pp. 3–87. Nauka, Moscow (1980) (in Russian). Reviews of Plasma Physics, vol. 10. Edited by Acad. Leontovich, M.A. Translated from Russian by Glebov, O. Translation editor: ter Haar, D., Department of Theoretical Physics, University of Oxford, Oxford, England. Published by Consultants Bureau, New York (1986)

26. Kovalev, V.F., Pustovalov, V.V.: Functional self-similarity in a problem of plasma theory with electron nonlinearity. Theor. Math. Phys. **81**, 1060–1071 (1990)

27. Shirkov, D.V.: Several topics on renorm-group theory. In: Shirkov, D.V., Priezzhev, V.B. (eds.) Renormalization Group '91, Proc. of Second Intern. Conf., Sept. 1991, Dubna, USSR, pp. 1–10. World Scientific, Singapore (1992)
Kovalev, V.F., Krivenko, S.V., Pustovalov, V.V.: The Renormalization group, method based on group analysis. In: Shirkov, D.V., Priezzhev, V.B. (eds.) Renormalization Group '91, Proc. of Second Intern. Conf., Sept. 1991, Dubna, USSR, pp. 300–314. World Scientific, Singapore (1992)

28. Kovalev, V.F., Pustovalov, V.V., Shirkov, D.V.: Group analysis and renormgroup symmetries. J. Math. Phys. **39**, 1170–1188 (1998). Preprint hep-th/9706056

29. Kovalev, V.F., Shirkov, D.V.: Bogoliubov renormalization group and symmetry of solution in mathematical physics. Phys. Rep. **352**(4–6), 219 (2001). hep-th/0001210

30. Kovalev, V.F., Shirkov, D.V.: The renormalization group symmetry for solution of integral equations. In: Nikitin, A.G. (ed.) Proc. of the 5th Intern. Conf. on Symmetry in Nonlinear Mathematical Physics, Kii'v, Ukraine, June 23–29, 2003. Proc. of the Inst. of Math. of the Natl. Acad. Sci. of Ukraine. Math. and its Appl., vol. 50, Pt. 2, pp. 850–861. Inst. of Math. of NAS Ukraine, Kiïv (2004)

31. Kovalev, V.F., Shirkov, D.V.: Renormgroup symmetry for functionals of boundary value problem solutions. J. Phys. A, Math. Gen. **39**, 8061–8073 (2006)

32. Kovalev, V.F., Shirkov, D.V.: Renormalization-group symmetries for solutions of nonlinear boundary value problems. Phys.-Usp. **51**(8), 815–830 (2008). Preprint arXiv:0812.4821 [math-ph]

33. Shirkov, D.V.: Renormalization group, invariance principle and functional self-similarity. Sov. Phys. Dokl. **27**, 197 (1982)

34. Ovsyannikov, L.V.: Group Analysis of Differential Equations. Academic Press, New York (1982)

35. Rudenko, O.V., Soluyan, S.I.: Theoretical Foundations of Nonlinear Acoustics. Consultants Bureau, New York (1977)

36. Pustovalov, V.V., Silin, V.P.: Nonlinear theory of the interaction of waves in a plasma. In: Proceedings of P.N. Lebedev Physical Institute, AN USSR, vol. 61, pp. 42–283. Nauka, Moscow (1972). English translation in: Skobel'tsyn, D.V. (ed.) Theory of Plasmas. Consultants Bureau, New York (1975)
37. Maksimchuk, A., Flippo, K., Krause, H., et al.: High-energy ion generation by short laser pulses. Plasma Phys. Rep. **30**(6), 473–495 (2004)
38. Kovalev, V.F., Bychenkov, V.Yu: Analytic solutions to the Vlasov equations for expanding plasmas. Phys. Rev. Lett. **90**(18), 185004 (2003) (4 pages)
39. Bychenkov, V.Yu, Kovalev, V.F.: Coulomb explosion in a cluster plasma. Plas. Phys. Rep. **31**(2), 178–183 (2005)
40. Bychenkov V.Yu., Kovalev, V.F.: On the maximum energy of ions in a disintegrating ultrathin foil irradiated by a high-power ultrashort laser pulse. Quantum Electron. **35**(12), 1143–1145 (2005)
41. Kovalev, V.F., Popov, K.I., Bychenkov, V.Yu., Rozmus, W.: Laser triggered Coulomb explosion of nanoscale symmetric targets. Phys. Plasmas **14**, 053103 (2007) (10 pages)
42. Kovalev, V.F., Pustovalov, V.V.: Group and renormgroup symmetry of a simple model for nonlinear phenomena in optics, gas dynamics and plasma theory. Math. Comput. Model. **25**, 165–179 (1997)
43. Akhmanov, S.A., Sukhorukov, A.P., Khokhlov, R.V.: On the self-focusing and self-chanelling of intense laser beams in nonlinear medium. Sov. Phys. JETP **23**(6), 1025–1033 (1966)
44. Murakami, M., Kang, Y.-G., Nishihara, K., et al.: Ion energy spectrum of expanding laser-plasma with limited mass. Phys. Plasmas **12**, 062706 (2005) (8 pages)

# Chapter 5
# Symmetries of Stochastic Differential Equations

Stochastic differential equations are often obtained by including random fluctuations in differential equations, which have been deduced from phenomenological or physical view. For example, the motion of a small particle suspended in a moving liquid is described by the differential equation

$$\frac{dx}{dt} = b(t, x),$$

where $b(t, x)$ is the velocity of the fluid at the point $x$ and at time $t$. The function $b(t, x)$ represents the resistance caused by the viscosity of the liquid. If a particle is randomly bombarded by molecules of the fluid, then this can be modeled by the equation

$$\frac{dX}{dt} = b(t, X) + \sigma(t, X)W_t, \qquad (5.0.1)$$

where $W_t$ denotes "white noise". The second term on the right hand side represents the large number of collisions of the pollen grain with the molecules of the liquid. Formally the white noise is written as $W_t = dB_t/dt$, and (5.0.1) is rewritten in the differential form

$$dX = b(t, X)\,dt + \sigma(t, X)\,dB_t. \qquad (5.0.2)$$

Here $B_t$ is a Brownian motion. Equation (5.0.2) is called a stochastic differential equation. A solution $X(t)$ of the stochastic differential equation (5.0.2) is a stochastic process $X(t)$ which has a stochastic differential (5.0.2). Stochastic differential equations of the type (5.0.2) have been used widely in other areas of the science.

There are many textbooks on stochastic differential equations. We just mention here some of them [2, 13, 20, 24, 27, 31].

In contrast to deterministic differential equations, only few attempts to apply group analysis to stochastic differential equations can be found in the literature [1, 10, 12, 19, 23, 29, 30, 32, 34]. This chapter deals with applications of the group analysis method to stochastic differential equations. It is also worth to note that this theory is still developing.

Y.N. Grigoriev et al., *Symmetries of Integro-Differential Equations*, 209
Lecture Notes in Physics 806,
DOI 10.1007/978-90-481-3797-8_5, © Springer Science+Business Media B.V. 2010

The chapter is organized as follows. Before defining an admitted symmetry for stochastic differential equations an introduction into stochastic differential equations is given. The introduction includes the discussion of a stochastic integration, a stochastic differential and a change of the variables (Itô formula) in stochastic differential equations. Applications of the Itô formula are considered in the next section which deals with the linearization problem. The Itô formula and the change of time in stochastic differential equations are the main tools of defining admitted transformations for stochastic differential equations. After introducing an admitted Lie group for SDEs and supporting material of the introduced definition, some examples of applications of the given definition are studied.

## 5.1 Stochastic Integration of Processes

This section is devoted to developing the tools for stochastic processes. In particular, it discusses stochastic integrals with respect to Brownian motion, martingales, alternative fields and changes of time.

### 5.1.1 Stochastic Processes

Let $\Omega$ be a given set of elementary events $\omega$, $\mathscr{F}$ a $\sigma$-algebra of subsets of $\Omega$ and $\mathscr{P}$ a probability (or probability measure) on $\mathscr{F}$. The triple $(\Omega, \mathscr{F}, \mathscr{P})$ is called a probability space. It is assumed that the $\sigma$-algebra $\mathscr{F}$ is generated by a family of $\sigma$-algebras $\mathscr{F}_t$ ($t \geq 0$) such that

$$\mathscr{F}_s \subset \mathscr{F}_t \subset \mathscr{F}, \quad \forall s \leq t, \quad s, t \in I,$$

where $I = [0, T]$, $T \in (0, \infty]$.

The nondecreasing family of $\sigma$-algebras $\mathscr{F}_t$ is also called a filtration and the $\sigma$-algebra $\mathscr{F}$ is denoted by $\mathscr{F} = (\mathscr{F}_t)_{t \geq 0}$. The triple $(\Omega, \mathscr{F}, \mathscr{P})$ is called a filtrated probability space.

Let $\mathscr{B}$ be the Borel $\sigma$-algebra on $R^N$. A mapping $X : \Omega \to R^N$ is called an $N$-dimensional random variable if for each $B \subset \mathscr{B}$ the set $X^{-1}(B) \in \mathscr{F}$. A collection $\{X(t)\}_{t \geq 0}$ of random variables on $(\Omega, \mathscr{F}, \mathscr{P})$ is called a stochastic process.[1] A process $\{X(t)\}_{t \geq 0}$ is said to be adapted to $(\mathscr{F}_t)_{t \geq 0}$ if $X(t)$ is $\mathscr{F}_t$-measurable for each $t$. A process $X$ is called measurable if $(t, \omega) \longmapsto X(t, \omega)$ is a $\mathscr{B} \otimes \mathscr{F}$-measurable mapping. The process $X$ is said to be continuous if the trajectories $t \longmapsto X(t, \omega)$ are continuous for almost all $\omega \in \Omega$. It is called progressively measurable if $X : [0, t] \times \Omega \longmapsto R$ is a $\mathscr{B}([0, t]) \otimes \mathscr{F}_t$-measurable mapping for each $0 \leq t < \infty$. Note that a progressively measurable process is measurable and adapted.

---

[1] A stochastic process depends on two variables $X = X(t, \omega)$; the second variable $\omega$ usually is omitted.

If $X$ is integrable, that is if $X$ or equivalently $|X|$ belongs to $L^1(\Omega, \mathscr{F}, P)$, then its expectation is defined as

$$E(X) = \int_\Omega X \, d\mathscr{P},$$

where the integral is of the Lebesgue type.

A scalar (standard) Brownian motion or Wiener–Lévy process is a stochastic process $B(t)$ satisfying the following properties:

(i) $\mathscr{P}(\{B(0) = 0\}) = 1$;
(ii) for any finite partition $\{t_i\}_{i=0}^n t_i < t_{i+1}$ of $I = [0, T]$, $T > 0$, the random variables $B(t_{i+1}) - B(t_i)$ are independent;
(iii) for all $t, s \in I$, the probability distribution $B(t) - B(s)$ is Gaussian with $E(B(t) - B(s)) = 0$ and $E([B(t) - B(s)]^2) = \mu^2|t - s|$, where $\mu$ is a nonzero constant.

An $M$-dimensional Brownian motion is a stochastic process $B(t) = (B_1(t), B_2(t), \ldots, B_M(t))$, where $B_i(t)$ ($i = 1, 2, \ldots, M$) are independent scalar Brownian motions.

As is usual, for ease of notation we will omit the stochastic variable $\omega$, switch freely between the notations $X_t$, $X(t)$ or $X(t, \omega)$, and make the convention that identities hold a.s. only.

## 5.1.2 The Itô Integral

From now on, unless stated otherwise, we let $\{B(t)\}_{t \geq 0}$ be a standard Brownian motion and $\mathscr{F}_t = \sigma(\{B(s); 0 \leq s \leq t\})$, $t \geq 0$. Let $0 = t_0 < t_1 < \cdots < t_n = T$ be a partition of $[0, T]$ and $Y_0, Y_1, \ldots, Y_{n-1}$ some random variables which are adapted to $\mathscr{F}_0, \mathscr{F}_{t_1}, \ldots, \mathscr{F}_{t_{n-1}}$ respectively and satisfy the conditions $E(Y_0^2), \ldots, E(Y_{n-1}^2) < \infty$. The process $\{X(t)\}_{t \geq 0}$ which is defined by

$$X(t) = Y_0 I_{\{0\}}(t) + \sum_{i=1}^n Y_{i-1} I_{(t_{i-1}, t_i]}(t), \quad t \in [0, T]$$

is called a simple process. Here, $I_S$ denotes the characteristic function of a set $S$. The set of simple processes forms the class $S_T$. In the case $T = \infty$, there is one more requirement for a simple process: $Y_{n-1} = 0$. For a process $X \in S_T$, the Itô integral of $\{X(t)\}_{t \in [0,T]}$ is defined by

$$\int_0^t X(s) \, dB(s) = X(t_m)(B(t) - B(t_m)) + \sum_{i=1}^m X(t_{i-1})(B(t_i) - B(t_{i-1})),$$

$t \in (t_m, t_{m+1}]$.

A stochastic process $\{X(t)\}_{t \geq 0}$ is said to belong to the class $E_T$, $T \in (0, \infty]$ if it is measurable and adapted to $(\mathscr{F}_t)_{t \geq 0}$ with

$$E\left(\int_0^T X^2(s)\,ds\right) < \infty.$$

For a stochastic process $X \in E_T$, the Itô integral of $\{X(t)\}_{t \in [0,T]}$ is defined in the sense of convergence in the mean

$$\int_0^t X(s)\,dB(s) = \lim_{n \to \infty} \int_0^t X_n(s)\,dB(s), \quad t \in [0, T],$$

where $\{X_n\}_{n=1}^{\infty}$ is a sequence of simple processes such that

$$\lim_{n \to \infty} \int_0^T E\left(X_n(s) - X(s)\right)\,ds = 0.$$

A stochastic process $\{X(t)\}_{t \geq 0}$ is said to belong to the class $P_T$ of predictable processes on $[0, T]$, $T \in (0, \infty]$ if it is measurable and adapted to $(\mathscr{F}_t)_{t \geq 0}$ with

$$\mathscr{P}\left\{\int_0^T X^2(s)\,ds < \infty\right\} = 1.$$

Note that $S_T \subset E_T \subset P_T$.

A stochastic process $X$ is a nonanticipating functional if it is measurable and adapted to $(\mathscr{F}_t)_{t \geq 0}$ with

$$\mathscr{P}\left\{\int_0^t X^2(s)\,ds < \infty, \; t \geq 0\right\} = 1.$$

For a process $X \in P_T$, the Itô integral of $\{X(t)\}_{t \in [0,T]}$ is defined in the sense of convergence in probability,

$$\int_0^t X(s)\,dB(s) = \lim_{n \to \infty} \int_0^t X_n(s)\,dB(s), \quad t \in [0, T],$$

where $\{X_n\}_{n=1}^{\infty}$ is a sequence of processes which belong to the class $E_T$ such that

$$\lim_{n \to \infty} \int_0^T \left(X_n(s) - X(s)\right)\,ds = 0,$$

with limit in the sense of convergence in probability.

In the literature on stochastic processes it is proven that processes $X, Y \in P_T$ satisfy the following properties:[2]

---

[2]The proofs can be found, for example, in [2].

(a) the Itô integral $\int_0^t X(s)\,dB(s)$ is well-defined for $0 \le t \le T$,

(b) $E((\int_0^t X(s)\,dB(s))^2) = \int_0^t E(X^2(s))\,ds$ for $0 \le t \le T$ (Itô isometry property),

(c) $E((\int_0^t X(s)\,dB(s))(\int_0^t Y(s)\,dB(s))) = \int_0^t E(X(s)Y(s))\,ds$ for $0 \le t \le T$,

(d) $\int_0^t (\alpha X(s) + \beta Y(s))\,dB(s) = \alpha \int_0^t X(s)\,dB(s) + \beta \int_0^t Y(s)\,dB(s)$   a.s.   for all $\alpha, \beta \in R$ and $0 \le t \le T$,

(e) $\int_0^t X(s)\,dB(s)$ is $\mathscr{F}_t$-measurable for $0 \le t \le T$,

(f) $(\int_0^t X(s)\,dB(s))_{t \ge 0}$ is continuous and progressively measurable, with probability one,

(g) $X(t)B(t) = \int_0^t X(s)\,dB(s) + \int_0^t B(s)\,dX(s)$ for $0 \le t \le T$.

### 5.1.3 The Itô Formula

A change of variables in the stochastic Itô integral is not according to the chain rule of classical calculus. An additional term appears and the resulting expression is called the Itô formula. This formula plays one of key roles in constructing an admitted Lie group for stochastic differential equations.

Let $\{f(t)\}_{t \ge 0}$ and $\{g(t)\}_{t \ge 0}$ be two stochastic processes, such that $\sqrt{|f|} \in P_T$, $g \in P_T$ and[3]

$$X_j(t, \omega) = X_j(0, \omega) + \int_0^t f_j(s, \omega)\,ds + \int_0^t g_{jk}(s, \omega)\,dB_k(s). \qquad (5.1.1)$$

The last relation written in a symbolical form

$$dX_j = f_j\,dt + g_{jk}\,dB_k,$$

or in the vector form

$$dX = f\,dt + g\,dB,$$

is called a stochastic differential. Here $g_{jk}$ are components of the matrix $g$, $f_j$ and $B_k$ are coordinates of the vectors $f$ and $B$, respectively. For each $\omega$ the first integral in (5.1.1) is a Riemann integral and the second term is an Itô integral. Assume that the function $U(t, x)$ has continuous partial derivatives $U_t$, $U_{x_j}$, $U_{x_j x_k}$. Then the stochastic process $Y(t, \omega) = U(t, X(t, \omega))$ satisfies the formula

$$\begin{aligned} Y(t, \omega) = Y(0, \omega) &+ \int_0^t \big(U_t(s, X(s, \omega)) + f_j(s, \omega)U_{x_j}(s, X(s, \omega)) \\ &+ \tfrac{1}{2}g_{ji}(s, \omega)g_{ki}(s, \omega)U_{x_j x_k}(s, X(s, \omega))\big)\,ds \\ &+ \int_0^t U_{x_j}(s, X(s, \omega))g_{jk}(s, \omega)\,dB_k(s, \omega). \end{aligned}$$

---

[3] Summation over the repeated indices is assumed.

This formula is called the Itô formula. In the differential form this formula is written as

$$dY = \left( U_t + f_j U_{x_j} + \frac{1}{2} g_{ji} g_{ki} U_{x_j x_k} \right) dt + (U_{x_j} g_{jk}) dB_k. \qquad (5.1.2)$$

### 5.1.4  Time Change in Stochastic Integrals

One of the mathematical tools required for defining the transformation of Brownian motion is the formula of time change in stochastic integrals.

Let $\eta(t, x, b)$ be a sufficiently many times continuously differentiable function, $\{X(t)\}_{t \geq 0}$ be a continuous and adapted stochastic process, and $\{B(t)\}_{t \geq 0}$ be a standard Brownian motion. Since $\eta(t, x, b)$ is continuous, $\eta^2(t, X(t, \omega), B(t, \omega))$ is also an adapted process. Define[4]

$$\beta_X(t) = \beta(t, X) = \int_0^t \eta^2(s, X(s), B(s)) \, ds, \quad t \geq 0. \qquad (5.1.3)$$

For brevity we write $\beta(t)$ instead of $\beta(t, X)$. The function $\beta(t)$ is called a random time change with time change rate $\eta^2(t, X(t, \omega), B(t, \omega))$. Note that $\beta(t)$ is an adapted process. Suppose now that $\eta(t, x, b) \neq 0$ for all $(t, x, b)$. Then for each $\omega$, the map $t \longmapsto \beta(t)$ is strictly increasing. Next define

$$\alpha_X(t) = \alpha(t, X) = \inf_{s \geq 0} \{s : \beta(s, X) > t\}, \qquad (5.1.4)$$

and for brevity, we write $\alpha(t)$ instead of $\alpha(t, X)$. For almost all $\omega$, the map $t \longmapsto \alpha(t)$ is nondecreasing and continuous. One easily shows that for almost all $\omega$, and for all $t \geq 0$,

$$\beta(\alpha(t)) = t = \alpha(\beta(t)). \qquad (5.1.5)$$

Since $\beta(t)$ is an $\mathscr{F}_t$-adapted process, one has

$$\{\omega : \alpha(t) \leq s\} = \{\omega : t \leq \beta(s)\} \in \mathscr{F}_s, \quad \forall t \geq 0, \ \forall s \geq 0.$$

Hence $t \longmapsto \alpha(t)$ is an $\mathscr{F}_s$-stopping time for each $t$.

The following theorem will be crucial for defining the transformation of a Brownian motion.[5]

**Theorem 5.1.1** *Let $\eta(t, x, b)$ and $\{X(t)\}_{t \geq 0}$ be as above and $\{B(t)\}_{t \geq 0}$ a standard Brownian motion. Define*

$$\bar{B}(t) = \int_0^t \eta(s, X(s, \omega), B(s, \omega)) \, dB(s), \quad t \geq 0. \qquad (5.1.6)$$

---

[4]Notice that the function $\eta(t, x, b)$ also depends on $x$ and $b$.

[5]The proof of the theorem is similar to [24] and can be found in [29].

Then $(\bar{B}_{\alpha(t)}, \mathscr{F}_{\alpha(t)})$ is a standard Brownian motion, where

$$\mathscr{F}_{\alpha(t)} = \{A \in \mathscr{F} : A \cap \{\omega : \alpha(t) \leq s\} \in \mathscr{F}_s, \ \forall s \geq 0\}.$$

## 5.2 Stochastic Ordinary Differential Equations

### 5.2.1 Itô Stochastic Differential Equations

The stochastic process

$$Y(t) = X(t_0) + \int_{t_0}^{t} f(s, X(s)) \, dt + \int_{t_0}^{t} g(s, X(s)) \, dB_s \qquad (5.2.1)$$

is called an Itô process, and formally it is defined by the stochastic differential

$$dY_t = f(t, X_t) dt + g(t, X_t) dB_t. \qquad (5.2.2)$$

The equation

$$dX_t = f(t, X_t) dt + g(t, X_t) dB_t. \qquad (5.2.3)$$

is called a stochastic differential equation. The vector function $f(t, x) = (f_i(t, x))$ $(i = 1, 2, \ldots, N)$ is called a drift vector, the matrix $g(t, x) = (g_{ij}(t, x))$ $(i = 1, 2, \ldots, N; \ j = 1, 2, \ldots, M)$ is called a diffusion matrix.

**Definition 5.2.1** A stochastic process $\{X(t)\}_{t \geq 0}$ which satisfies (5.2.3) is called a strong solution.

A solution is called a strong solution, because it is a solution of (5.2.3) with a Brownian motion given in advance. If only the functions $f(t, x)$ and $g(t, x)$ are given and one is asked for a pair of processes $(\hat{X}_t, \hat{B})$ on a probability space $(\Omega, \mathscr{F}, \mathscr{P})$ such that (5.2.4) holds, then $\hat{X}_t$ (or more precisely $(\hat{X}_t, \hat{B})$) is called a weak solution.

**Definition 5.2.2** A stochastic process $\{\hat{X}(t)\}_{t \geq 0}$ which satisfy (5.2.3) for some Brownian motion $\{\hat{B}(t)\}_{t \geq 0}$

$$d\hat{X}_t = f(t, \hat{X}_t) \, dt + g(t, \hat{X}_t) \, d\hat{B}_t \qquad (5.2.4)$$

is called a weak solution.

A strong solution is of course also a weak solution.

Any solution of a diffusion type stochastic differential equation is called a diffusion process.

Existence of solutions of stochastic differential equations can be guaranteed by the following theorem.

**Theorem 5.2.1** (Existence and Uniqueness [24]) *Suppose that the coefficients f and g of equations*

$$X_i(t, \omega) = X_i(0, \omega) + \int_0^t f_i(s, X(s, \omega)) \, ds$$

$$+ \int_0^t g_{ik}(s, X(s, \omega)) \, dB_k(s), \tag{5.2.5}$$

*satisfy a space-variable Lipschitz condition*

$$|f(t, x) - f(t, y)|^2 + |g(t, x) - g(t, y)|^2 \le k|x - y|^2,$$

*and the spatial growth condition*

$$|f(t, x)|^2 + |g(t, x)|^2 \le k(1 + |x|^2),$$

*for some positive constant k. Then there exists a continuous adapted solution $X_t$ of (5.2.5) that is uniformly bounded in $L^2$: $\sup_{0 \le t \le T} E(X_t^2) < \infty$. Moreover, if $X_t$ and $Y_t$ are both continuous $L^2$ bounded solutions of (5.6.11) then*

$$\mathscr{P}(X_t = Y_t \text{ for all } t \in [0, T]) = 1.$$

In the theorem some conditions for the initial random variable $X(0, \omega)$ are also required. For the sake of simplicity these conditions are omitted here.

Further we assume that solutions of all stochastic differential equations exist locally, that is for $t \in [0, \varepsilon)$.

### 5.2.2 Stratonovich Stochastic Differential Equations

There is another interpretation of the white noise in (5.0.1). For example, in some applications stochastic differential equations are formulated in terms of Stratonovich stochastic differential equations

$$dX_t = \tilde{f}(t, X_t)dt + g(t, X_t) \circ dB_t. \tag{5.2.6}$$

The notation "∘" denotes the Stratonovich integral

$$\int_{t_0}^t g(s, X(s)) \circ dB_s.$$

The Stratonovich stochastic differential equations (5.2.6) can be expressed in terms of the Itô stochastic differential equations (5.2.4): a solution of a Stratonovich stochastic differential equation (5.2.6) is a solution of the Itô stochastic differential equation (5.2.4) with the modified drift coefficients

$$f_i = \tilde{f}_i + \frac{1}{2} \frac{\partial g_{ij}}{\partial x_k} g_{kj} \quad (i = 1, 2, \ldots, N).$$

One of the benefits of the Stratonovich form of stochastic differential equations is that a change of variables in the stochastic differential equations occurs according to the chain rule of classical calculus.

Since the Stratonovich form of stochastic differential equations (5.2.6) can be converted into Itô stochastic differential equations (5.2.4), further we will deal with the Itô interpretation of stochastic differential equations.

### 5.2.3 Kolmogorov Equations

Instead of considering trajectories of a stochastic processes $\{X(t)\}_{t \geq 0}$ Kolmogorov studied the properties of the transition probabilities

$$P(s, x; t, A) = P(X_t \in A \mid X_s = x),$$

i.e., the probability of the event that a trajectory of $X$ arrives in the set $A$ at time $t$, provided that $X_s = x$ at time $s$. This leads to the following backward parabolic differential equation

$$-\frac{\partial p}{\partial s} = f_i(s, x)\frac{\partial p}{\partial x_i} + \frac{1}{2}\sigma_{ij}(s, x)\frac{\partial^2 p}{\partial x_i \partial x_j}, \qquad (5.2.7)$$

and the forward parabolic equation

$$\frac{\partial p}{\partial t} = -\frac{\partial}{\partial y_i}(f_i(t, y)p) + \frac{1}{2}\frac{\partial^2}{\partial y_i \partial y_j}(\sigma_{ij}(t, y)p), \qquad (5.2.8)$$

where $\sigma_{ij} = g_{ik}g_{jk}$, and the function $p(s, x; t, y)$ is defined by the probability $P(s, x; t, (-\infty, y])$ [27]. Equation (5.2.8) is also called the Fokker–Planck equation.

Notice that the one-dimensional backward Kolmogorov equation

$$u_t + fu_x + \frac{g^2}{2}u_{xx} = 0 \qquad (5.2.9)$$

is equivalent with respect to a change of the dependent and independent variables to the heat equation

$$v_\tau = v_{yy},$$

if and only if it satisfies the condition[6]

$$H_{xxx}g^2 + 2H_{tx} + 2H_{xx}(g_x g + f) + 2H_x(f_x - 2H) = 0, \qquad (5.2.10)$$

where

$$H = (2g)^{-1}(g_{xx}g^2 + 2(g_t - f_x g + g_x f)). \qquad (5.2.11)$$

In particular, the backward Kolmogorov equation is equivalent to the heat equation for all stochastic differential equations with $H_x = 0$.

---

[6]Criteria for a linear parabolic equation to be equivalent to one of the Lie canonical equations can be found in [14, 16, 18].

## 5.3  Linearization of First-Order Stochastic ODE

**Definition 5.3.1** If the functions $f(t, x)$ and $g(t, x)$ are linear with respect to the variable $x$, i.e.,

$$f(t, x) = a_1(t)x + a_0(t), \quad g(t, x) = b_1(t)x + b_0(t),$$

then the stochastic differential equation

$$dX_t = (a_1(t)X_t + a_0(t))\, dt + (b_1(t)X_t + b_0(t))\, dB_t, \tag{5.3.1}$$

is called a linear stochastic differential equation. A linear stochastic differential equation with $b_1 = 0$ is called a linear stochastic differential equation in the narrow-sense [25].

Linear stochastic ordinary differential equations play a similar role as linear ordinary differential equations play in the classical theory of ordinary differential equations. For example, as in the classical theory of ordinary differential equations the reduction of a stochastic ordinary differential equation to a linear stochastic ordinary differential equation allow constructing an exact solution of the original equation [13, 25, 26]. Hence, one can state the linearization problem for a stochastic ordinary differential equations: find a change of variables such that a transformed equation becomes a linear equation.

Recall that in the general case the change of variables in stochastic differential equations differs from the change of variables in ordinary differential equations owing to the necessity of using the Itô formula instead of the formula for differentiation of a composite function (5.1.2): if $\varphi(t, x)$ is a continuous function with continuous derivatives $\varphi_t, \varphi_x, \varphi_{xx}$, and a stochastic process $\{X_t\}_{t \geq 0}$ is a solution of the stochastic differential equation

$$dX = f(t, X)\, dt + g(t, X)\, dB_t, \tag{5.3.2}$$

then the process $\varphi(t, X)$ has the stochastic differential

$$d\varphi(t, X) = \left(\varphi_t + f\varphi_x + kt^{2k-1}g^2\varphi_{xx}\right)(t, X)\, dt + (g\varphi_x)(t, X)\, dB_t.$$

Here $f(t, x)$ and $g(t, x)$ are deterministic functions.

In [5, 13, 25] the Itô formula was applied for solving the linearization problem of a scalar first-order stochastic ordinary differential equation

$$dX = f(t, X)\, dt + g(t, X)\, dW. \tag{5.3.3}$$

Particular criteria of the existence of a change of the dependent variable

$$y = \varphi(t, x) \tag{5.3.4}$$

such that (5.3.3) becomes the linear equation

$$dY = (a_1(t)Y + a_0(t))\, dt + (b_1(t)Y + b_0(t))\, dW, \tag{5.3.5}$$

are presented in [5, 13, 25, 33].[7] In [13] linearization criteria for autonomous stochastic ordinary differential equations ($f = f(x)$ and $g = g(x)$) were found. It was also obtained the reducibility conditions to the equation with $a_1 = 0$ and $b_1 = 0$. Many examples of stochastic ordinary differential equations satisfying these criteria are given in [13, 25]. On the base of this analysis the author [5] developed a MAPLE package containing routines which return explicit solutions of stochastic differential equations. Necessary and sufficient conditions for the linearization of the one-dimensional Itô stochastic differential equations driven by fractional Brownian motion are given in [33].

Since transformation (5.3.4) does not change the Brownian motion, the linearization by (5.3.4) can be called a strong linearization to contrast a reducibility to a linear differential equation with a different Brownian motion.

### 5.3.1  Weak Linearization Problem

Similar to the definition of a strong or weak solutions the reduction of a stochastic differential equation to a linear differential equation with a changed Brownian motion can be called a weak linearization. For example, the time change $t$ including the (random) time change $t$ with the time change rate $h^2(t, \omega)$:

$$y = x, \quad \tau(t, \omega) = \int_0^t h^2(s, \omega)\, ds, \qquad (5.3.6)$$

leads to the change of the Brownian motion by the formula [24]

$$\tilde{B}_t = \int_0^{\alpha(t,\omega)} h(s, \omega)\, dB_s, \qquad (5.3.7)$$

where $\tau(\alpha(t, \omega), \omega) = t$. This transformation can further simplify the coefficients of a linear stochastic differential equation (5.3.5). For example, one of the functions $b_1(t)$ or $b_0(t)$ of a diffusion coefficient can be reduced to one. Transformations of the type (5.3.7) were used in [3, 4, 11, 12, 22] for defining a fiber preserving admitted Lie group of stochastic differential equations.

In [30] it was proven that the generalization of (5.3.7)[8]

$$\tau(t) = \int_0^t \eta^2(s, X(s))\, ds, \quad \tilde{B}_t = \int_0^{\alpha(t)} \eta(s, X(s))\, dB_s, \qquad (5.3.8)$$

---

[7]Details and references can be found therein.

[8]Considering $h = h(t, x, b)$ this generalization can be extended including the Brownian motion $B_t$ into the transformation.

also gives a transformation of the Brownian motion $B_t$ into the Brownian motion $\tilde{B}_t$. Recall that any first-order (deterministic) ordinary differential equation can be mapped into the simplest equation $y' = 0$ by a suitable change of the dependent and independent variables. By virtue of its generality this transformation is not applicable to the linearization problem. Similar to first-order (deterministic) ordinary differential equations, one can find a transformation (5.3.7), which maps a stochastic ordinary differential equation to the simplest equation

$$dY_t = d\tilde{B}_t.$$

Among other weak transformations one also can mention an application of the Girsanov theorems.

## 5.3.2 Strong Linearization of First-Order SODE

This section gives a complete[9] study of the linearization problem by using (5.3.4).

Let be given a function $y = \varphi(t, x)$. Assume that $\varphi_x(t, x) \neq 0$, then by virtue of the inverse function theorem there is a local inverse function $x = \psi(t, y)$ such that $y \equiv \varphi(t, \psi(t, y))$ and $x \equiv \psi(t, \varphi(t, x))$.

Applying the Itô formula to a solution of (5.3.2) with the function $\varphi(t, x)$, one obtains that the stochastic process $Y(t, \omega) = \varphi(t, X(t, \omega))$ satisfies the equation

$$dY_t = \bar{f}(t, Y_t) dt + \bar{g}(t, Y_t) dB_t, \qquad (5.3.9)$$

where

$$\bar{f}(t, y) = \left( \varphi_t + f\varphi_x + \frac{1}{2}g^2\varphi_{xx} \right)(t, \psi(t, y)), \quad \bar{g}(t, y) = (g\varphi_x)(t, \psi(t, y)).$$

The linearization problem is to find a substitution $y = \varphi(t, x)$ such that (5.3.2)

$$dX_t = f(t, X_t) dt + g(t, X_t) dB_t, \qquad (5.3.10)$$

is reduced to a linear stochastic differential equation, i.e., (5.3.9) is the linear stochastic differential equation

$$dY_t = (\alpha_1(t)Y_t + \alpha_0(t)) dt + (\beta_1(t)Y_t + \beta_0(t)) dB_t. \qquad (5.3.11)$$

A stochastic differential equation (5.3.10) which has this property is called a reducible equation [25].

The problem of finding reducible stochastic differential equations using the Itô formula can be solved if one can find a function $\varphi(t, x)$ which satisfies the conditions

$$\varphi_t + f\varphi_x + \frac{1}{2}g^2\varphi_{xx} = \alpha_1\varphi + \alpha_0, \quad g\varphi_x = \beta_1\varphi + \beta_0. \qquad (5.3.12)$$

---

[9]In [5, 13, 25] only particular cases were studied. In [33] one of sufficient conditions is missing. The present section complete this niche.

Notice that $\beta_1^2 + \beta_0^2 \neq 0$.

From these equations one finds the derivatives

$$\varphi_x = (\beta_0 + \beta_1 \varphi)/g, \tag{5.3.13}$$

$$\varphi_t = (\varphi \beta_1 + \beta_0)\left(\frac{1}{2}(g_x - \beta_1) - g^{-1}f\right) + (\alpha_1 \varphi + \alpha_0). \tag{5.3.14}$$

Considering the mixed derivatives $(\varphi_x)_t = (\varphi_t)_x$, one obtains

$$\varphi(\beta_{1t} - \beta_1 N) + \beta_{0t} - \beta_0(N + \alpha_1) + \alpha_0 \beta_1 = 0, \tag{5.3.15}$$

where

$$N = g^{-1}\left(g_t + f g_x + \frac{g^2}{2}g_{xx} - g f_x\right). \tag{5.3.16}$$

**Remark 5.3.1** For the equation

$$p_t + f p_x + \frac{1}{2}g^2 p_{xx} = 0 \tag{5.3.17}$$

one has

$$H = N,$$

where $H$ is defined by formula (5.2.11).

Assuming that the coefficient of $\varphi$ in (5.3.15) is equal to zero, one finds

$$\beta_{1t} = \beta_1 N, \qquad \beta_{0t} = -\alpha_0 \beta_1 + \alpha_1 \beta_0 + \beta_0 N. \tag{5.3.18}$$

Since $\varphi_x \neq 0$, one has $\beta_1^2 + \beta_0^2 \neq 0$, that leads to

$$N_x = 0.$$

Notice also that excluding $N$ from (5.3.18), the coefficients $\alpha_0(t)$, $\alpha_1(t)$, $\beta_0(t)$ and $\beta_1(t)$ has to satisfy the equation

$$\beta_{1t}\beta_0 - \beta_1(\beta_{0t} + \alpha_0 \beta_1 - \alpha_1 \beta_0) = 0. \tag{5.3.19}$$

Stochastic differential equations satisfying the condition $N_x = 0$ were studied in [25]. Many examples of reducible equations of such type of equations were given in [5, 25]. The authors of [25] considered $\beta_1 = 0$, $\alpha_1 = 0$. Moreover without loss of generality one can also assume that $\alpha_0 = 0$: for satisfying (5.3.18) one can choose $\beta_1 = 0$, $\alpha_1 = 0$ and $\alpha_0 = 0$. Notice that by virtue of $N_x = 0$ the equation (5.3.17) satisfies the criterion (5.2.10): the equation (5.3.17) is equivalent to the heat equation

$$u_t = u_{xx}.$$

**Theorem 5.3.1** *Let the coefficients of a stochastic differential equation*

$$dX_t = f(t, X_t)\,dt + g(t, X_t)\,dB_t \tag{5.3.20}$$

satisfy the condition $N_x = 0$. Then the backward Kolmogorov equation corresponding to (5.3.20) is equivalent to the heat equation, and (5.3.20) is reducible to the linear stochastic differential equation

$$dY_t = e^J \, dB_t,$$

where

$$J(t) = \int N \, dt, \tag{5.3.21}$$

and the transition function $y = \varphi(t, x)$ is found by integrating the compatible system of partial differential equations

$$\varphi_t = e^J \, (g_x/2 - f/g), \quad \varphi_x = e^J/g.$$

Let us consider the case which was not studied in [25]: assume that the coefficient of $\varphi$ in (5.3.15) is not equal to zero

$$J_0 = \beta_{1t} - \beta_1 N \neq 0,$$

which also means that in this case $\beta_1 \neq 0$. From (5.3.15) one can define the function $\varphi$

$$\varphi = -\frac{\beta_{0t} - \beta_0(N + \alpha_1) + \alpha_0\beta_1}{(\beta_{1t} - \beta_1 N)}.$$

Substituting $\varphi$ into the right-hand side of (5.3.13), one has

$$\varphi_x = \frac{\beta_{1t}\beta_0 - \beta_{0t}\beta_1 - \alpha_0\beta_1^2 + \alpha_1\beta_0\beta_1}{g(\beta_{1t} - \beta_1 N)}. \tag{5.3.22}$$

Since it is assumed that $\varphi_x \neq 0$, the numerator of the last equation satisfies the condition

$$\beta_{1t}\beta_0 - \beta_1(\beta_{0t} + \alpha_0\beta_1 - \alpha_1\beta_0) \neq 0. \tag{5.3.23}$$

Substituting $\varphi$ into (5.3.22), one obtains

$$\beta_{1t} - \beta_1 N = N_x g. \tag{5.3.24}$$

By virtue of the assumption $J_0 \neq 0$, this means that $N_x \neq 0$.

Differentiating (5.3.24) with respect to $x$, one gets

$$(N_x g)_x + \beta_1 N_x = 0$$

Because of the assumption $N_x \neq 0$, one can find $\beta_1$:

$$\beta_1 = -\frac{(N_x g)_x}{N_x}.$$

Since the function $\beta_1$ does not depend on $x$, differentiating the last equation with respect to $x$, one obtains

$$\frac{\partial}{\partial x}\left(\frac{(N_x g)_x}{N_x}\right) = 0$$

or

$$N_{xxx} = (N_x g)^{-1}(N_{xx}^2 g - N_x(g_{xx}N_x + g_x N_{xx})).$$ (5.3.25)

Substituting $\beta_1$ into (5.3.24), one has the condition

$$\frac{\partial}{\partial t}\left(\frac{(N_x g)_x}{N_x}\right) - N\frac{(N_x g)_x}{N_x} + N_x g = 0$$

or

$$N_{txx} = (N_x g)^{-1}(N_{xx}(N_{tx}g + N_x g N - g_t N_x) - g_{tx}N_x^2 + g_x N_x^2 N - N_x^3 g).$$ (5.3.26)

Equation (5.3.14) becomes

$$\beta_0'' - (\mu_1 - \tfrac{1}{2}\beta_1^2 + 2\alpha_1)\beta_0' - (\mu_2 - \alpha_1\mu_1 + \alpha_1' - \alpha_1^2 + \tfrac{1}{2}\beta_1(\beta_1' + \alpha_1\beta_1))\beta_0$$
$$+ \alpha_0(2\beta_1' + \beta_1(\tfrac{1}{2}\beta_1^2 - \mu_1 - \alpha_1)) + \beta_1\alpha_0' = 0$$ (5.3.27)

where

$$\mu_1 = \frac{g_t + g_x f + g N}{g} + \frac{N_{tx} + N_{xx} f}{N_x} - \frac{1}{2}g_x\left(g_x + g\frac{N_{xx}}{N_x}\right),$$

$$\mu_2 = N_t + N_x\left(f - \frac{1}{2}g_x g\right) + N^2 - N\mu_1.$$

By virtue of the conditions for the function $N$, one obtains $\mu_{1x} = 0$, $\mu_{2x} = 0$.
If a stochastic differential equation with $N_x \neq 0$ is linearizable, then without loss of generality one can choose, for example, $\alpha_1 = \beta_0 = 0$. Hence, one gets

$$\alpha_0 = e^{J(t)}, \qquad \varphi = -\frac{\beta_1 e^{J(t)}}{\beta_{1t} - \beta_1 N},$$

where

$$J(t) = \int q\, dt, \qquad q = \mu_1 - 2\frac{\beta_1'}{\beta_1}.$$ (5.3.28)

Notice also that in this case the backward Kolmogorov equation is not equivalent to the classical heat equation. In fact, substituting (5.3.25) into (5.2.10), one finds

$$N_{tx} = -(2N_x g)^{-1}\left(N_{xx}^2 g^3 + N_{xx}N_x g(2f + g_x g) + 2N_x^2(g_t + g_x f - 3g N)\right).$$ (5.3.29)

Then (5.3.26) and (5.3.29) give the contradiction $N_x = 0$.

**Theorem 5.3.2** *Assume that $N_x \neq 0$, and the function $N$ satisfies (5.3.25), (5.3.26). Then the change*

$$y = -\frac{\beta_1 e^J}{\beta_{1t} - \beta_1 N}$$

*transforms a solution of the equation*

$$dX_t = f(t, X_t)\, dt + g(t, X_t)\, dB_t,$$

*into a solution of the linear stochastic differential equation*

$$dY_t = e^J\, dt + \beta_1 Y_t\, dB_t.$$

*Here* $\beta_1 = -N_x^{-1}(N_x g)_x$, $J(t) = \int q\, dt$, *and*

$$q = \beta_1(g_x - \beta_1)/2 + (g_t - f\beta_1)/g + N_{tx}/N_x - 2N_x g/\beta_1 - N.$$

**Remark 5.3.2** For a stochastic differential equation with fractional Brownian motion (fBm) $B^h$ with the Hurst parameter $h \in (0, 1)$:

$$dX = f(t, X)\, dt + g(t, X)\, dB^h$$

the linearization conditions are similar to the function

$$N = g^{-1}\left(g_t + fg_x + ht^{2h-1}g_{xx}g^2 - gf_x\right).$$

The criteria for the linearization of a stochastic differential equation with fBm were obtained in [33].[10]

**Remark 5.3.3** Considering stochastic ordinary differential equation (5.3.20) as the system of stochastic differential equations

$$dX_t = f(t, X_t)\, dt + g(t, X_t)\, dB_t,$$
$$dZ = dB_t,$$

one can include Brownian motion in the linearizing transformation

$$Y = \varphi(t, X, Z).$$

In this case it is required for the function $\varphi(t, x, z)$ to satisfy the deterministic system of equations

$$\varphi_t + f\varphi_x + \tfrac{1}{2}g^2\varphi_{xx} + g\varphi_{xz} + \tfrac{1}{2}\varphi_{zz} = \alpha_1\varphi + \alpha_0, \qquad (5.3.30)$$
$$g\varphi_x + \varphi_z = \beta_1\varphi + \beta_0.$$

These equations are obtained similarly to (5.3.12) (using the Itô formula for two stochastic processes). Analysis of the last system of equations (5.3.30) shows that the extension of the transformation does not extend the set of linearizable stochastic differential equations (5.3.20).

---

[10]Unfortunately one of sufficient conditions is missing in [33].

## 5.4 Strongly Linearizable Second-Order SODE

Many mathematical models in chemistry and physics are based on the second-order equation

$$\ddot{X} = f(t, X, \dot{X}) + g(t, X, \dot{X}) \dot{B}_t, \tag{5.4.1}$$

where $\dot{B}_t$ is a white noise. For example, the second-order Langevin equation

$$\ddot{X} = f(t, X, \dot{X}) + \sigma \dot{B}_t \tag{5.4.2}$$

describes the motion of a particle in a noise-perturbed force field. In particular, for the case of the harmonic oscillator,

$$f(t, x, \dot{x}) = -v^2 x - \beta \dot{x},$$

and (5.4.2) is a linear second-order stochastic differential equation.The Langevin equation is also used in lasers, chemical kinetics, population dynamics.

While solving problems connected with nonlinear ordinary differential equations (deterministic) it is often expedient to simplify the equations by a suitable change of variables. The simplest form of a second order ordinary differential equation $\ddot{x} = f(t, x, \dot{x})$ is a linear form. S. Lie [17][11] showed that a second-order ordinary differential equation $\ddot{x} = f(t, x, \dot{x})$ is linearizable by a change of the independent and dependent variables if, and only if, it is a polynomial of third degree with respect to the first-order derivative:

$$\ddot{x} + a\dot{x}^3 + b\dot{x}^2 + c\dot{x} + d = 0,$$

where the coefficients $a(t, x)$, $b(t, x)$, $c(t, x)$ and $d(t, x)$ satisfy the conditions

$$L_1 = 3a_{tt} - 2b_{tx} + c_{xx} - 3a_t c + 3a_x d + 2b_t b - 3c_t a - c_x b + 6d_x a = 0,$$
$$L_2 = b_{tt} - 2c_{tx} + 3d_{xx} - 6a_t d + b_t c + 3b_x d - 2c_x c - 3d_t a + 3d_x b = 0.$$

### 5.4.1 Linearization Conditions

A scalar second-order stochastic ordinary differential equation (5.4.1) in differential form is written as the following equation

$$d\dot{X} = f(t, X, \dot{X}) dt + g(t, X, \dot{X}) dB_t. \tag{5.4.3}$$

Here the first term in the right-hand side corresponds to a Riemann integral, and the second term corresponds to an Itô integral. The problem is to find a change of the dependent variable $y = \varphi(t, x)$ such that the transformed equation is a linear stochastic differential equation

$$d\dot{Y} = \left(a_1(t)Y + b_1(t)\dot{Y} + c_1(t)\right) dt + \left(a_2(t)Y + b_2(t)\dot{Y} + c_2(t)\right) dB_t. \tag{5.4.4}$$

Here we just formulate the final result.

---

[11]Details and references one can find in [21].

**Theorem 5.4.1** *A second-order stochastic differential equation*

$$\ddot{X} = f(t, X, \dot{X}) + g(t, X, \dot{X})\dot{B}_t$$

*is linearizable by a change of the dependent variables if and only if*

$$f = -(p^2 b + pc + d), \quad g = pb_2 + \psi,$$

*where the functions* $b(t, x)$, $c(t, x)$, $d(t, x)$, $g(t, x)$, $\psi(t, x)$ *and* $b_2(t)$ *satisfy the conditions*:

$$c_x = 2b_t, \quad b_{tt} + d_{xx} + b_t c + b_x d + d_x b = 0,$$
$$(\psi_x + b\psi)_x - b_t b_2 = 0.$$

**Theorem 5.4.2** *If a second-order stochastic differential equation*

$$\ddot{X} = f(t, X, \dot{X}) + g(t, X, \dot{X})\dot{B}_t,$$

*is linearizable, then the deterministic second-order ordinary differential equation*

$$\ddot{x} = f(t, x, \dot{x})$$

*is also linearizable.*

## 5.5 Transformations of Autonomous Stochastic First-Order ODEs

The first step in the study of admitted Lie group of transformations consists of a discussion of defining admitted transformations. Recall that according to Itô's formula: if $\varphi(t, x)$ is a continuous function with continuous derivatives $\varphi_t$, $\varphi_x$, $\varphi_{xx}$, and a stochastic process $\{X_t\}_{t \geq 0}$ is a solution of the stochastic differential equation

$$dX = f(t, X)\,dt + g(t, X)\,dB_t, \tag{5.5.1}$$

then the process $\varphi(t, X)$ has the stochastic differential

$$d\varphi(t, X) = \left(\varphi_t + f\varphi_x + \frac{g^2}{2}\varphi_{xx}\right)(t, X)\,dt + (g\varphi_x)(t, X)\,dB_t.$$

Here $f = f(t, x)$ and $g = g(t, x)$ are measurable deterministic functions.

### 5.5.1 Admitted Transformations

Let $\eta(t, x) \neq 0$ be a continuous in $t$ function defining a random time change with time change rate $\eta^2(t, X(t, \omega))$:

$$\beta_X(t) = \beta(t, X) = \int_0^t \eta^2(s, X(s))\,ds, \quad t \geq 0 \tag{5.5.2}$$

with the inverse

$$\alpha_X(t) = \alpha(t, X) = \inf_{s \geq 0}\{s : \beta_X(s) > t\}, \tag{5.5.3}$$

$$\beta_X(\alpha_X(t)) = t = \alpha_X(\beta_X(t)), \quad t \geq 0. \tag{5.5.4}$$

Using the function $\varphi(t, x)$ and the random time change, one can define a transformation $\bar{X}(\bar{t})$ of the stochastic process $X(t)$ by

$$\bar{X}(\bar{t}) = \varphi\big(\alpha(\bar{t}, X), X(\alpha(\bar{t}, X))\big). \tag{5.5.5}$$

Setting $\psi(t) = \varphi(t, X(t))$, it can be shown [29, 30] that for almost all $\omega$, there is the relation

$$\bar{X}(\beta_X(t)) = \psi(t). \tag{5.5.6}$$

Due to the Itô formula one has

$$\psi(t) = \psi(0) + \int_0^t \left(\varphi_t + f\varphi_x + \frac{g^2}{2}\varphi_x\right)(s, X(s))\, ds$$

$$+ \int_0^t g\varphi_x(s, X(s))\, dB(s). \tag{5.5.7}$$

Because $X(t)$ is a solution of (5.5.1) and $\varphi_x(t, x)$ is a continuous function, the process $\varphi_x(t, X(t))g(t, X(t))$ is a continuous process and $g\varphi_x$ is a nonanticipating functional. According to the time change formula for Itô integrals [20], a nonanticipating functional $Y$ with

$$\mathscr{P}\left(\int_0^t Y^2\, ds + \int_0^t \eta^2\, ds < \infty, \, t \geq 0\right) = 1$$

satisfies the formula

$$\int_0^{\alpha_X(t)} Y(s)\, dB(s) = \int_0^t Y(\alpha_X(s)) \frac{1}{\eta(\alpha_X(s), X(\alpha_X(s)))}\, d\bar{B}(s). \tag{5.5.8}$$

Correspondingly, the last term of (5.5.7) changes to

$$\int_0^{\beta_X(t)} \big(\eta^{-1}g\varphi_x\big)(\alpha_X(s), X(\alpha_X(s)))\, d\bar{B}(s).$$

Since $\beta_X(t) = \int_0^t \eta^2(s, X(s))\, ds$ and $\beta(\alpha_X(\bar{t}, X) = \bar{t}$, then

$$\eta^2\big(\alpha_X, X(\alpha_X)\big)\alpha_X'(\bar{t}) = 1,$$

and hence (5.5.7) becomes

$$\psi(t) = \psi(0) + \int_0^{\beta_X(t)} \left[ \eta^{-2} \left( \varphi_t + f \varphi_x + \tfrac{g^2}{2} \varphi_{xx} \right) \right] (\alpha_X(s), X(\alpha_X(s))) \, ds$$

$$+ \int_0^{\beta_X(t)} \left( \eta^{-1} g \varphi_x \right) (\alpha_X(s), X(\alpha_X(s))) \, d\bar{B}(s). \tag{5.5.9}$$

On the other hand, requiring that $\bar{X}(\bar{t})$ is a weak solution of (5.5.1), one obtains

$$\bar{X}(\bar{t}) = \bar{X}(0) + \int_0^{\bar{t}} f(s, \bar{X}(s)) \, ds + \int_0^{\bar{t}} g(s, \bar{X}(s)) \, d\bar{B}(s).$$

Substituting[12] $\bar{t} = \beta_X(t)$ into this equation, one gets

$$\bar{X}(\beta_X(t)) = \bar{X}(0) + \int_0^{\beta_X(t)} f(s, \bar{X}(s)) \, ds + \int_0^{\beta_X(t)} g(s, \bar{X}(s)) \, d\bar{B}(s).$$

$$\tag{5.5.10}$$

Equations (5.5.9) and (5.5.10) will certainly be equal if the integrands of the two Riemann integrals as well those of the Itô integrals coincide. Comparing the Riemann and Itô integrands, respectively, one comes to the equalities

$$\left( \varphi_t + f \varphi_x + \frac{g^2}{2} \varphi_{xx} \right) (\alpha_X(t), X(\alpha_X(t))) = f(t, \bar{X}(t)) \eta^2 (\alpha_X(t), X(\alpha_X(t))),$$

$$g \varphi_x (\alpha_X(t), X(\alpha_X(t))) = g(t, \bar{X}(t)) \eta(\alpha_X(t), X(\alpha_X(t))). \tag{5.5.11}$$

Replacing $t$ by $\bar{t} = \beta_X(t)$ and using (5.5.4), these two equations become

$$\left( \varphi_t + f \varphi_x + \frac{g^2}{2} \varphi_{xx} \right) (t, X(t)) = f(\beta_X(t), \bar{X}(\beta_X(t))) \eta^2(t, X(t)),$$

$$g \varphi_x (t, X(t)) = g(\beta_X(t), \bar{X}(\beta_X(t))) \eta(t, X(t)). \tag{5.5.12}$$

Using (5.5.6), this pair of equations can be rewritten as

$$\left( \varphi_t + f \varphi_x + \tfrac{g^2}{2} \varphi_{xx} \right) (t, X(t)) = f(\beta_X(t), \varphi(t, X(t))) \eta^2(t, X(t)),$$

$$(g \varphi_x)(t, X(t)) = g(\beta_X(t), \varphi(t, X(t))) \eta(t, X(t)). \tag{5.5.13}$$

Notice that by virtue of the presence of $\beta_X(t)$ in (5.5.13), these equations are still functional equations. If the functions $f$ and $g$ do not depend on time $t$ or $\eta_x = 0$, then the problem for (5.5.13) to be functional is overcome.

From here onwards it is assumed that (5.5.1) is autonomous:

$$dX_t = f(X_t) \, dt + g(X_t) \, dB_t. \tag{5.5.14}$$

---

[12] These considerations are similar to the constructions applied in [29, 30]. It should be noted that there is no change of variables in the integrands as the authors of [8] misleadingly state.

The assumption $f = f(x)$ and $g = g(x)$ allows us to consider (5.5.13) as the deterministic equations for the functions $\varphi_x(t, x)$ and $\eta(t, x)$:

$$\varphi_t(t, x) + f(x)\varphi_x(t, x) + \frac{g^2(x)}{2}\varphi_{xx}(t, x) = f(\varphi(t, x))\eta^2(t, x), \quad (5.5.15)$$
$$g(x)\varphi_x(t, x) = g(\varphi(t, x))\eta(t, x).$$

Considering the second equation of (5.5.15) as an equation defining the function

$$\eta(t, x) = \frac{g(x)\varphi_x(t, x)}{g(\varphi(t, x))}, \quad (5.5.16)$$

the first equation becomes the parabolic nonlinear equation

$$\varphi_t(t, x) + f(x)\varphi_x(t, x) + \frac{g^2(x)}{2}\varphi_{xx}(t, x)$$
$$= f(\varphi(t, x))\left(\frac{g(x)\varphi_x(t, x)}{g(\varphi(t, x))}\right)^2. \quad (5.5.17)$$

Thus, if the function $\varphi(t, x)$ is a solution of (5.5.17), and $X(t)$ is a solution of (5.5.1), then $\varphi(t, X(t))$ is also a solution of (5.5.1). In the case that

$$\eta(t, x) = \frac{g(x)\varphi_x(t, x)}{g(\varphi(t, x))} = 1 \quad (5.5.18)$$

the solution $\varphi(t, X(t))$ is a strong solution. In this case we call the transformation $y = \varphi(t, x)$ a strong admitted transformation. Otherwise the transformation is called a weak admitted transformation.

Let us study some examples of stochastic ordinary differential equations.

### 5.5.1.1 Geometric Brownian Motion

As an example, consider the autonomous stochastic ordinary differential equation

$$dX(t) = \mu X(t)dt + \sigma X(t)dB(t), \quad (5.5.19)$$

where $\mu > 0$ and $\sigma > 0$ are constant. The solution of (5.5.19) with the initial condition $X(0) = X_0$ is called geometric Brownian motion.

Since the solution of (5.5.17) and (5.5.18) is trivial: $\varphi = kx$, where $k$ is constant, there are no nontrivial strong admitted transformations.

For obtaining weak admitted transformations one has to solve (5.5.17) which is

$$\varphi_t + \mu x\varphi_x + \frac{\sigma^2 x^2}{2}\varphi_{xx} = \mu \frac{x^2\varphi_x^2}{\varphi}. \quad (5.5.20)$$

Notice that if $\mu = \sigma^2/2$, then using the substitution $u = \varphi^2$, the nonlinear equation (5.5.20) is reduced to the linear backward Kolmogorov equation (5.2.9):

$$u_t + \mu x u_x + \frac{\sigma^2 x^2}{2}u_{xx} = 0.$$

For this equation the function $H$ defined by formula (5.2.11) vanishes: $H = 0$. This means that according to the criteria (5.2.10) for $\mu = \sigma^2/2$ the nonlinear equation

(5.5.20) is reduced to the classical heat equation. Below we consider the general case of $\mu$ and $\sigma$.

A simple check shows that this equation, considered as a deterministic equation, admits the Lie group with the generators

$$X_1 = x\partial_x, \quad X_2 = \varphi\partial_\varphi, \quad X_3 = \partial_t.$$

These generators compose an Abelian Lie algebra.[13] Invariant solutions constructed on the basis of this Lie algebra are exhausted by two classes. One class of invariant solutions is based on the generator

$$X_1 - kX_2 = x\partial_x - k\varphi\partial_\varphi$$

while the second class is based on the generator

$$2X_3 - \lambda\sigma^2 X_1 + 2kX_2 = 2\partial_t - \lambda\sigma^2 x\partial_x + 2k\varphi\partial_\varphi,$$

where $k$ and $\lambda$ are arbitrary[14] constants.

The first class of invariant solutions has the representation

$$\varphi = x^k v(t).$$

Substituting this representation into (5.5.20), one gets

$$v' - k(k-1)\left(\mu - \frac{\sigma^2}{2}\right)v = 0. \tag{5.5.21}$$

Hence, the transformation is

$$\varphi = Cx^k \exp\left(tk(k-1)\left(\mu - \frac{\sigma^2}{2}\right)\right). \tag{5.5.22}$$

The second class has the representation

$$\varphi = e^{kt}v(z), \quad z = xe^{\lambda\sigma^2 t/2}.$$

Substituting this representation into (5.5.20), one gets

$$\sigma^2 z^2 vv'' + z(\lambda\sigma^2 + 2\mu)vv' - 2\mu z^2 v'^2 + 2kv^2 = 0.$$

This equation is linearizable. Introducing $y = \ln(z)$, similar to the linear Euler equation it can be reduced to the equation

$$vv'' + k_1 vv' - (\gamma + 1)v'^2 + k_3 v^2 = 0, \tag{5.5.23}$$

where the constants are

$$\gamma = 2\frac{\mu}{\sigma^2} - 1, \quad k_1 = \gamma + \lambda, \quad k_3 = 2\frac{k}{\sigma^2}.$$

This equation can be mapped into the free particle equation $u''(\tau) = 0$ by the change of variables

$$u = h(y)v^{-\gamma}, \quad \tau = q(y),$$

---

[13] The admitted Lie algebra of (5.5.20) is infinite dimensional.

[14] Here signs and scaling these constants are chosen for further convenience.

where

$$q' = h^2 e^{-k_1 y}, \quad h' = \psi h, \quad \psi' = \psi^2 - k_1 \psi - k_3 \gamma. \tag{5.5.24}$$

Thus, the general solution of (5.5.23) is

$$h(y)v^{-\gamma} = \tilde{c}_1 q(y) + \tilde{c}_0.$$

Solutions of (5.5.24) corresponding to $\varphi = v(z)$ $(k = 0)$ depend on $k_1$. If $k_1 = 0$, then

$$\psi = -\frac{1}{y}, \quad h = \frac{1}{y}, \quad q = -\frac{1}{y},$$

$$\varphi = (c_1 y + c_0)^{-1/\gamma}, \quad y = \ln(x) - t\gamma\sigma^2/2. \tag{5.5.25}$$

If $k_1 \neq 0$, then

$$\psi = \frac{k_1}{1 + e^{k_1 y}}, \quad h = \frac{e^{k_1 y}}{1 + e^{k_1 y}}, \quad q = \frac{1}{k_1} \frac{e^{k_1 y}}{1 + e^{k_1 y}}, \quad y = \ln(x) + \lambda t\sigma^2/2,$$

$$\varphi = \left( c_1 + c_o x^{-k_1} e^{-k_1 \lambda t\sigma^2/2} \right)^{-1/\gamma}. \tag{5.5.26}$$

## 5.5.2 Autonomous Systems of Stochastic First-Order ODEs

In this section the approach developed above is applied to a system of autonomous stochastic first-order ordinary differential equations. For simplicity, we illustrate this with a system of two equations

$$\begin{aligned} dX_1 &= f_1(X_1, X_2)\, dt + g_1(X_1, X_2)\, dB_t, \\ dX_2 &= f_2(X_1, X_2)\, dt + g_2(X_1, X_2)\, dB_t. \end{aligned} \tag{5.5.27}$$

Let us make a transformation of the dependent variables,

$$y_1 = \varphi_1(t, x_1, x_2), \quad y_2 = \varphi_2(t, x_1, x_2). \tag{5.5.28}$$

Comparison of integrands leads to the equations

$$\eta^{-2}\left( \varphi_{1t} + f_1\varphi_{1x_1} + f_2\varphi_{1x_2} + \frac{g_1^2}{2}\varphi_{1x_1x_1} + g_1 g_2 \varphi_{1x_1x_2} + \frac{g_2^2}{2}\varphi_{1x_2x_2} \right) = f_1(\varphi_1, \varphi_2),$$

$$\eta^{-2}\left( \varphi_{2t} + f_1\varphi_{2x_1} + f_2\varphi_{2x_2} + \frac{g_1^2}{2}\varphi_{2x_1x_1} + g_1 g_2 \varphi_{2x_1x_2} + \frac{g_2^2}{2}\varphi_{2x_2x_2} \right) = f_2(\varphi_1, \varphi_2),$$

$$\eta^{-1}(g_1\varphi_{1x_1} + g_2\varphi_{1x_2}) = g_1(\varphi_1, \varphi_2),$$

$$\eta^{-1}(g_1\varphi_{2x_1} + g_2\varphi_{2x_2}) = g_2(\varphi_1, \varphi_2). \tag{5.5.29}$$

Here on the left hand sides, $\eta = \eta(t, x_1, x_2)$, $f_i = f_i(t, x_1, x_2)$, and $g_i = g_i(t, x_1, x_2)$, $(i = 1, 2)$.

Next two applications are considered: SODE of order grater than one and deterministic change of Brownian motion.

### 5.5.2.1 Second-Order SODE

Notice that a second-order stochastic ordinary differential equation

$$dX = f(X, \dot{X}) dt + g(X, \dot{X}) dB_t, \tag{5.5.30}$$

can be rewritten as the system of first-order stochastic ordinary differential equations

$$\begin{aligned} dX_1 &= X_2\, dt, \\ dX_2 &= f(X_1, X_2)\, dt + g(X_1, X_2)\, dB_t, \end{aligned} \tag{5.5.31}$$

where usual for deterministic differential equations $X = X_1$, $\dot{X} = X_2$ are applied. Equations (5.5.29) become

$$\eta^{-2}(\varphi_{1t} + x_2\varphi_{1x_1}) = \varphi_2,$$
$$\eta^{-2}\left(\varphi_{2t} + x_2\varphi_{2x_1} + f\varphi_{2x_2} + \tfrac{g^2}{2}\varphi_{2x_2x_2}\right) = f(\varphi_1, \varphi_2), \tag{5.5.32}$$
$$\varphi_{1x_2} = 0, \quad \eta^{-1}g\varphi_{2x_2} = g(\varphi_1, \varphi_2).$$

We observe that the first equation is similar to the prolongation formula if one takes into account the change of time

$$\beta(t, X, \dot{X}) = \int_0^t \eta^2(s, X(s), \dot{X}(s))\, ds. \tag{5.5.33}$$

From the last equation of (5.5.32) one finds

$$\eta(t, x_1, x_2) = \frac{g(x_1, x_2)\varphi_{2x_2}(t, x_1, x_2)}{g\left(\varphi_1(t, x_1, x_2), \varphi_2(t, x_1, x_2)\right)}. \tag{5.5.34}$$

Substituting (5.5.34) into the remaining equations of (5.5.32), one obtains the overdetermined system of partial differential equations for the functions $\varphi_1(t, x_1, x_2)$ and $\varphi_2(t, x_1, x_2)$:

$$\varphi_{1x_2} = 0, \quad \varphi_{1t} + x_2\varphi_{1x_1} = \varphi_2\left(g\frac{\varphi_{2x_2}}{g(\varphi_1,\varphi_2)}\right)^2,$$
$$\varphi_{2t} + x_2\varphi_{2x_1} + f\varphi_{2x_2} + \tfrac{g^2}{2}\varphi_{2x_2x_2} = f(\varphi_1, \varphi_2)\left(g\frac{\varphi_{2x_2}}{g(\varphi_1,\varphi_2)}\right)^2. \tag{5.5.35}$$

For example, for the Ornstein–Uhlenbeck equation

$$d\dot{X} = -b\dot{X}\, dt + \sigma\, dB_t, \tag{5.5.36}$$

where $b > 0$ is the friction coefficient, and $\sigma \neq 0$ is the diffusion coefficient, (5.5.35) become

$$\varphi_{1x_2} = 0, \quad \varphi_{1t} + x_2\varphi_{1x_1} = \varphi_2\varphi_{2x_2}^2,$$
$$\varphi_{2t} + x_2\varphi_{2x_1} - bx_2\varphi_{2x_2} + \tfrac{\sigma^2}{2}\varphi_{2x_2x_2} = -b\varphi_2\varphi_{2x_2}^2. \tag{5.5.37}$$

### 5.5.2.2 Deterministic Change of Brownian Motion

Considering (5.5.14) as a system of equations,

$$dX_1 = f(X_1)dt + g(X_1)dB_t,$$
$$dX_2 = dB_t,$$

where we have set $X_1 = X$, one can include Brownian motion into the transformation (5.5.5):

$$y_1 = \varphi_1(t, x_1, x_2), \quad y_2 = \varphi_2(t, x_1, x_2). \tag{5.5.38}$$

In this case $f_1 = f_1(x_1)$, $g_1 = g_1(x_1)$, $f_2 = 0$ and $g_2 = 1$, and (5.5.27) become

$$\varphi_{1t} + f_1\varphi_{1x_1} + \tfrac{g_1^2}{2}\varphi_{1x_1x_1} + g_1\varphi_{1x_1x_2} + \tfrac{1}{2}\varphi_{1x_2x_2} = f_1(\varphi_1)(g_1\varphi_{2x_1} + \varphi_{2x_2})^2,$$
$$g_1\varphi_{1x_1} + \varphi_{1x_2} = g_1(\varphi_1)(g_1\varphi_{2x_1} + \varphi_{2x_2}),$$
$$\varphi_{2t} + f_1\varphi_{2x_1} + \tfrac{g_1^2}{2}\varphi_{2x_1x_1} + g_1\varphi_{2x_1x_2} + \tfrac{1}{2}\varphi_{2x_2x_2} = 0,$$
$$\eta = g_1\varphi_{2x_1} + \varphi_{2x_2}.$$

For example, for the geometric Brownian motion

$$\varphi_{1t} + \mu x_1\varphi_{1x_1} + \tfrac{\sigma^2 x_1^2}{2}\varphi_{1x_1x_1} + \sigma x_1\varphi_{1x_1x_2} + \tfrac{1}{2}\varphi_{1x_2x_2} = \mu\varphi_1(\sigma x_1\varphi_{2x_1} + \varphi_{2x_2})^2,$$
$$\sigma x_1\varphi_{1x_1} + \varphi_{1x_2} = \sigma\varphi_1(\sigma x_1\varphi_{2x_1} + \varphi_{2x_2}),$$
$$\varphi_{2t} + \mu x_1\varphi_{2x_1} + \tfrac{\sigma^2 x_1^2}{2}\varphi_{2x_1x_1} + \sigma x_1\varphi_{2x_1x_2} + \tfrac{1}{2}\varphi_{2x_2x_2} = 0,$$
$$\eta = \sigma x_1\varphi_{2x_1} + \varphi_{2x_2}.$$

One of solutions of these equations is

$$\varphi_1 = e^{\sigma(k_1x_2+k_2)}q(z), \quad \varphi_2 = k_1x_2 + k_2 + k_3z^{1-\gamma},$$

where $k_i$ $(i = 1, 2, 3)$ are arbitrary constants, $z = x_1e^{-\sigma x_2}$, and the function $q(z)$ is a solution of the linear second-order Euler-type equation

$$z^2q'' + \gamma zq' - \gamma k_1^2q = 0.$$

For this solution the function

$$\eta = k_1.$$

In particular, if $k_1 = 1$, $k_2 = k_3 = 0$, this transformation becomes

$$\varphi_1 = c_1x_1 + c_2x_1^{-\gamma}e^{\sigma(\gamma+1)x_2}, \quad \varphi_2 = x_2,$$

where $c_i$ $(i = 1, 2)$ are arbitrary constants.

## 5.6 Lie Group of Transformations for Stochastic Processes

### 5.6.1 Short Historical Review

While symmetry techniques have found a wide range of applications in the analysis of ordinary and partial differential equations, there have been only few and recent attempts to extend these techniques to stochastic differential equations. The main obstacle which one encounters here is the non-differentiability of stochastic processes, which makes it difficult to include change of time in the symmetry transformations. Another difficulty is how to match the time change in the Lie groups of transformations with the time change in the stochastic processes.

Symmetries of stochastic differential equations are usually considered for scalar equations or systems of stochastic ordinary differential equations

$$dX_i = f_i(t, X)\, dt + g_{ik}(t, X)\, dB_k \quad (i = 1, 2, \ldots, n),$$

where $B_k$ $(k = 1, 2, \ldots, r)$ are standard Brownian motions. Most approaches fall into two general groups as follows.

The first approach [1, 3, 4, 7–9, 11, 12, 19, 22, 23, 32] employs fiber-preserving transformations only,

$$\bar{x} = \varphi(t, x, a), \quad \bar{t} = H(t, a),$$

and thus avoids the problem of how to include the dependent variables in the time change. Here $x$ is the vector of dependent, that is spatial variables, $t$ is the independent variable, usually time, and $a$ is the group parameter.

Misawa [23] and Albeverio & Fei [1] considered $H(t, a) = t$. Gaeta & Quinter [12] and Gaeta [9] allowed time to be changed, but did not apply the time change to Brownian motion. Gaeta later [11] extended the approach developed in [9, 12] to include Brownian motion in the transformation. Mahomed & Wafo Soh [19] and Ünal [32] used an infinitesimal transformation for Brownian motion,

$$d\bar{B} = dB + \frac{1}{2}\varepsilon \left( \tau_t + f\tau_x + \frac{1}{2}g_{xx}^2 \tau_{xx} \right) dB,$$

where $\tau(t, x) = \frac{\partial H}{\partial a}(t, x, 0)$ is the coefficient of the infinitesimal generator of the Lie group. Fredericks & Mahomed [7, 8] tried to reconcile [19] and [32]. Melnick [22] and Alexandrova [3, 4] also include Brownian motion in the transformation of the dependent variables.

In general, the change of variables in stochastic differential equations differs from the change of variables in ordinary differential equations, as the Itô formula takes the place of the chain rule of differentiation. Exploiting the Itô formula and the requirement that a solution of a stochastic differential equation is mapped into a solution of the same equation, the determining equations of an admitted Lie group can be obtained. This approach has been applied to stochastic dynamical systems [1, 3, 4, 23], to the Fokker–Plank equation [11, 12, 15, 32, 34], to scalar second-order stochastic ordinary differential equations [19], and to the Hamiltonian–Stratonovich dynamical control system [34]. It has also been applied to stochastic partial differential equations [22].

The second approach [29, 30] includes the dependent variables in the transformation of time as well,

$$\bar{x} = \varphi(t, x, a), \quad \bar{t} = H(t, x, a).$$

In particular, the transformation of Brownian motion is defined through the transformation of the dependent and independent variables. Generalizing the change of time formula [24], it was proven in [29] that the transformed Brownian motion

$$\bar{B}(t) = \int\limits_0^t \eta\big(s, X(s), a\big) \, dB(s)$$

(where $\eta(t, x, a) \neq 0$) satisfies again the properties of Brownian motion. This transformation of Brownian motion is a logical generalization of the time change in the Itô integral to the case where the stochastic process is included in the change. Exploiting the Itô formula, this transformation of Brownian motion and the requirement that a solution of the stochastic differential equation is mapped into a solution of the same equation, and finally equating the Riemann and Itô integrands, the determining equations of an admitted Lie group were obtained. The definition of an admitted Lie group for stochastic differential equations was given using these determining equations.

It is worth to note that if $H = H(t, a)$, then these determining equations coincide with those obtained in [8].

In spite of its greater generality, the definition of an admitted Lie group for stochastic differential equations given in [29, 30] has some weaknesses, as we explain now. First, the relation of the function $\eta$ defined in [29, 30] by the formula

$$\eta^2(t, x, a) = H_t(t, x, a)$$

restricts the set of transformations substantially. Second, the determining equations defined in [29, 30] only give necessary conditions for the transformed function to be a solution of the original equations, as they are obtained by equating integrands. Compare this to deterministic equations, where the determining equations are obtained by differentiating the original equations with the transformed solution substituted into them, and hence give also sufficient conditions. Some of these difficulties will be overcome for autonomous equations in the next section.

In this section the approach [29, 30] is described. We construct a Lie group of transformations, involving both the dependent variables and the Brownian motion in the transformations. Mapping Brownian motions into Brownian motions allow obtaining a correct generalization of application of group analysis to stochastic differential equations.

### 5.6.2 Admitted Lie Group and Determining Equations

Let us consider a system of stochastic ordinary differential equations:

$$dX_i = f_i(t, X)dt + g_{ik}(t, X) \, dB_k \quad (i = 1, \ldots, N; \; k = 1, \ldots, M). \quad (5.6.1)$$

### 5.6.2.1  Lie Group of Transformations

Assume that the set of transformations

$$\bar{t} = H(t, x, a), \quad \bar{x}_i = \varphi_i(t, x, a) \tag{5.6.2}$$

composes a Lie group with the infinitesimal generator

$$h(t, x)\partial_t + \xi_i(t, x)\partial_{x_i}.$$

Since the initial point in the Riemann and Itô integrals is fixed ($t = 0$), then it has to be invariant of admitted transformations. This gives

$$h(0, x) = 0. \tag{5.6.3}$$

This requirement can be omitted if one allows changes of the initial point in the integrals.

According to Lie's theorem, the functions $H(t, x, a)$ and $\varphi(t, x, a)$ satisfy the Lie equations

$$\frac{\partial H}{\partial a} = h(H, \varphi), \quad \frac{\partial \varphi_i}{\partial a} = \xi_i(H, \varphi) \tag{5.6.4}$$

and the initial conditions for $a = 0$:

$$H(t, x, 0) = t, \quad \varphi_i(t, x, 0) = x_i. \tag{5.6.5}$$

Using the functions $\varphi_i(t, x, a)$, one can define a transformation $\bar{X}(\bar{t})$ of a stochastic process $X(t)$ by

$$\bar{X}(\bar{t}) = \varphi\big(\alpha_X(\bar{t}), X(\alpha_X(\bar{t})), a\big), \tag{5.6.6}$$

where the functions $\alpha_X(\bar{t})$ is as in formulae (5.5.3). This gives an action of Lie group (5.6.2) on the set of stochastic processes.

In contrast to deterministic differential equations one notices that the function $H(t, x, a)$ is not involved in the definition of the transformed stochastic process (5.6.6). By analogy with deterministic differential equations one can relate the function $\eta(t, x, a)$ with the Lie group. For deterministic differential equations the function $\eta^2$ plays the role of the total derivative of the function $H(t, x, a)$ with respect to $t$: $\eta^2 = H_{,t} + H_{,i}\dot{x}_i$. Since the stochastic process $X(t, \omega)$ is not differentiable, then the following relation can be considered instead,

$$H_{,t} = \eta^2. \tag{5.6.7}$$

This relation was used in [29, 30]. For deterministic differential equations $\dot{x}_i = f_i$, one may also choose the following relation for stochastic differential equations

$$H_{,t} + H_{,i} f_i = \eta^2. \tag{5.6.8}$$

Recall that for stochastic differential equations the Itô formula plays the role of the total derivative. Hence, one further relation between Lie group and the function $\eta(t, x, a)$ can be considered

$$H_{,t} + H_{,i} f_i + \frac{1}{2} g_{jk} g_{lk} H_{,jl} = \eta^2. \tag{5.6.9}$$

Since $H_{,t}(t, x, 0) = 1$ and $H_{,x_i}(t, x, 0) = 0$, then all these choices do not contradict positivity of the right hand sides.

Recall that a one-parameter Lie group of transformations is related to the generator

$$h(t, x)\partial_t + \xi(t, x)\partial_{x_i},$$

where

$$h(t, x) = \frac{\partial H}{\partial a}(t, x, 0), \quad \xi_i(t, x) = \frac{\partial \varphi_i}{\partial a}(t, x, 0).$$

In calculations of an admitted Lie group of transformations it is useful to introduce the function

$$\tau(t, x) = \frac{\partial \eta}{\partial a}(t, x, 0).$$

If one assumes any of the relations (5.6.7)–(5.6.9), then similar to deterministic differential equations, the functions $\tau(t, x)$ and $\xi_i(t, x)$ define a Lie group of transformations for stochastic processes. In fact, if the function $\tau(t, x)$ is given, then the function $h(t, x)$ is the unique solution of the Cauchy problems, respectively,

$$h_{,t} = 2\tau, \qquad h(0, x) = 0;$$

$$h_{,t} + h_{,i} f_i(t, x) = 2\tau, \qquad h(0, x) = 0; \qquad (5.6.10)$$

$$h_{,t} + h_{,i} f_i + \frac{1}{2} g_{jk} g_{lk} h_{,jl} = 2\tau, \qquad h(0, x) = 0.$$

Integrating the Lie equations

$$\frac{\partial H}{\partial a}(t, x, a) = h(H(t, x), \varphi(t, x)), \qquad \frac{\partial \varphi_i}{\partial a}(t, x, a) = \xi_i(H(t, x), \varphi(t, x))$$

with the initial conditions

$$H(t, x, 0) = t, \qquad \varphi(t, x, 0) = x$$

one then finds the functions $H(t, x, a)$ and $\varphi(t, x, a)$. Notice that each of these Cauchy problems have an unique solution.

### 5.6.2.2 Determining Equations

The problem which remains is to find sufficient conditions which guarantee that the transformed process (5.6.6) is again a solution of the system (5.6.1). In the following we will overcome the difficulties encountered in [29, 30] for autonomous equations of form (5.5.14).[15] Our approach is to use the admissibility condition for transformations (5.5.17) to define admitted Lie groups of transformations. In particular, the stochastic time change (5.5.2) need not be directly related with the time change (5.6.2) by the admitted Lie group.

---

[15]In [8] the authors also tried to correct [29, 30]. Their attempt led to the strong restriction: all possible admitted transformations are fiber preserving.

Assume that the deterministic functions $\varphi(t, x, a)$ and $H(t, x, a)$ compose a Lie group of transformations $G^1$ with the generator

$$Y = h(t, x)\partial_t + \xi(t, x)\partial_x.$$

Let $X(t)$ be a continuous and adapted stochastic process satisfying (5.6.1)

$$X_i(t) = X_i(0) + \int_0^t f_i(s, X(s))\, ds + \int_0^t g_{ik}(s, X(s))\, dB_k(s), \quad (5.6.11)$$

where the drift vector $f = (f_1, \ldots, f_N)$ and the diffusion matrix $g = (g_{ik})_{N \times M}$ are given functions, $B = (B_1, \ldots, B_M)$ is multi-Brownian motion, $\int_0^t f(s, X(s))\, dt$ is a Riemann integral and $\int_0^t g(s, X(s))\, dB(s)$ is an Itô integral.

Requiring that $\bar{X}(\bar{t})$ determined by (5.6.6) is a weak solution of (5.6.11), and applying the approach developed for constructing (5.5.13), one obtains the equations

$$\left( \varphi_{i,t} + \varphi_{i,j} f_j + \frac{1}{2} \varphi_{i,jl} g_{jk} g_{lk} \right)(t, X(t), a)$$

$$= f_i(\beta_X(t), \varphi(t, X(t), a))\eta^2(t, X(t), a), \quad (5.6.12)$$

$$\varphi_{i,j} g_{jk}(t, X(t), a) = g_{ik}(\beta_X(t), \varphi(t, X(t), a))\eta(t, X(t), a) \quad (5.6.13)$$

$$(i = 1, \ldots, N; \ k = 1, \ldots, M),$$

where the function $\beta_X(t)$ is as in formula (5.5.2).

**Definition 5.6.1** A Lie group of transformations (5.6.2) is called admitted by stochastic differential equations (5.6.11), if for any solution $X(t)$ of (5.6.11) the functions $\eta(t, x, a)$ and $\varphi(t, x, a)$ satisfy (5.6.12) and (5.6.13).

This definition of admitted Lie group for stochastic ordinary differential equations is similar to one of definitions of admitted Lie group for deterministic equations.

In analogy with deterministic differential equations one can obtain so-called determining equations by differentiating (5.6.12) and (5.6.13) with respect to the parameter $a$, and substituting $a = 0$:

$$\xi_{i,t}(t, X(t)) + \xi_{i,j} f_j(t, X(t)) + \tfrac{1}{2}\xi_{i,jl} g_{jk} g_{lk}(t, X(t))$$

$$- 2 f_{i,t}(t, X(t)) \int_0^t \tau(s, X(s))\, ds$$

$$- f_{i,j}\xi_j(t, X(t)) - 2 f_i \tau(t, X(t)) = 0, \quad (5.6.14)$$

$$\xi_{i,j} g_{jk}(t, X(t)) - 2 g_{ik,t}(t, X(t)) \int_0^t \tau(s, X(s))\, ds$$

$$- g_{ik}\tau(t, X(t)) - g_{ik,j}\xi_j(t, X(t)) = 0 \quad (5.6.15)$$

$$(i = 1, \ldots, N; \ k = 1, \ldots, M).$$

Here we used the relation

$$\frac{\partial \beta_X}{\partial a}\bigg|_{a=0} = 2 \int_0^t \frac{\partial \eta}{\partial a}\bigg|_{a=0} ds = 2 \int_0^t \tau(s, X(s))\, ds$$

which is obtained after differentiation $\beta_X(t, a) = \int_0^t \eta^2(s, X(s), a)\, ds$ with respect to the group parameter $a$ and substitution $a = 0$ into it.

Equations (5.6.14), (5.6.15) are integro-differential equations for the functions $\tau(t, x)$ and $\xi(t, x)$. These equations have to be satisfied for any solution $X(t, \omega)$ of the stochastic differential equation (5.6.11).

The determining equations (5.6.14), (5.6.15) only give necessary conditions for the transformed function to be a solution of the original equations, as they are obtained by equating integrands. Compare this to deterministic equations, where the determining equations are obtained by differentiating the original equations with the transformed solution substituted into them, and hence give also sufficient conditions. Some of these difficulties will be overcome for autonomous equations in the next section.

It is worth to note that if $H = H(t, a)$, then these determining equations coincide with the determining equations obtained in [8]. Coincidence of the determining equations also occurs in the case of autonomous stochastic ordinary differential equations which are considered in Sect. 5.7.

## 5.7 Admitted Lie Groups of Transformations of Autonomous SODEs

Autonomy of stochastic ordinary differential equations allows one to reduce the definition of an admitted Lie group of transformations (5.6.12) and (5.6.13) to solving a system of deterministic differential equations. For the sake of simplicity a one-dimensional ($N = 1$) autonomous stochastic ordinary differential equation with a single Brownian motion ($M = 1$) is considered in this section.

Assume that the deterministic functions $\varphi(t, x, a)$ and $H(t, x, a)$ compose a Lie group of transformations $G^1$ with the generator

$$Y = h(t, x)\partial_t + \xi(t, x)\partial_x.$$

Hence, the functions $\varphi(t, x, a)$ and $H(t, x, a)$ satisfy the Lie equations

$$\frac{d\varphi}{da}(t, x, a) = \xi\big(H(t, x, a), \varphi(t, x, a)\big),$$

$$\frac{dH}{da}(t, x, a) = h\big(H(t, x, a), \varphi(t, x, a)\big),$$

(5.7.1)

and the initial conditions

$$\varphi(t, x, 0) = x, \qquad H(t, x, 0) = t.$$

Assuming that $\varphi(t, x, a)$ is admitted, the function $\varphi(t, x, a)$ has to satisfy the deterministic equation (5.5.17):

$$S(t, x, a) \equiv \varphi_t(t, x, a) + f(x)\varphi_x(t, x, a) + \frac{g^2(x)}{2}\varphi_{xx}(t, x, a)$$

$$- f(\varphi(t, x, a))\left(\frac{g(x)\varphi_x(t, x, a)}{g(\varphi(t, x, a))}\right)^2 = 0. \qquad (5.7.2)$$

**Definition 5.7.1** A Lie group $G^1$ is admitted by the autonomous stochastic differential equation

$$dX = f(X)\,dt + g(X)\,dB_t \qquad (5.7.3)$$

if it satisfies (5.7.2).

This definition overcomes the weaknesses described in Sect. 5.6.

## 5.7.1 Determining Equations

For finding admitted Lie group one can directly solve (5.7.2). A solution $\varphi(t, x, a)$ of (5.7.2) determines the coefficient

$$\xi(t, x) = \frac{\partial \varphi}{\partial a}(t, x, 0).$$

If $\xi_t = 0$, then, by virtue of the uniqueness of the solution of the Cauchy problem for the Lie equations, one has that $\varphi = \varphi(x, a)$. Hence, $\varphi(x, a)$ is a solution of the second-order nonlinear ordinary differential equation

$$f(x)\varphi_x(x, a) + \frac{g^2(x)}{2}\varphi_{xx}(x, a) - f(\varphi(x, a))\left(\frac{g(x)\varphi_x(x, a)}{g(\varphi(x, a))}\right)^2 = 0. \qquad (5.7.4)$$

In this case the function could be chosen as the solution of the Cauchy problem

$$\frac{\partial H}{\partial a}(t, x, a) = h(H(t, x, a), \varphi(t, x, a)), \quad H(t, x, 0) = t.$$

Here the function $h(t, x)$ is an arbitrary function. The choice of the function $h(t, x)$ can depend on additional conditions. If there are no any additional conditions, then one can choose $h(t, x) = 1$, which gives $H(t, x, a) = t$.

If $\xi_t \neq 0$, then the first equation of the Lie equations (5.7.1) defines the function $H(t, x, a)$.

As an alternative for finding an admitted Lie group, one can also use determining equations obtained by expanding the left hand side of (5.7.2) with respect to the group parameter $a$,

$$S(t, x, a) = a S_a(t, x, 0) + \frac{a^2}{2!} S_{aa}(t, x, 0) + \frac{a^3}{3!} S_{aaa}(t, x, 0) + \cdots.$$

By virtue of the Lie equations (5.7.1) the coefficients of this expansion $\frac{\partial^k S}{\partial a^k}(t, x, 0)$ can be written through the coefficients of the infinitesimal generator. For example, the first coefficient of the expansion[16] is

$$S_a(t, x, 0) = \xi_t - f(x)\xi_x + \frac{g^2(x)}{2}\xi_{xx} + \left(2f(x)\frac{g'(x)}{g(x)} - f'(x)\right)\xi.$$

Necessary conditions for $S(t, x, a) = 0$ are $\frac{\partial^k S}{\partial a^k}(t, x, 0) = 0$. In the case that $\xi_t = 0$ the equation $S_a(t, x, 0) = 0$ is a linear second-order ordinary differential equation, in contrast to the nonlinear equation (5.7.4). A solution of this equation also gives sufficient conditions for $S(t, x, a) = 0$. Examples considered below also show that $\frac{\partial^k S}{\partial a^k}(t, x, 0) = 0$ ($k = 1, 2, 3$) are sufficient for $S(t, x, a) = 0$.

**Remark 5.7.1** There is the problem of finding the minimal number $N$ of the terms of the expansion $\frac{\partial^k S}{\partial a^k}(t, x, 0) = 0$ ($k = 1, 2, \ldots, N$) which guarantee that $S(t, x, a) = 0$. Compare with deterministic differential equations where it is sufficient to solve just determining equations for $k = 1$.

### 5.7.2 Admitted Lie Group of Geometric Brownian Motion

For geometric Brownian motion (5.5.19)

$$f = \mu x, \quad g = \sigma x \quad (\mu > 0, \ \sigma > 0),$$

and (5.7.2) is

$$\varphi_t + \mu x \varphi_x + \frac{\sigma^2 x^2}{2}\varphi_{xx} - \mu x^2 \frac{\varphi_x^2}{\varphi} = 0.$$

Particular solutions of this equation given in the previous section are (5.5.22), (5.5.25) and (5.5.26).

$$\varphi_t + \mu x \varphi_x + \frac{\sigma^2 x^2}{2}\varphi_{xx} = \mu \frac{x^2 \varphi_x^2}{\varphi}. \tag{5.7.5}$$

Let us consider the solution (5.5.22):

$$\varphi = C x^k \exp\left(t k(k-1)\gamma\sigma^2/2\right). \tag{5.7.6}$$

If $k = 1$, then setting $C = e^a$, one obtains $\xi = x$. Because $\xi_t = 0$, one has $H = t$. Hence, the admitted generator is

$$x \partial_x. \tag{5.7.7}$$

This generator was also obtained in [28–30].

---

[16]Other terms of the expansion are cumbersome and not presented here.

If $k \neq 1$, then setting $k = e^a$, one obtains

$$\xi = x\big(\ln(x) + t\gamma\sigma^2/2\big).$$

Since $\xi_t \neq 0$, one also finds

$$H(t, x, a) = te^{2a}.$$

Hence, the admitted generator is

$$2t\partial_t + x\big(\ln(x) + t\gamma\sigma^2/2\big)\partial_x. \tag{5.7.8}$$

This generator was also obtained in [8].

Solutions corresponding to (5.5.25) do not satisfy the property

$$\varphi(t, x, 0) = x.$$

Let us study a possibility to compose Lie group for solutions corresponding to (5.5.26)

$$\varphi = \big(c_1 + c_o x^{-k_1} e^{-k_1\lambda t\sigma^2/2}\big)^{-1/\gamma}, \quad k_1 = \gamma + \lambda. \tag{5.7.9}$$

Assume that the constants $c_1 = c_1(a)$, $c_o = c_o(a)$, and $\lambda = \lambda(a)$. Since for a Lie group $\varphi(t, x, 0) = x$, then

$$c_1(0) = 0, \quad c_o(0) = 1, \quad \lambda(0) = 0.$$

The coefficient of the infinitesimal generator is obtained by differentiating (5.7.9) with respect to the group parameter $a$ and setting it to zero:

$$\xi = \frac{\lambda'(0)}{\gamma} x\big(\ln(x) + t\gamma\sigma^2/2\big) - \frac{c_o'(0)}{\gamma} x - \frac{c_1'(0)}{\gamma} x^{\gamma+1}.$$

Forming a linear combination with the generators (5.7.7) and (5.7.8) one obtains the only generator

$$x^{\gamma+1}\partial_x. \tag{5.7.10}$$

It is worth to notice that since solutions (5.7.5), (5.5.25) and (5.5.26) are particular solutions of (5.7.5), the generators (5.7.7), (5.7.8) and (5.7.10) do not exhaust the set of admitted generators.

Let us employ the determining equations $\frac{\partial^k S}{\partial a^k}(t, x, 0) = 0$ ($k = 1, 2, \ldots$) for finding admitted Lie group. The first determining equation is

$$2\xi_t + \sigma^2\big(x^2\xi_{xx} - (\gamma + 1)x\xi_x + (\gamma + 1)\xi\big) = 0. \tag{5.7.11}$$

Recall that $2\mu = \sigma^2(\gamma + 1)$.

If $\xi_t = 0$, then the general solution of this equation is

$$\xi = C_1 x + C_2 x^{\gamma+1}.$$

If $\xi_t \neq 0$, then from the equations $S_a = 0$ and $S_{aa} = 0$ one can find $\xi_t$ and $h_t$. Substituting them into $S_{aaa} = 0$, one obtains the equation

$$\frac{\partial^5 \xi}{\partial x^5} hh_x + \Phi\left(x, \xi, \xi_x, \ldots, \frac{\partial^4 \xi}{\partial x^4}, h, h_x, \ldots, \frac{\partial^3 h}{\partial x^3}\right) = 0. \tag{5.7.12}$$

Further study depends on quantity of $h_x$.

If $h_x = 0$, then

$$h_t = 2(\xi_x - \xi/x).$$

The general solution of this equation is

$$\xi = \frac{x}{2}(h_t \ln(x) + 2h_1),$$

where $h_1 = h_1(t)$ is an arbitrary function. Substituting this solution into (5.7.11), and splitting it with respect to $x$, one finds

$$h_{tt} = 0, \quad h_1' = (\gamma + 1)\sigma^2/4.$$

This gives that

$$h = c_1 t + c_0, \quad h_1 = t(\gamma + 1)\sigma^2/4 + c_2,$$

where $c_0$, $c_1$ and $c_2$ are arbitrary constants. The requirement[17] (5.6.3) forces the constant $c_0$ to vanish and one thus obtains the generators (5.7.7) and (5.7.8).

If $h_x \neq 0$, then (5.7.12) gives $\frac{\partial^5 \xi}{\partial x^5}$. The equation

$$\frac{\partial^5}{\partial x^5}\left(\frac{\partial \xi}{\partial t}\right) = \frac{\partial}{\partial t}\left(\frac{\partial^5 \xi}{\partial x^5}\right)$$

and $\frac{\partial^k S}{\partial a^k}(t, x, 0) = 0$ $(k = 1, 2, \ldots)$ are satisfied.[18] Since, there are no other equations for the function $h(t, x)$, there is an infinite number of admitted generators.

**Remark 5.7.2** In [29] the admitted Lie group

$$\varphi(t, x, a) = (a + x^{-\gamma})^{-1/\gamma}, \quad H(t, x, a) = t(1 + ax^\gamma)^{-2}$$

for geometric Brownian motion was presented as an example. In the context of the present study, this group arises from transformation (5.7.9) with $\lambda = c_o = 0$ and the additional relation (5.6.7).

Another example considered in [29] is the equation

$$dX_t = \mu dt + dB_t,$$

where $\mu > 0$ is constant. Here (5.7.2) becomes

$$\varphi_t + \mu\varphi_x + \frac{1}{2}\varphi_{xx} - \mu\varphi_x^2 = 0.$$

One easily verifies that the function obtained in [29],

$$\varphi(t, x, a) = x - \frac{1}{2\mu}\ln(1 - 2\mu a e^{2\mu x})$$

---

[17]In [8] the constant $c_0$ is mistakenly kept.

[18]Using symbolic calculations on computer we checked equations $\frac{\partial^k S}{\partial a^k}(t, x, 0) = 0$ $(k \leq 7)$. It is likely that this identity holds for all large $k$.

solves this equation. The additional relation (5.6.7) then yields

$$H(t, x, a) = t\left(1 - 2\mu a e^{2\mu x}\right)^{-2}$$

as already obtained in [29]. This confirms that the examples of transformations considered in [29] are correct, contrary to what [8] claims.

## 5.8  Lie Groups of Transformations of Some Stochastic ODEs

According to the above discussion an alternative way for finding an admitted Lie group of transformations is the way of solving determining equations. The present section demonstrates this method for obtaining an admitted Lie group of some stochastic ordinary differential equations. The method considered in this section consists of solving determining equations (5.6.14) and (5.6.15). Following [28–30] we assume that the stochastic time change (5.5.2) is related to the time change of an admitted Lie group of transformations (5.6.2) by the relation (5.6.7). Recall that (5.6.7) gives that the functions $\tau(t, x)$ and $h(t, x)$ have to satisfy the Cauchy problem (5.6.10).

### 5.8.1  Geometric Brownian Motion

Let us consider ones more the equation describing geometric Brownian motion (5.5.19):

$$dX(t) = \mu X(t)dt + \sigma X(t)dB(t), \tag{5.8.1}$$

where $\mu$ and $\sigma > 0$ are constant. The system of determining equations becomes

$$\xi_t + \mu x \xi_x + \tfrac{1}{2}\sigma^2 x^2 \xi_{xx} - \mu \xi - 2\mu x \tau = 0, \quad \tau = \xi_x - \xi/x. \tag{5.8.2}$$

Substituting $\tau$ into the first equation of (5.8.2), one obtains that the function $\xi(t, x)$ has to satisfy the linear parabolic partial differential equation

$$\xi_t - \mu x \xi_x + \frac{1}{2}\sigma^2 x^2 \xi_{xx} + \mu \xi = 0. \tag{5.8.3}$$

Using the change

$$\bar{t} = -t, \quad y = \frac{\sqrt{2}}{\sigma}\ln(x), \quad \xi = u e^{ky + t(\mu - k^2)} \quad \left(k = \frac{\sqrt{2}(\sigma^2 + 2\mu)}{4\sigma}\right),$$

(5.8.3) is reduced to the heat equation

$$u_{\bar{t}} - u_{yy} = 0.$$

Any solution of the heat equation provides a solution of the determining equations (5.8.2). For the sake of simplicity we study the particular class of solutions of (5.8.3) defined by the additional assumption $\xi = \xi(x)$. In this case the determining equation (5.8.3) becomes

$$x^2\xi'' - (\gamma + 1)(x\xi' - \xi) = 0,$$

where $\gamma = 2\mu\sigma^{-2} - 1$. The general solution of the last equation is

$$\xi = C_1 x + C_2 x^{\gamma+1}, \tag{5.8.4}$$

and, hence, $\tau = C_2\gamma x^\gamma$. The solution of the Cauchy problem (5.6.10) is $h(t, x) = 2C_2\gamma t x^\gamma$. Thus, a basis of admitted generators consists of the generators

$$x\partial_x, \quad x^\gamma(x\partial_x + 2\gamma t\partial_t).$$

The first admitted generator coincides with (5.7.7). Taking into account Remark 5.7.2, one can note that the second generator also coincides with (5.7.10).

For the first generator the transformations are $\bar{x} = xe^a$, $\bar{t} = t$ and the function $\eta = 1$.

Let us construct the Lie group of transformations corresponding to the second admitted generator

$$x^\gamma(\gamma^{-1}x\partial_x + 2t\partial_t).$$

The solution of the Cauchy problem

$$\begin{cases} \frac{dH}{da} = h(H, \varphi), & \frac{d\varphi}{da} = \xi(H, \varphi), \\ H(t, x, 0) = t, & \varphi(t, x, 0) = x \end{cases}$$

is

$$\bar{t} = H = t(1 + ax^\gamma)^{-2}, \quad \bar{x} = \varphi = (a + x^{-\gamma})^{-1/\gamma}.$$

Then $\eta^2 = H_t = (1 + ax^\gamma)^{-2}$.

## 5.8.2 Narrow-Sense Linear System

Let $\mu$, $\nu$ and $\sigma \neq 0$ be constant. Consider the system of equations discussed in [13]

$$dX(t) = Y(t)dt,$$
$$dY(t) = -\left(\nu^2 X(t) + \mu Y(t)\right)dt + \sigma dB(t). \tag{5.8.5}$$

This system of equations is called the narrow-sense linear system. The system of determining equations (5.6.14), (5.6.15) for (5.8.5) becomes

$$\xi_{1,t} + y\xi_{1,x} - (\nu^2 x + \mu y)\xi_{1,y} + \tfrac{1}{2}\sigma^2\xi_{1,yy} - 2y\tau - \xi_2 = 0,$$
$$\xi_{2,t} + y\xi_{2,x} - (\nu^2 x + \mu y)\xi_{2,y} + \tfrac{1}{2}\sigma^2\xi_{2,yy} + 2(\nu^2 x + \mu y)\tau + \nu^2\xi_1 + \mu\xi_2 = 0,$$
$$\xi_{1,y} = 0, \quad \xi_{2,y} - \tau = 0. \tag{5.8.6}$$

From the last two equations one finds

$$\tau = \xi_{2,y}, \quad \xi_1 = \xi_1(t, x).$$

Substituting these into the remaining equations of (5.8.6), one obtains that the functions $\xi_1$ and $\xi_2$ have to satisfy the equations

$$\xi_{1,t} + y\xi_{1,x} - 2y\xi_{2,y} - \xi_2 = 0,$$
$$\xi_{2,t} + y\xi_{2,x} + (v^2x + \mu y)\xi_{2,y} + \tfrac{1}{2}\sigma^2\xi_{2,yy} + v^2\xi_1 + \mu\xi_2 = 0. \qquad (5.8.7)$$

Differentiating the first equation in (5.8.7) with respect to $y$, one gets

$$\xi_{1,x} - 2y\xi_{2,yy} - 3\xi_{2,y} = 0. \qquad (5.8.8)$$

Differentiating (5.8.8) with respect to $y$ again, one finds

$$2y\xi_{2,yyy} + 5\xi_{2,yy} = 0.$$

The general solution of this equation is

$$\xi_2 = H_1 + \frac{1}{\sqrt{y}}H_2 + yH_3,$$

where $H_1(t, x)$, $H_2(t, x)$ and $H_3(t, x)$ are arbitrary functions. Substituting $\xi_2$ into the second equation of (5.8.7), one obtains

$$\tfrac{3}{8}\sigma^2 y^{-5/2}H_2 - \tfrac{1}{2}y^{-3/2}v^2 x H_2 + y^{-1/2}(\tfrac{1}{2}\mu H_2 + H_{2,t}) + y^{1/2}H_{2,x}$$
$$+ y(H_{3,t} + H_{1,x} + 2\mu H_3) + y^2 H_{3,x} + H_{1,t} + vx^2 H_3 + v^2\xi_1 + \mu H_1 = 0.$$

Splitting the last equation with respect to $y$, one has $H_2 = 0$, $H_3 = f(t)$, and

$$H_{1,t} + \mu H_1 + vx^2 f + v^2\xi_1 = 0, \quad H_{1,x} + f' + 2\mu f = 0. \qquad (5.8.9)$$

Substituting $\xi_2$ into (5.8.8), and solving it, one finds

$$\xi_1 = 3xf(t) + g(t).$$

Then the first equation of (5.8.7) gives

$$H_1 = 3xf' + g'.$$

Substituting $\xi_1$ and $H_1$ into the first equation of (5.8.9), one finds

$$vx^2 f + x(3f'' + 3f(v^2 + \mu)) + g'' + \mu g' + v^2 g = 0.$$

Splitting the last equation with respect to $x$, one has $f(t) = 0$, and

$$g'' + \mu g' + v^2 g = 0.$$

The general solution of the last equation depends on the relation between $v$ and $\mu$:

$$g = \begin{cases} e^{-\mu t/2}(C_1 e^{\gamma_1 t} + C_2 e^{-\gamma_1 t}), & \mu^2 > 4v^2; \\ e^{-\mu t/2}(C_1 + tC_2), & \mu^2 = 4v^2; \\ e^{-\mu t/2}(C_1 \sin(\gamma_2 t) + C_2 \cos(\gamma_2 t)), & \mu^2 < 4v^2, \end{cases}$$

where

$$\gamma_1 = \sqrt{\mu^2/4 - v^2}, \quad \gamma_2 = \sqrt{v^2 - \mu^2/4}.$$

For brevity we only proceed with the first case: $\mu^2 > 4v^2$. In this case $\tau = 0$, $h = 0$, and

$$\xi_1 = e^{-\mu t/2}(C_1 e^{\gamma_1 t} + C_2 e^{-\gamma_1 t}),$$

$$\xi_2 = e^{-\mu t/2}(C_1(\gamma_1 - \mu/2)e^{\gamma_1 t} - C_2(\gamma_1 + \mu/2)e^{-\gamma_1 t}). \tag{5.8.10}$$

The basis of admitted generators consists of the generators

$$e^{(\gamma_1 - \mu/2)t}(\partial_x + (\gamma_1 - \mu/2)\partial_y), \quad e^{-(\gamma_1 + \mu/2)t}(\partial_x - (\gamma_1 + \mu/2)\partial_y).$$

A Lie group of transformations based on these generators does not change time $t$. Applying the Itô formula one can show that the transformations corresponding to these generators map a solution of (5.8.5) into a solution of the same equations.

## 5.8.3 Black and Scholes Market

Consider the system of equations discussed in [24],

$$dX(t) = \rho X(t)dt,$$

$$dY(t) = \mu Y(t)dt + \sigma Y(t)dB(t), \tag{5.8.11}$$

where $\rho$, $\mu$ and $\sigma$ are nonzero constants. For (5.8.11) the corresponding system of determining equations becomes

$$\xi_{1,t} + \rho x \xi_{1,x} + \mu y \xi_{1,y} + \tfrac{1}{2}\sigma^2 y^2 \xi_{1,yy} - 2\rho x \tau - \rho \xi_1 = 0,$$

$$\xi_{2,t} + \rho x \xi_{2,x} + \mu y \xi_{2,y} + \tfrac{1}{2}\sigma^2 y^2 \xi_{2,yy} - 2\mu y \tau - \mu \xi_2 = 0, \tag{5.8.12}$$

$$y\xi_{1,y} = 0, \quad y\xi_{2,y} - y\tau - \xi_2 = 0.$$

From the last pair of equations of (5.8.12), one finds

$$\xi_1 = \xi_1(t, x), \quad \tau = \xi_{2,y} - \frac{\xi_2}{y}.$$

Substituting these into the remaining equations of (5.8.12), one obtains that the functions $\xi_1$ and $\xi_2$ have to satisfy the system of equations

$$\xi_{1,t} + \rho x \xi_{1,x} - 2\rho x(\xi_{2,y} - \tfrac{\xi_2}{y}) - \rho \xi_1 = 0,$$

$$\xi_{2,t} + \rho x \xi_{2,x} - \mu y \xi_{2,y} + \tfrac{1}{2}\sigma^2 y^2 \xi_{2,yy} + \mu \xi_2 = 0. \tag{5.8.13}$$

Differentiating the first equation of (5.8.13) with respect to $y$, one gets

$$\xi_{2,yy} - \frac{1}{y}\xi_{2,y} + \frac{1}{y^2}\xi_2 = 0.$$

The general solution of the last equation is

$$\xi_2 = yF(t, x) + yG(t, x)\ln y.$$

Substituting this $\xi_2$ into the second equation of (5.8.13), and splitting it with respect to $y$, one obtains

$$G_t + \rho x G_x = 0, \quad F_t + \rho x F_x - \left(\mu - \frac{1}{2}\sigma^2\right)G = 0.$$

Thus

$$G = F_1, \quad F = \gamma \ln x F_1 + F_2,$$

where $\gamma = \rho^{-1}(\mu - \sigma^2/2)$, and $F_1 = F_1(z)$, $F_2 = F_2(z)$ are arbitrary functions of the argument $z = t - \rho^{-1}\ln x$. Substituting $\xi_2$ into the first equation of (5.8.13), one obtains

$$\xi_{1,t} + \rho x \xi_{1,x} - \rho \xi_1 - 2\rho x F_1 = 0.$$

Hence,

$$\xi_1 = 2x \ln x F_1 + x F_3, \quad \xi_2 = (y \ln y + \gamma y \ln x)F_1 + y F_2, \quad \tau = F_1, \tag{5.8.14}$$

where $F_3 = F_3(z)$. A basis of admitted generators corresponding to (5.8.14) is

$$y F_2(z)\partial_y, \quad x F_3(z)\partial_x, \quad F_1(z)(2x \ln x \partial_x + y(\ln y + \gamma \ln x)\partial_y) + h\partial_t,$$

where $z = t - \rho^{-1}\ln x$, and $h = 2\int_0^t F_1(s - \frac{\ln x}{\rho})ds$.

### 5.8.4 Nonlinear Itô System

Let $\mu_1$ and $\mu_2$ be constants. Consider the system of equations discussed in [12],

$$dX(t) = \frac{\mu_1}{X(t)}dt + dB_1(t),$$
$$dY(t) = \mu_2 dt + dB_2(t). \tag{5.8.15}$$

The associated Fokker–Planck equation is

$$u_t = \frac{1}{2}(u_{xx} + u_{yy}) + \frac{\mu_1}{x^2}u - \frac{\mu_1}{x}u_x - \mu_2 u_y,$$

which has been studied by [6]. For (5.8.15) the corresponding system of determining equations for (5.6.14), (5.6.15) becomes

$$\xi_{1,t} + \frac{\mu_1}{x}\xi_{1,x} + \mu_2\xi_{1,y} + \frac{1}{2}\xi_{1,xx} + \frac{1}{2}\xi_{1,yy} - 2\frac{\mu_1}{x}\tau + \frac{\mu_1}{x^2}\xi_1 = 0,$$
$$\xi_{2,t} + \frac{\mu_1}{x}\xi_{2,x} + \mu_2\xi_{2,y} + \frac{1}{2}\xi_{2,xx} + \frac{1}{2}\xi_{2,yy} - 2\mu_2\tau = 0, \tag{5.8.16}$$
$$\xi_{1,x} - \tau = 0, \quad \xi_{2,y} - \tau = 0, \quad \xi_{1,y} = 0, \quad \xi_{2,x} = 0.$$

From the last equations of (5.8.16) one can find

$$\xi_1 = \xi_1(t, x), \quad \xi_2 = \xi_2(t, y), \quad \tau = \xi_{2,y}.$$

Substituting these relations into the remaining equations of (5.8.16), one obtains that the function $\xi_1$ and $\xi_2$ have to satisfy the system of equations

$$\xi_{1,t} + \frac{\mu_1}{x}\xi_{1,x} + \frac{1}{2}\xi_{1,xx} - 2\frac{\mu_1}{x}\xi_{2,y} + \frac{\mu_1}{x^2}\xi_1 = 0,$$
$$\xi_{2,t} - \mu_2\xi_{2,y} + \frac{1}{2}\xi_{2,yy} = 0, \qquad (5.8.17)$$
$$\xi_{1,x} - \xi_{2,y} = 0.$$

Differentiating the third equation of (5.8.17) with respect to $x$ and $y$, respectively, one obtains

$$\xi_{1,xx} = 0, \quad \xi_{2,yy} = 0,$$

thus

$$\xi_1 = h_1(t)x + h_2(t), \quad \xi_2 = h_3(t)y + h_4(t).$$

Substituting them into other equations of (5.8.17), and splitting it with respect to $x$, one obtains

$$h_1 - h_3 = 0, \quad h_2 = 0, \quad h'_3 = 0, \quad h'_4 - \mu_2 h_3 = 0.$$

Hence

$$\xi_1 = C_1 x, \quad \xi_2 = C_1(y + \mu_2 t) + C_2, \quad \tau = C_1, \qquad (5.8.18)$$

and $h = 2C_1 t$. Thus, a basis of generators corresponding to (5.8.18) is

$$\partial_y, \quad x\partial_x + (y + \mu_2 t)\partial_y + 2t\partial_t.$$

# References

1. Albeverio, S., Fei, S.: Remark on symmetry of stochastic dynamical systems and their conserved quantities. J. Phys. A, Math. Gen. **28**, 6363–6371 (1995)
2. Albin, P.: Stochastic Calculus. Chalmers University of Technology and Gothenburg University, Gothenburg (2001). www.math.chalmers.se/~palbin/sc-notes.ps
3. Alexandrova, O.V.: Group analysis of a two-dimensional Itô stochastic equation. Vestn. DonNABA **49**(3/4), 255–280 (2006)
4. Alexandrova, O.V.: Group analysis of the Itô stochastic system. Diff. Equ. Dyn. Syst. **14**(3/4), 255–280 (2006)
5. Cyganowski, S.: Solving stochastic differential equations with maple. Maple-Tech Newsl. **3**(2), 38–40 (1996)
6. Finkel, F.: Symmetries of the Fokker–Planck equation with a constant diffusion matrix in $2+1$ dimensions. J. Phys. A, Math. Theor. **32**, 2671–2684 (1999)
7. Fredericks, E., Mahomed, F.M.: Symmetries of first-order stochastic ordinary differential equations revisited. Math. Methods Appl. Sci. **30**, 2013–2025 (2007)
8. Fredericks, E., Mahomed, F.M.: A formal approach for handling Lie point symmetries of scalar first-order Itô stochastic ordinary differential equations. J. Nonlinear Math. Phys. **15**, 44–59 (2008)
9. Gaeta, G.: Lie-point symmetries and stochastic differential equations. J. Phys. A, Math. Theor. **33**, 4883–4902 (2000)
10. Gaeta, G.: Symmetry of stochastic equations. arXiv:nlin.SI/0305044, pp. 1–18 (2004)
11. Gaeta, G.: Symmetry of stochastic equations. In: Samoilenko, A.M. (ed.) Proceedings of the Institute of Mathematics of NAS of Ukraine, SYMMETRY in Nonlinear Mathematical Physics, pp. 98–109. Institute of Mathematics of NAS of Ukraine, Kyiv (2004)

12. Gaeta, G., Quinter, N.R.: Lie-point symmetries and differential equations. J. Phys. A, Math. Gen. **32**, 8485–8505 (1999)
13. Gard, T.C.: Introduction to Stochastic Differential Equations. Dekker, New York and Basel (1988)
14. Ibragimov, N.H., Meleshko, S.V.: A solution to the problem of invariants for parabolic equations. Commun. Nonlinear Sci. Numer. Simul. **14**, 2551–2558 (2009)
15. Ibragimov, N.H., Ünal, G., Jogréus, C.: Approximate symmetries and conservation laws for Itô and Stratonovich dynamical systems. J. Math. Anal. Appl. **297**, 152–168 (2004)
16. Johnpillai, I.K., Mahomed, F.: Singular invariant equation for the $(1+1)$ Fokker–Planck equation. J. Phys. A, Math. Gen. **34**, 11,033–11,051 (2001)
17. Lie, S.: Klassifikation und Integration von gewöhnlichen Differentialgleichungen zwischen $x, y$. Die eine Gruppe von Transformationen gestatten. III. Arch. Mat. Naturvidenskab **8**(4), 371–427 (1883). Reprinted in Lie's Gessammelte Abhandlungen, 1924, 5, paper XIY, pp. 362–427
18. Mahomed, F.M.: Complete invariant characterization of scalar linear $(1+1)$ parabolic equations. J. Nonlinear Math. Phys. **15**, 112–123 (2008)
19. Mahomed, F.M., Wafo Soh, C.: Integration of stochastic ordinary differential equations from a symmetry standpoint. J. Phys. A, Math. Gen. **34**, 777–782 (2001)
20. McKean, H.P.: Stochastic Integrals. Academic Press, New York (1969)
21. Meleshko, S.V.: Methods for Constructing Exact Solutions of Partial Differential Equations. Springer, New York (2005)
22. Melnick, S.A.: The group analysis of stochastic partial differential equations. Theory Stoch. Process. **9(25)**(1–2), 99–107 (2003)
23. Misawa, T.: New conserved quantities form symmetry for stochastic dynamical systems. J. Phys. A, Math. Gen. **27**, 177–192 (1994)
24. Øksendal, B.: Stochastic Ordinary Differential Equations. An Introduction with Applications. 5th edn. Springer, Berlin (1998).
25. Platen, E.P.K.: Numerical Solution of Stochastic Differential Equations. Springer, Berlin (1992)
26. Rao, M.M.: Higher order stochastic differential equation. In: Rao, M.M. (ed.) Real and Stochastic Analysis. Recent Advances, pp. 225–302. CRC Press, Boca Raton, New York (1977)
27. Shiryaev, A.N.: Essentials of Stochastic Finance. Facts, Models, Theory. World Scientific, Singapore (1999).
28. Srihirun, B.: Application of group analysis to stochastic differential equations. Ph.D. thesis, School of Applied Mathematics, Suranaree University of Technology, Nakhon Ratchasima, Thailand (2005)
29. Srihirun, B., Meleshko, S.V., Schulz, E.: On the definition of an admitted Lie group for stochastic differential equations with multi-Brownian motion. J. Phys. A, Math. Theor. **39**, 13,951–13,966 (2006)
30. Srihirun, B., Meleshko, S.V., Schulz, E.: On the definition of an admitted Lie group for stochastic differential equations. Commun. Nonlinear Sci. Numer. Simul. **12**(8), 1379–1389 (2007)
31. Steele, J.M.: Stochastic Calculus and Financial Applications. Springer, New York (2001)
32. Ünal, G.: Symmetries of Ito and Stratonovich dynamical systems and their conserved quantities. Nonlinear Dyn. **32**, 417–426 (2003)
33. Ünal, G., Dinler, A.: Exact linearization of one dimensional Itô equations driven by fBm: Analytical and numerical solutions. Nonlinear Dyn. **53**, 251–259 (2008)
34. Ünal, G., Sun, J.Q.: Symmetries conserved quantities of stochastic dynamical control systems. Nonlinear Dyn. **36**, 107–122 (2004)

# Chapter 6
# Delay Differential Equations

Many mathematical models in biology, physics and engineering, where there is a time lag or aftereffect, are described by retarded differential equations. Retarded differential equations are similar to ordinary differential equations, but their evolution involves past values of the state variable. The solution of retarded differential equations therefore requires knowledge of not only the current state, but also of the state at previous moments. Although this type of equations plays a key role in many branches, the theory of retarded differential equations is still being developed.

In this chapter, applications of group analysis to delay differential equations (DDE) are considered. For the sake of completeness short introduction into the theory[1] of delay differential equations is also presented here.

## 6.1 Delay Differential Equations in Mathematical Modeling

Delay differential equations appear in problems with delaying links where certain information processing is needed, for example, in population dynamics and bioscience problems, in control problems, electrical networks containing lossless transmission lines. Many examples of applications of delay differential equations in mathematical modeling one can find in [2, 9, 12, 15, 23]. Here we present one illustrative example.

Let us consider simple models describing a change of population. One of such models is proposed by T.R. Malthus in 1798. It has the form

$$\dot{N}(t) = \lambda N(t), \quad N(t_o) = N_o > 0. \tag{6.1.1}$$

Here $N(t)$ is the population at time $t \geq t_o$ and $\lambda > 0$ is the growth coefficient. According to this model the population is growing exponentially, which is not realistic. This model does not take into account overcrowding and food shortage. Preventing

---

[1] The theory and applications of functional differential equations can be found in many books, for example, in [2, 9, 12, 15, 23] and in many others.

Y.N. Grigoriev et al., *Symmetries of Integro-Differential Equations*, 251
Lecture Notes in Physics 806,
DOI 10.1007/978-90-481-3797-8_6, © Springer Science+Business Media B.V. 2010

weakness of the Malthus model, P.F. Verhulst published in 1838 the logistic equation:

$$\dot{N}(t) = \lambda N(t) \left( 1 - \frac{N(t)}{K} \right), \quad N(t_o) = N_o > 0. \tag{6.1.2}$$

The logistic equation can be integrated exactly, and has the solution

$$N(t) = \frac{K}{1 + CKe^{-\lambda(t-t_o)}},$$

where $C = 1/N_o - 1/K$. Since $N(t) \to K$ when $t \to \infty$, this solution indicates that the population growth tends to the finite number $K$, which is the maximum number of individuals that the environment can support. The term $(1 - \frac{N(t)}{K})$ in (6.1.2) manages the rate of the population growth: the population is growing slower if it tends to the stationary number $K$. The model described by (6.1.2) is more realistic than the model described by (6.1.1), but it has some defects as well. One of them lies in the fact that the population growth is monotone, whereas in the reality the population is oscillating near the stationary solution. Another defect is related to the fact that the population immediately reacts on the change of the population. The model proposed by G.E. Hutchinson in 1948 corrects these defects and has the form

$$\dot{N}(t) = \lambda N(t) \left( 1 - \frac{N(t-\tau)}{K} \right). \tag{6.1.3}$$

Here the reaction of the population is considered with the delay $\tau$. This delay can arise from variety of causes, for example, slow replacement of food supplies.

The presence of the delay in an equation cannot be considered separately from setting the initial value problem. For (6.1.3), the initial value problem, in contrast to ordinary differential equations, is set on the interval $[t_o - \tau, t_o]$:

$$N(t_o - s) = N_0(s), \quad s \in [t_o - \tau, t_o]. \tag{6.1.4}$$

The set of solutions of the initial value problem (6.1.3), (6.1.4) is richer and their behavior is closer to the reality than solutions of the previous models. In particular, for some parameters solutions demonstrate oscillating behavior in tending to their stationary state.

## 6.2 Mathematical Background of Delay Ordinary Differential Equations

One of the main keys to solving determining equations of an admitted Lie group of partial differential equations is an existence theorem of a local solution of the Cauchy problem. For delay differential equations the existence of a local solution plays a similar role. Since the theory of existence for delay ordinary differential equations is well-developed, in this section we give a short review of results of this theory. We use the terminology accepted in the theory of delay differential equations.[2] The reader familiar with delay differential equations can skip this section.

---

[2]See, for example, [6].

### 6.2.1 Definitions and Theorems

Let us consider a system of ordinary delay differential equations

$$\dot{x}(t) = G(t, x(g_1(t)), x(g_2(t)), \ldots, x(g_n(t))), \tag{6.2.1}$$

where $x \in R^m$, $g_j(t) \in [t - r, t]$, $\forall t \geq t_0$ $(j = 1, 2, \ldots, n)$, for some constant $r \geq 0$. For example, for $g_1(t) = t - 1$, $g_2(t) = t - 2$, $n = 2$, $m = 1$, the equation (6.2.1) is written as

$$\dot{x}(t) = 3x(t - 1) - x(t - 2).$$

**Definition 6.2.1** If a vector-function $\chi(t)$ is defined at least on $[t - r, t]$, then a new function $\chi_t : [-r, 0] \to R^m$ is defined by

$$\chi_t(s) = \chi(t + s), \quad s \in [-r, 0].$$

We will denote the set of continuous on $[-r, 0]$ functions with values in $D$ by

$$\mathcal{D}_D = \{\chi \in C([-r, 0]) \mid \chi(t) \in D \subset R^m, \ \forall t \in [-r, 0]\}.$$

**Definition 6.2.2** The equation

$$\dot{x}(t) = F(t, x_t) \tag{6.2.2}$$

with the functional $F : J \times \mathcal{D}_D \to R^m$ is called a functional differential equation. Here $J$ is an interval $(\alpha, \beta) \in R$.

For any function $\chi \in \mathcal{D}_D$ we define the value

$$\|\chi\|_r = \sup_{-r \leq s \leq 0} |\chi(s)|,$$

which can be considered as a norm of the Banach space $\mathcal{D}_{R^m}$ containing the space of functions $\chi \in \mathcal{D}_D$.

**Example 6.2.1** For system of equations (6.2.1) the functional $F(t, x_t)$ is defined as

$$F(t, x_t) = G(t, x_t(g_1(t) - t), x_t(g_2(t) - t), \ldots, x_t(g_n(t) - t)).$$

**Example 6.2.2** If the functional $F$ is $F(t, \chi) = \int_{-r}^{0} \chi(s)\, ds$, then the functional differential equation has the form

$$\dot{x}(t) = \int_{-r}^{0} x_t(s)\, ds = \int_{t-r}^{t} x(s)\, ds.$$

In studying existence and uniqueness for a system of ordinary differential equations we usually consider continuity and Lipschitz conditions. In the case of the functional $F(t, x_t)$ these conditions are replaced in the following way.

**Definition 6.2.3** A continuity condition is satisfied if for each given continuous function $\chi : [t_0 - r, \beta) \to D$, the function $F(t, \chi_t)$ is a continuous function with respect to $t \in [t_0, \beta)$.

Let $\mathscr{Q}$ be a subset of $J \times \mathscr{Q}_D$.

**Definition 6.2.4** The functional $F : J \times \mathscr{Q}_D \to R^m$ satisfies the Lipschitz condition on $\mathscr{Q}$ (or $F$ is Lipschitzian on $\mathscr{Q}$) if there exists a positive constant $L$ such that

$$|F(t, \chi) - F(t, \hat{\chi})| \le L \|\chi - \hat{\chi}\|_r,$$

for arbitrary $(t, \chi)$ and $(t, \hat{\chi})$ from $\mathscr{Q}$.

**Definition 6.2.5** The functional $F : J \times \mathscr{Q}_D \to R^m$ is said to be locally Lipschitzian if for each given $(t_0, \chi_0) \in J \times \mathscr{Q}_D$ there exist numbers $a > 0$ and $b > 0$ such that $F$ is Lipschitzian on the subset

$$\mathscr{Q} = \{(t, \chi) \in J \times \mathscr{Q}_D \mid t \in [t_0 - a, t_0 + a] \cap J, \ \chi \in \mathscr{Q}_D, \ \|\chi - \chi_0\|_r \le b\}.$$

Notice that if a function $F : J \times D^n \to R^m$ satisfies the Lipschitz condition (as a function), then it satisfies the Lipschitz condition as a functional $F : J \times \mathscr{Q}_D \to R^m$.

**Definition 6.2.6** The problem of finding a solution of system (6.2.2), with the initial data

$$x_{t_0}(s) = \psi(s), \quad s \in [t_0 - r, t_0] \tag{6.2.3}$$

is called an initial value problem.

One has the following definitions and theorems like their counterparts for ordinary differential equations.

**Theorem 6.2.1** *Let the functional $F : [t_0, \beta) \times \mathscr{Q}_D \to R^m$ satisfy the continuity condition and be locally Lipschitzian. Then the initial value problem (6.2.2), (6.2.3) with $\psi \in \mathscr{Q}_D$ has at most one solution on $[t_0 - r, \beta_1)$, for any $\beta_1 \in (t_0, \beta]$.*

**Theorem 6.2.2** *Let the functional $F : [t_0, \beta) \times \mathscr{Q}_D \to R^m$ satisfy continuity condition with respect to $t$ in $[t_0, \beta)$ and let it be locally Lipschitzian. Then the initial value problem (6.2.2), (6.2.3) with $\psi \in \mathscr{Q}_D$ has a unique solution on $[t_0 - r, t_0 + \Delta)$, for some $\Delta > 0$.*

**Definition 6.2.7** Let $\phi(t)$ on $[t_0 - r, \beta_1)$ and $\hat{\phi}(t)$ on $[t_0 - r, \beta_2)$ both be solutions of problem (6.2.2), (6.2.3). If $\beta_2 > \beta_1$, then the solution $\hat{\phi}(t)$ is called an extension of $\phi(t)$ to $[t_0 - r, \beta_2)$. A solution $\phi(t)$ of (6.2.2), (6.2.3) is nonextendable if it has no extension.

**Theorem 6.2.3** *Let $F : [t_0, \beta) \times \mathscr{Q}_D \to R^m$ satisfy the continuity condition, and let it be locally Lipschitzian. Then for each $\psi \in \mathscr{Q}_D$, problem (6.2.2), (6.2.3) has a unique nonextendable solution.*

## 6.3 On the Definition of Admitted Lie Group for DDE

This section is devoted to constructing determining equations of an admitted Lie group for delay differential equations. We start with an example which relates definitions of determining equations of an admitted Lie group for integro-differential equations and definitions of determining equations of an admitted Lie group for delay differential equations.

### 6.3.1 Example

Let us construct a determining equation of an admitted Lie group for the integro-differential equation

$$x'(t) = \int_{-r}^{0} x(t+s)\, ds. \tag{6.3.1}$$

According to the definition of the admitted Lie group with the generator $X = \tau(t, x)\partial_t + \eta(t, x)\partial_x$, the determining equation is written as

$$\eta_t(t, x(t)) + x'(t)\eta_x(t, x(t)) - x''(t)\tau(t, x(t))$$
$$- x'(t)\left(\tau_t(t, x(t)) + x'(t)\tau_x(t, x(t))\right)$$
$$= \int_{-r}^{0} \left(\eta(t+s, x(t+s)) - x'(t+s)\tau(t+s, x(t+s))\right)\, ds. \tag{6.3.2}$$

Equation (6.3.2) should be satisfied for any solution $x = x(t)$ of (6.3.1). The determining equation of an admitted Lie group is still an integro-differential equation, and it is not easy to split it. However, differentiating (6.3.1), one obtains the delay differential equation

$$x''(t) = x(t) - x(t - r). \tag{6.3.3}$$

The Cauchy problem for this equation is posed by adding the initial conditions [15]

$$x'(t_o) = x_1, \quad x_{t_o}(s) = \psi(s), \quad s \in [t_o - r, t_o], \tag{6.3.4}$$

with an arbitrary value $x_1$ and an arbitrary continuous function $\psi$.

### 6.3.2 Admitted Lie Group of DODE

Relations between integro-differential equation and delay differential equation considered in the example above give an idea of an application of the definition of the admitted Lie group, developed initially for integro-differential equations, to delay differential equations.[3] For the sake of simplicity the method of constructing the determining equations of an admitted Lie group is explained for a single first-order delay differential equation ($x \in R^1$) with one delay

$$\Phi(t, x(t)) = \dot{x}(t) - F(t, x(t), x(g_1(t))) = 0, \tag{6.3.5}$$

and the function $F$ that depends on the independent variables $(t, x, y)$. Extension to a multi-dimensional case, multiple delays, and higher order is made without any additional obstacles.

Suppose a one-parameter Lie group $G^1(X)$ of the transformations

$$\bar{t} = f^t(t, x, a), \quad \bar{x} = f^x(t, x, a) \tag{6.3.6}$$

with the generator

$$X = \tau(t, x)\partial_t + \eta(t, x)\partial_x$$

is given. Formally[4] the determining equations of an admitted Lie group for delay differential equations is constructed similarly to those for integro-differential equations. Assume that the Lie group $G^1(X)$ transforms a solution $x_0(t)$ of (6.3.5)

$$\Phi(t, x(t)) = \dot{x}(t) - F(t, x(t), x(g_1(t))) = 0,$$

into the solution $x_a(t)$ of the same equation. The transformed function $x_a(t)$ is

$$x_a(\bar{t}) = f^x(t, x_0(t), a)$$

with the expression $t = \psi^t(\bar{t}; a)$ substituted, which is found from the relation $\bar{t} = f^t(t, x_0(t), a)$. Differentiating $\Phi(\bar{t}, x_a(\bar{t}))$ with respect to the group parameter $a$, and considering this equation for the value $a = 0$, one obtains the determining equation of an admitted Lie group

$$\zeta^{\dot{x}}(t, x_o(t), \dot{x}_o(t), \ddot{x}_o(t)) - \zeta^x(t, x_o(t), \dot{x}_o(t)) F_x(t, x_o(t), x_o(g_1(t)))$$
$$- \zeta^x(g_1(t), x_o(g_1(t)), \dot{x}_o(g_1(t))) F_y(t, x_o(t), x_o(g_1(t))) = 0, \tag{6.3.7}$$

where

$$\zeta^x(t, x, \dot{x}) = \eta(t, x) - \dot{x}\tau(t, x), \quad \zeta^{\dot{x}}(t, x, \dot{x}, \ddot{x}) = D_t\zeta^x(t, x, \dot{x}),$$

and $D_t$ is the operator of the total derivative with respect to $t$:

$$D_t = \frac{\partial}{\partial t} + \dot{x}\frac{\partial}{\partial x} + \ddot{x}\frac{\partial}{\partial \dot{x}} + \cdots.$$

Equation (6.3.7) can be rewritten in the form

$$(\bar{X}\Phi)_{|x=x_o(t)} = 0, \tag{6.3.8}$$

---

[3]This idea was realized in [21, 22], see also [13].

[4]The reason to consider this procedure as a formal construction was discussed in the section devoted to integro-differential equations.

where the operator $\bar{X}$ is the prolongation of the canonical Lie–Bäcklund operator $\tilde{X} = \zeta^x \partial_x$ equivalent to the generator $X$ extended by $\zeta^y \partial_y$ for the delay term $y = x(g_1(t))$:

$$\bar{X} = \zeta^x \partial_x + \zeta^y \partial_y + \zeta^{\dot{x}} \partial_{\dot{x}}.$$

The coefficient $\zeta^y$ is equal to the coefficient $\zeta^x$ considered at the delayed point:

$$\zeta^y(t, x(t), \dot{x}(t)) = \zeta^x(g_1(t), x(g_1(t)), \dot{x}(g_1(t))).$$

**Definition 6.3.1** A one-parameter Lie group $G^1$ of transformations (6.3.6) is a symmetry group admitted by (6.3.5) if $G^1$ satisfies the determining equation (6.3.8) for any solution $x(t)$ of (6.3.5).

The main features of the determining equation in the given definition is that they should be valid for any solution of (6.3.5). This allows splitting the determining equation with respect to arbitrary elements. Since arbitrary elements of delay differential equations are contained in the determining equations of an admitted Lie group similarly to the case of differential equations, the process of solving the determining equations for delay differential equations is analogous to finding the solutions of the determining equations for differential equations. This will be demonstrated in examples.

Notice that the given definition is free from the requirement that the admitted Lie group should transform a solution into a solution, and also it can be applied for finding an equivalence group, contact and Lie–Bäcklund transformations for functional differential equations.

### 6.3.3 Symmetries of a Model Equation

The determining equation of an admitted Lie group for (6.3.3) considered at the point $(t_o + 0)$ (right hand side limit), after substitution into it the derivatives

$$\ddot{x}(t_o) = x(t_o) - x(t_o - r), \quad x'''(t_o) = x_1 - \dot{x}(t_o - r),$$

becomes

$$2\eta_{tx}(t_o, x_o)x_1 + \eta_{tt}(t_o, x_o) + \eta_{xx}(t_o, x_o)x_1^2 + \eta_x(t_o, x_o)x_o - \eta_x(t_o, x_o)x_2$$
$$- 2\tau_{tx}(t_o, x_o)x_1^2 - \tau_{tt}(t_o, x_o)x_1 - 2\tau_t(t_o, x_o)x_o + 2\tau_t(t_o, x_o)x_2 - \tau_{xx}(t_o, x_o)x_1^3$$
$$- 3\tau_x(t_o, x_o)x_ox_1 + 3\tau_x(t_o, x_o)x_2x_1 - \eta(t_o, x_o) + \eta(t_o - r, x_2)$$
$$+ (\tau(t_o, x_o) - \tau(t_o - r, x_2))x_3 = 0, \tag{6.3.9}$$

where $x_o = \psi(t_o)$, $x_2 = \psi(t_o - r)$, $x_3 = \psi'(t_o - r)$. By virtue of the theorem of the existence of a solution of the Cauchy problem (6.3.3), (6.3.4) with an arbitrary function $\psi(s)$ and an arbitrary value $x_1$, the values $t_o, x_o, x_1, x_2$ and $x_3$ can be assigned arbitrarily. The arbitrariness of these variables allows one to split the determining equation, and then to find the general solution of the determining equation. Further analysis is similar to the classical analysis of solving determining equations of an admitted Lie group for partial differential equations. Since the values $t_o$ and $x_o$ are arbitrary in the determining equation, they are written as $t$ and $x$, respectively.

The determining equation (6.3.9) can be split with respect to $x_1$ and $x_3$ into the five equations

$$2\eta_{tx}(t, x) - \tau_{tt}(t, x) - 3\tau_x(t, x)x + 3\tau_x(t, x)x_2 = 0,$$

$$\eta_{xx}(t, x) - 2\tau_{tx}(t, x) = 0, \quad \tau_{xx}(t, x) = 0, \quad \tau(t, x) - \tau(t - r, x_2) = 0,$$

(6.3.10)

$$\eta_{tt}(t, x) + \eta_x(t, x)x - \eta_x(t, x)x_2 - 2\tau_t(t, x)x + 2\tau_t(t, x)x_2$$

$$- \eta(t, x) + \eta(t - r, x_2) = 0.$$

The first equation can be also split with respect to $x_2$:

$$2\eta_{tx}(t, x) - \tau_{tt}(t, x) = 0, \quad \tau_x(t, x) = 0.$$

Hence, the first four equations of (6.3.10) give

$$2\eta(t, x) = x(c_1 + \tau'(t)) + \varphi(t), \quad \tau(t) = \tau(t - r),$$

where $c_1$ is constant. Substituting this representation into the last equation of (6.3.10), and splitting it with respect to $x_o$ and $x_2$, one obtains

$$\tau'''(t) - 4\tau'(t) = 0, \quad 3\tau'(t) + \tau'(t - r) = 0,$$

$$\varphi''(t) - \varphi(t) + \varphi(t - r) = 0.$$

The general solution of the first two equations is $\tau = c_2$, where $c_2$ is constant. Then the infinitesimal generator corresponding to the admitted Lie group is

$$X = c_2\partial_t + c_1x\partial_x + \varphi(t)\partial_x,$$

where the function $\varphi(t)$ is an arbitrary solution of (6.3.3).

### 6.3.4 Differential-Difference Equations

Delay differential equations with a constant delay can be considered as a particular case of differential-difference equations. Differential-difference equations appear, for example, as numerical models of differential equations. It is known that the same system of partial differential equations can be approximated by different numerical schemes. Inheritance of symmetry properties of differential equations in a numerical scheme can be a criterion for choosing the scheme. Several approaches are being developed for application of group analysis to differential-difference equations.[5] Since the discussion of the results of such applications would require extension of our study to numerical schemes which are beyond the scope of the present book, the reader is referred to [3][6] where this subject is systematically developed.

---

[5] See a review of results in [11] (vol. 1).

[6] And references therein.

## 6.4 Group Classification of Second-Order DODEs

This section deals with a second-order delay ordinary differential equation

$$y'' = f(x, y, y_\tau, y', y'_\tau) \tag{6.4.1}$$

where $y_\tau = y(x - \tau)$, $y'_\tau = y'(x - \tau)$. The Cauchy problem for (6.4.1) is formulated as follows. The initial conditions are

$$y(x) = \chi(x), \quad x \in (x_0 - \tau, x_0), \quad y'(x_0) = y_1. \tag{6.4.2}$$

Assuming some continuity conditions of the functions $f$ and $\chi$, the initial value problem (6.4.1)–(6.4.2) has a solution. Here $x_0$, $y_1$ and $\chi$ are arbitrary.

The purpose of this section is to give a complete group classification of second-order DODEs with respect to admitted Lie groups.[7]

**Remark 6.4.1** For an ordinary differential equation the presence of an admitted Lie group allows one to reduce the order of the equation. The use of the admitted Lie group for DODEs is still an open problem.

**Remark 6.4.2** A symmetry classification of second-order difference equations was studied in [4, 5].

### 6.4.1 Introduction into the Problem

The group classification problem of differential equations was formulated by S. Lie. He proved that any non-singular invariant system of differential equations can be expressed in terms of differential invariants of the corresponding symmetry group.

The group classification of an ordinary differential equation is based upon the enumeration of all possible nonequivalent Lie algebras of operators admitted by the chosen type of equations. S. Lie gave the classification of all dissimilar Lie algebras (under complex change of variables) in two complex variables. In 1992 A. Gonzalez-Lopez et al. ordered the Lie classification of realizations of complex Lie algebras [7] and extended it to the real case [8]. The mentioned works do not exhaust all papers devoted to realizations of Lie algebras. We use here Table B.1 that appears in [16].

Recall that for second-order ordinary differential equations with one dependent variable the group classification was obtained using the following strategy.[8] First, all Lie algebras on the plane being nonequivalent with respect to a change of the variables were constructed. Differential invariants of operator of these Lie algebras

---

[7]This result was obtained in [19, 20].

[8]The study of the problem of group classification for second-order ordinary differential equations in complex domain was carried out by S. Lie and is reviewed in [11].

prolonged up to second-order were obtained. Using these invariants, the representation of second-order ordinary differential equations were found. These equations compose a group classification of second-order ordinary differential equations. This classification is presented in Table 3 [11].[9]

## 6.4.2 Determining Equations

Let a one-parameter Lie group $G^1$ of transformations

$$\tilde{x} = \varphi(x, y, a), \quad \tilde{y} = \psi(x, y, a) \tag{6.4.3}$$

has the generator

$$X = \xi(x, y)\partial_x + \eta(x, y)\partial_y. \tag{6.4.4}$$

According to the definition, the one-parameter Lie group $G^1$ of transformations (6.4.3) is admitted by (6.4.1) if the coefficients of the generator satisfy the determining equation

$$\bar{X}\big(y'' - f(x, y, y_\tau, y', y'_\tau)\big)\big|_{(6.4.1)} = 0, \tag{6.4.5}$$

where the operator $\bar{X}$ is the prolongation of the canonical Lie–Bäcklund operator equivalent to the generator $X$:

$$\tilde{X} = \zeta^y \partial_y + \zeta^{y'} \partial_{y'} + \zeta^{y''} \partial_{y''}, \tag{6.4.6}$$

extended for the delay terms:

$$\bar{X} = \tilde{X} + \zeta^{y_\tau} \partial_{y_\tau} + \zeta^{y'_\tau} \partial_{y'_\tau}.$$

Here $D$ is the operator of the total derivative with respect to $x$,

$$\zeta^y(x, y, y') = \eta(x, y) - y'\xi(x, y),$$
$$\zeta^{y_\tau}(x, y_\tau, y'_\tau) = \zeta^y(x - \tau, y_\tau, y'_\tau)$$
$$= \eta(x - \tau, y_\tau) - y'_\tau \xi(x - \tau, y_\tau),$$
$$\zeta^{y'}(x, y, y', y'') = D(\zeta^y)$$
$$= \eta_x(x, y) + [\eta_y(x, y) - \xi_x(x, y)]y'$$
$$- \xi_y(x, y)y'^2 - \xi(x, y)y'',$$
$$\zeta^{y'_\tau}(x, y_\tau, y'_\tau, y''_\tau) = \zeta^{y'}(x - \tau, y_\tau, y'_\tau, y''_\tau)$$
$$= \eta_x(x - \tau, y_\tau) + [\eta_y(x - \tau, y_\tau) - \xi_x(x - \tau, y_\tau)]y'_\tau$$
$$- \xi_y(x - \tau, y_\tau)y'^2_\tau - \xi(x - \tau, y_\tau)y''_\tau,$$
$$\zeta^{y''}(x, y, y', y'', y''') = D(\zeta^{y'})$$

---

[9]See vol. 3, page 201.

$$= \eta_{xx}(x, y) + [2\eta_{xy}(x, y) - \xi_{xx}(x, y)]y'$$
$$+ [\eta_{yy}(x, y) - 2\xi_{xy}(x, y)]y'^2 - \xi_{yy}(x, y)y'^3$$
$$+ [\eta_y(x, y) - 2\xi_x(x, y)]y'' - 3\xi_y(x, y)y'y'' - \xi(x, y)y'''.$$

Equation (6.4.5) holds for any solution of (6.4.1), hence the derivatives $y''$, $y''_\tau$ and $y'''$ are:

$$y'' = f, \quad y''_\tau = f_\tau,$$
$$y''' = Df = f_x + y'f_y + y'_\tau f_{y_\tau} + ff_{y'} + f_\tau f_{y'_\tau},$$

where $f_\tau = f(x - \tau, y_\tau, y_{2\tau}, y'_\tau, y'_{2\tau})$, $y_{2\tau} = y(x - 2\tau)$ and $y'_{2\tau} = y'(x - 2\tau)$. Substituting these derivatives into the determining equation (6.4.5), it becomes

$$-\xi_{yy}y'^3 + (\eta_{yy} - 2\xi_{xy} + \xi_y f_{y'})y'^2 + \xi^\tau_{y_\tau} f_{y_\tau} y'^2_\tau + (2\eta_{xy} - \xi_{xx})y'$$
$$+ (\xi_x - \eta_y)f_{y'}y' - 3\xi_y fy' + \eta_{xx} - \eta_x f_{y'} + (\eta_y - 2\xi_x)f - \eta^\tau_x f_{y_\tau}$$
$$+ (\xi^\tau_x - \eta^\tau_{y_\tau})f_{y'_\tau}y'_\tau - f_x\xi - f_y\eta - \eta^\tau f_{y_\tau}$$
$$+ (\xi^\tau - \xi)f_{y_\tau}y'_\tau + (\xi^\tau - \xi)f_\tau f_{y'_\tau} = 0. \tag{6.4.7}$$

### 6.4.3 Properties of Admitted Generators

By virtue of the Cauchy problem $x$, $y$, $y_\tau$, $y'$, $y'_\tau$, $y_{2\tau}$ and $y'_{2\tau}$ in (6.4.7) can be considered as arbitrary variables thus allowing splitting the determining equation with respect to these variables.

If $f_{y'_\tau} \neq 0$, then splitting the determining equation (6.4.7) with respect to $y'_{2\tau}$ implies $\xi = \xi^\tau$. If $f_{y'_\tau} = 0$, then the assumption of DODE implies that $f$ must depend on the delay terms, i.e. $f_{y_\tau} \neq 0$. Splitting (6.4.7) with respect to $y'_\tau$, we also get $\xi = \xi^\tau$. This shows the periodic property of $\xi$, i.e.,

$$\xi(x, y) = \xi(x - \tau, y_\tau). \tag{6.4.8}$$

Since this property is valid for any solution of the Cauchy problem, then (6.4.8) implies that the function $\xi$ does not depend on $y$, i.e., $\xi_y = 0$. Moreover, property (6.4.8) allows rewriting the determining equation (6.4.5) in the form

$$\hat{X}\left(y'' - f(x, y, y_\tau, y', y'_\tau)\right)\big|_{(6.4.1)} = 0, \tag{6.4.9}$$

where the operator

$$\hat{X} = \bar{X} + \xi D$$
$$= \xi\partial_x + \eta^y\partial_y + \eta^{y_\tau}\partial_{y_\tau} + \eta^{y'}\partial_{y'} + \eta^{y'_\tau}\partial_{y'_\tau} + \eta^{y''}\partial_{y''} \tag{6.4.10}$$

has the coefficients

$$\eta^y(x, y) = \eta(x, y),$$

$$\eta^{y_\tau}(x, y_\tau) = \eta(x - \tau, y_\tau),$$

$$\eta^{y'}(x, y, y') = \eta_x(x, y) + [\eta_y(x, y) - \xi_x(x, y)]y' - \xi_y(x, y)y'^2,$$

$$\eta^{y'_\tau}(x, y_\tau, y'_\tau) = \eta^{y'}(x - \tau, y_\tau, y'_\tau)$$

$$= \eta_x(x - \tau, y_\tau) + [\eta_y(x - \tau, y_\tau) - \xi_x(x - \tau, y_\tau)]y'_\tau$$

$$- \xi_y(x - \tau, y_\tau)y'^2_\tau,$$

$$\eta^{y''}(x, y, y', y'') = \eta_{xx}(x, y) + [2\eta_{xy}(x, y) - \xi_{xx}(x, y)]y'$$

$$+ [\eta_{yy}(x, y) - 2\xi_{xy}(x, y)]y'^2 - \xi_{yy}(x, y)y'^3$$

$$+ [\eta_y(x, y) - 2\xi_x(x, y)]y'' - 3\xi_y(x, y)y'y''.$$

The difference between the generators $\hat{X}$ and $\bar{X}$ is the following. The generator $\hat{X}$ acts in the space of variables $(x, y, y_\tau, y', y'_\tau, y'')$, whereas the coefficients of the operator $\bar{X}$ also include the derivatives $y''_\tau$ and $y'''$.

Equation (6.4.9) means that the manifold defined by (6.4.1) is an invariant manifold of the generator $\hat{X}$. Since any nonsingular invariant manifold can be represented in terms of invariants of the generator $\hat{X}$, then for describing equations admitting the generator $X$, one needs to find all invariants of the generator $\hat{X}$.

Another property of admitted generators, which allows developing a method for classifying all second-order delay ordinary differential equations is the following. Direct calculations show that if two generators $X_1$ and $X_2$ are admitted by (6.4.1), then their commutator $[X_1, X_2]$ is also admitted by (6.4.1). This property allows stating that the set of infinitesimal generators admitted by (6.4.1) composes a Lie algebra on the real plane. Notice that in contrast to differential equations this property is not necessary for delay differential equations. Since the set of all finite-dimensional nonequivalent Lie algebras acting on the real plane is known,[10] then the Lie algebra admitted by (6.4.1) is equivalent to one of this set.

As it was mentioned, realizations of Lie algebras in vector fields on the real plane is given up to local diffeomorphisms, i.e., up to a change of the dependent and independent variables. DODEs do not posses an equivalence transformation related with the change of the dependent and independent variables. Hence, for finding invariant second-order DODE one needs to consider the representation of a Lie algebra in the form which is equivalent to one of the Lie algebras of Table 1 [16]. The equivalence is considered with respect to a change of the dependent and independent variables.

Consider an invertible (nonsingular) change of variables:

$$x = \bar{h}(\bar{x}, \bar{y}), \qquad y = \bar{g}(\bar{x}, \bar{y}), \tag{6.4.11}$$

and its inverse

$$\bar{x} = h(x, y), \qquad \bar{y} = g(x, y), \tag{6.4.12}$$

with the Jacobian $\Delta = h_x g_y - g_x h_y \neq 0$.

---

[10]For example, Table 1 [16].

The operator $X$ (6.4.4) in the variables $(\bar{x}, \bar{y})$ becomes

$$X = \bar{\xi}(\bar{x}, \bar{y})\frac{\partial}{\partial \bar{x}} + \bar{\eta}(\bar{x}, \bar{y})\frac{\partial}{\partial \bar{y}}, \tag{6.4.13}$$

where the coefficients are related by the formulae:

$$\begin{aligned}
\xi(x, y) &= X(\bar{h}(\bar{x}, \bar{y}))\big|_{\bar{x}=h(x,y), \bar{y}=g(x,y)} \\
&= \big[\bar{\xi}(h(x, y), g(x, y))\bar{h}_{\bar{x}}(h(x, y), g(x, y)) \\
&\quad + \bar{\eta}(h(x, y), g(x, y))\bar{h}_{\bar{y}}(h(x, y), g(x, y))\big],
\end{aligned}$$

$$\begin{aligned}
\eta(x, y) &= X(\bar{g}(\bar{x}, \bar{y}))\big|_{\bar{x}=h(x,y), \bar{y}=g(x,y)} \\
&= \big[\bar{\xi}(h(x, y), g(x, y))\bar{g}_{\bar{x}}(h(x, y), g(x, y)) \\
&\quad + \bar{\eta}(h(x, y), g(x, y))\bar{g}_{\bar{y}}(h(x, y), g(x, y))\big].
\end{aligned}$$

The derivatives

$$\bar{h}_{\bar{x}}(h(x, y), g(x, y)), \quad \bar{h}_{\bar{y}}(h(x, y), g(x, y)),$$

and

$$\bar{g}_{\bar{x}}(h(x, y), g(x, y)), \quad \bar{g}_{\bar{y}}(h(x, y), g(x, y))$$

can be found by differentiating the identities

$$x = \bar{h}(h(x, y), g(x, y)), \quad y = \bar{g}(h(x, y), g(x, y))$$

with respect to $x$ and $y$:

$$\begin{aligned}
1 &= \bar{h}_{\bar{x}}(h(x, y), g(x, y))h_x(x, y) + \bar{h}_{\bar{y}}(h(x, y), g(x, y))g_x(x, y), \\
0 &= \bar{h}_{\bar{x}}(h(x, y), g(x, y))h_y(x, y) + \bar{h}_{\bar{y}}(h(x, y), g(x, y))g_y(x, y), \\
0 &= \bar{g}_{\bar{x}}(h(x, y), g(x, y))h_x(x, y) + \bar{g}_{\bar{y}}(h(x, y), g(x, y))g_x(x, y), \\
1 &= \bar{g}_{\bar{x}}(h(x, y), g(x, y))h_y(x, y) + \bar{g}_{\bar{y}}(h(x, y), g(x, y))g_y(x, y).
\end{aligned}$$

Solving these equations, one finds

$$\begin{aligned}
\bar{h}_{\bar{x}}(h(x, y), g(x, y)) &= \Delta^{-1}(x, y)g_y(x, y), \\
\bar{h}_{\bar{y}}(h(x, y), g(x, y)) &= -\Delta^{-1}h_y(x, y), \\
\bar{g}_{\bar{x}}(h(x, y), g(x, y)) &= -\Delta^{-1}g_x(x, y), \\
\bar{g}_{\bar{y}}(h(x, y), g(x, y)) &= \Delta^{-1}h_x(x, y).
\end{aligned} \tag{6.4.14}$$

Thus,

$$\xi(x, y) = \Delta^{-1}(x, y)\big[\bar{\xi}(h(x, y), g(x, y))g_y(x, y) - \bar{\eta}(h(x, y), g(x, y))h_y(x, y)\big],$$

$$\eta(x, y) = \Delta^{-1}(x, y)\big[\bar{\eta}(h(x, y), g(x, y))h_x(x, y) - \bar{\xi}(h(x, y), g(x, y))g_x(x, y)\big].$$

$$\tag{6.4.15}$$

## 6.4.4 Strategy for Obtaining a Complete Classification of DODEs

Since the set of generators admitted by a second-order DODE composes a Lie algebra, this algebra is equivalent to one of 56 Lie algebras of Table 1 [16]. Thus, for complete group classification of second-order DODEs one needs to carry out the following steps for each class of 56 Lie algebras:

(a) select a Lie algebra[11] from Table B.1,
(b) change the variables original variables $x$ and $y$ to the variables $\bar{x}$ and $\bar{y}$ (6.4.12),
(c) find invariants of the Lie algebra in the space of the changed variables $(x, y, y_\tau,$
    $y', y'_\tau, y'')$,
(d) use the found invariants to form a second-order DODE.

Applying this strategy one obtains representations of all second-order DODEs admitting a Lie group.

## 6.4.5 Illustrative Examples

This subsection gives examples which illustrate an application of the above strategy. Complete results of the classification are presented in Table B.2.

The construction of all equations (6.4.1) admitting a Lie algebra $L$ requires the solution of the determining equation (6.4.7) for known coordinates $\bar{\xi}_i(\bar{x}, \bar{y})$ and $\bar{\eta}_i(\bar{x}, \bar{y})$ $(i = 1, 2, \ldots, k)$ of basis operators of a Lie algebra $L_j^n$. Here the notation $L_j^n$ denotes the $n$-dimensional Lie algebra of the number $j$ from Table B.1.

**Example 6.4.1** Let us consider the three-dimensional Lie algebra $L_{10}^3$, which is defined by the generators

$$X_1 = \partial_{\bar{x}}, \quad X_2 = \partial_{\bar{y}}, \quad X_3 = \bar{x}\partial_{\bar{x}} + (\bar{x} + \bar{y})\partial_{\bar{y}}. \tag{6.4.16}$$

Changing the variables $\bar{x} = h(x, y)$, $\bar{y} = g(x, y)$, and using (6.4.15), the components $\xi_i$ become:

$$\xi_1 = \Delta^{-1}g_y, \quad \xi_2 = -\Delta^{-1}h_y, \quad \xi_3 = \Delta^{-1}(hg_y - (h+g)h_y).$$

Conditions (6.4.8) imply that $(\xi_i)_y = 0$ $(i = 1, 2, 3)$. Equations $(\xi_2)_y = 0$ and $(\xi_3)_y = 0$ lead us to the restrictions $h_y = 0$ and

$$h(x) - h(x - \tau) = c, \tag{6.4.17}$$

where $c$ is an arbitrary constant. Then the Jacobian is $\Delta = h_x g_y$. Generators (6.4.16) become

$$X_1 = \frac{1}{h_x}\partial_x - \frac{g_x}{h_x g_y}\partial_y, \quad X_2 = \frac{1}{g_y}\partial_y, \quad X_3 = \frac{h}{h_x}\partial_x + \frac{(h+g)h_x - hg_x}{h_x g_y}\partial_y.$$

---

[11]The selected Lie algebra is considered in the variables $(\bar{x}, \bar{y})$.

For finding invariants, we consequently solve the equations

$$\hat{X}_1 J = 0, \quad \hat{X}_2^{(2)} J = 0, \quad \hat{X}_3^{(2)} J = 0,$$

where $J = J(x, y, y_\tau, y', y'_\tau, y'')$, $\hat{X}_i$ is the operator (6.4.10) of the generators $X_i$ ($i = 1, 2, 3$).

To find invariants with respect to $X_1$ we have to solve the equation

$$\hat{X}_1 J = 0, \tag{6.4.18}$$

where

$$\hat{X}_1 = \xi_1(x)\partial_x + \eta_1(x, y)\partial_y + \eta_1^{y'}(x, y, y')\partial_{y'} + \eta_1^{y''}(x, y, y', y'')\partial_{y''}$$
$$+ \eta_1(x - \tau, y_\tau)\partial_{y_\tau} + \eta_1^{y'_\tau}(x - \tau, y_\tau, y'_\tau)\partial_{y'_\tau}.$$

For integrating (6.4.18) one has to solve the characteristic system of equations

$$\frac{dx}{\xi_1} = \frac{dy}{\eta_1} = \frac{dy'}{\eta_1^{y'}} = \frac{dy''}{\eta_1^{y''}} = \frac{dy_\tau}{\eta_1^\tau} = \frac{dy'_\tau}{\eta_1^{y'_\tau}}.$$

Since the coefficients of this characteristic system are cumbersome, it is difficult to find its invariants. However, one may note that the first part of this system (without last two equations containing the variables related to delay) is equivalent to the system which corresponds to the prolongation of the original generator $X_1$ with the variables $(\bar{x}, \bar{y}, \bar{y}', \bar{y}'')$:

$$\frac{d\bar{x}}{1} = \frac{d\bar{y}}{0} = \frac{d\bar{y}'}{0} = \frac{d\bar{y}''}{0}.$$

Differential invariants of the last system are easily obtained: $\bar{y}, \bar{y}', \bar{y}''$. Hence, three invariants of (6.4.18) are:[12]

$$J_1(x, y) = g(x, y), \quad J_2(x, y, y') = \frac{Dg(x, y)}{Dh(x)},$$
$$J_3(x, y, y', y'') = \frac{DJ_2(x, y, y')}{Dh(x)}. \tag{6.4.19}$$

The other two invariants are chosen as follows

$$J_1^\tau = J_1(x - \tau, y_\tau), \quad J_2^\tau = J_2(x - \tau, y_\tau, y'_\tau). \tag{6.4.20}$$

Direct calculations show that (6.4.19)–(6.4.20) compose the universal differential invariant of the generator $\hat{X}_1$. Hence, the general solution of (6.4.18) is

$$J = \Phi(J_1, J_1^\tau, J_2, J_2^\tau, J_3). \tag{6.4.21}$$

At this step, the function $\Phi(y_1, y_2, y_3, y_4, y_5)$ is an arbitrary function.

For solving the other two equations

$$\hat{X}_2 J = 0, \quad \hat{X}_3 J = 0, \tag{6.4.22}$$

---

[12]These invariants are obtained by changing the dependent and independent variables in the invariants $(\bar{x}, \bar{y}, \bar{y}', \bar{y}'')$.

one has to find the function $\Phi(y_1, y_2, y_3, y_4, y_5)$ which satisfies the equations

$$\hat{X}_2 \Phi(J_1, J_1^\tau, J_2, J_2^\tau, J_3) = 0, \tag{6.4.23}$$

$$\hat{X}_3 \Phi(J_1, J_1^\tau, J_2, J_2^\tau, J_3) = 0. \tag{6.4.24}$$

Equation (6.4.23) becomes[13]

$$\Phi_{y_1} + \Phi_{y_2} = 0.$$

The general solution of this equation is

$$\Phi = \psi(y_1 - y_2, y_3, y_4, y_5)$$

where the function $\psi(z_1, z_2, z_3, z_4)$ is arbitrary.

For solving (6.4.24), we have to find the function $\psi(z_1, z_2, z_3, z_4)$ which satisfies the equation

$$\psi_{z_2} + \psi_{z_3} + z_1 \psi_{z_1} - z_4 \psi_{z_4} = 0.$$

This equation was obtained by substituting $J = \psi(J_1 - J_1^\tau, J_2, J_2^\tau, J_3)$ into (6.4.24). The general solution of this equation is

$$\psi = H(z_2 - z_3, z_1 e^{-z_2}, z_4 e^{z_2}),$$

where $H$ is an arbitrary function.

Thus, the universal invariant of the Lie algebra $L_{10}^3$ consists of the invariants

$$J_2 - J_2^\tau, \quad (J_1 - J_1^\tau)e^{-J_2}, \quad J_3 e^{J_2}.$$

The set of equations admitting the Lie algebra $L_{10}^3$ has the following form

$$J_3 = e^{-J_2} f(J_2 - J_2^\tau, (J_1 - J_1^\tau)e^{-J_2}).$$

Because of the meaning the functions $J_1, J_1^\tau, J_2, J_2^\tau$ and $J_3$, we represent this equation in Table B.2 as

$$y'' = e^{-y'} f(y' - y_\tau', (y - y_\tau)e^{-y'}).$$

**Example 6.4.2** Let us consider the Lie algebra determined by the generators

$$X_1 = \partial_{\bar{x}}, \quad X_2 = \eta_1(\bar{x})\partial_{\bar{y}}, \quad X_3 = \eta_2(\bar{x})\partial_{\bar{y}}, \quad \ldots, \quad X_{r+1} = \eta_r(\bar{x})\partial_{\bar{y}}$$

where the functions $\eta_1, \eta_2, \eta_3, \ldots, \eta_r$ form a fundamental system of solutions of an $r$-order ordinary differential equation with constant coefficients

$$\eta^{(r)} + c_1 \eta^{(r-1)} + \cdots + c_{r-1} \eta' + c_r \eta = 0.$$

These Lie algebras are $L_8^3, L_9^3, L_{11}^3, L_{15}^3, L_{17}^3, L_{26}^4, L_{27}^4, L_{28}^4, L_{31}^4, L_{32}^4, L_{33}^4, L_{34}^4, L_{36}^4, L_{37}^4, L_{50}^5$ of Table B.1.

---

[13] The computer system of symbolic calculations Reduce [10] was used for obtaining these substitutions.

**Case** $r = 2$. The Lie algebra is defined by the generators

$$X_1 = \partial_{\bar{x}}, \quad X_2 = \eta_1(\bar{x})\partial_{\bar{y}}, \quad X_3 = \eta_2(\bar{x})\partial_{\bar{y}},$$

where $\eta_1$, $\eta_2$ satisfy the equation

$$\eta'' = -(c_1\eta' + c_2\eta). \tag{6.4.25}$$

In Table B.1 these Lie algebras are $L_8^3$, $L_9^3$, $L_{11}^3$, $L_{15}^3$ and $L_{17}^3$. Changing the variables (6.4.12), the generators become

$$X_1 = \frac{1}{h_x}\partial_x - \frac{g_x}{h_x g_y}\partial_y, \quad X_2 = \frac{\eta_1(h)}{g_y}\partial_y, \quad X_3 = \frac{\eta_2(h)}{g_y}\partial_y.$$

The general solution of the equation $\hat{X}_1 J = 0$ is obtained similar to (6.4.21). Applying the generators $\hat{X}_2$, $\hat{X}_3$ to the function $J = \Psi(y_1, y_2, y_3, y_4, y_5, y_6)$ with

$$y_1 = J_1, \quad y_2 = J_1^\tau, \quad y_3 = J_2, \quad y_4 = J_2^\tau, \quad y_5 = J_3, \quad y_6 = J_3^\tau,$$

where $J_3^\tau = J_3(x - \tau, y_\tau, y_\tau', y_\tau'')$, one obtains the system of differential equations

$$\eta_1\Psi_{y_1} + \eta_1'\Psi_{y_3} + \eta_1''\Psi_{y_5} + \eta_1^\tau\Psi_{y_2} + \eta'^\tau_1\Psi_{y_4} + \eta''^\tau_1\Psi_{y_6} = 0,$$
$$\eta_2\Psi_{y_1} + \eta_2'\Psi_{y_3} + \eta_2''\Psi_{y_5} + \eta_2^\tau\Psi_{y_2} + \eta'^\tau_2\Psi_{y_4} + \eta''^\tau_2\Psi_{y_6} = 0, \tag{6.4.26}$$

where $\eta_i^\tau = \eta_i(h(x - \tau))$, $\eta'{}_i^\tau = \eta_i'(h(x - \tau))$, $\eta''{}_i^\tau = \eta_i''(h(x - \tau))$ $(i = 1, 2)$. Notice that for second-order delay differential equations of the form (6.4.1) one has $\Psi_{y_6} = 0$. Substituting $\eta_i''$ and $\eta''{}_i^\tau$ found from (6.4.25) into (6.4.26), they become

$$\eta_1\Psi_{y_1} + \eta_1'\Psi_{y_3} - (c_1\eta_1' + c_2\eta_1)\Psi_{y_5} + \eta_1^\tau\Psi_{y_2} + \eta'^\tau_1\Psi_{y_4} - (c_1\eta'^\tau_1 + c_2\eta_1^\tau)\Psi_{y_6} = 0,$$
$$\eta_2\Psi_{y_1} + \eta_2'\Psi_{y_3} - (c_1\eta_2' + c_2\eta_2)\Psi_{y_5} + \eta_2^\tau\Psi_{y_2} + \eta'^\tau_2\Psi_{y_4} - (c_1\eta'^\tau_2 + c_2\eta_2^\tau)\Psi_{y_6} = 0.$$

In matrix form, these equations can be rewritten as

$$\Phi\mathbf{z} - \Psi_{y_5}\Phi\mathbf{c} + \Phi^\tau\mathbf{z}^\tau = 0. \tag{6.4.27}$$

Here

$$\Phi = \begin{bmatrix} \eta_1 & \eta_1' \\ \eta_2 & \eta_2' \end{bmatrix}, \quad \mathbf{c} = \begin{bmatrix} c_2 \\ c_1 \end{bmatrix}, \quad \mathbf{z} = \begin{bmatrix} \Psi_{y_1} \\ \Psi_{y_3} \end{bmatrix}, \quad \mathbf{z}^\tau = \begin{bmatrix} \Psi_{y_2} \\ \Psi_{y_4} \end{bmatrix}, \quad \Phi^\tau = \Phi(h(x - \tau)).$$

Since $\eta_i$ composes a fundamental system of solutions of (6.4.25), $\Phi$ is a fundamental matrix, which has the properties $\Phi(v - c) = \Phi(v)C$ and $\det\Phi \neq 0$ with a nonsingular matrix $C = [c_{ij}]_{2\times2}$ [18]. By virtue of (6.4.17) one has that $\Phi(h(x - \tau)) = \Phi(h(x))C$. Multiplying (6.4.27) by $\Phi^{-1}$, system (6.4.27) is rewritten as

$$\mathbf{z} - \Psi_{y_5}\mathbf{c} + C\mathbf{z}^\tau = 0. \tag{6.4.28}$$

Hence, these equations are

$$\Psi_{y_1} - c_2\Psi_{y_5} + c_{11}\Psi_{y_2} + c_{12}\Psi_{y_4} = 0,$$
$$\Psi_{y_3} - c_1\Psi_{y_5} + c_{21}\Psi_{y_2} + c_{22}\Psi_{y_4} = 0.$$

Because the coefficients of these equations are constant, one easily obtains the universal invariant

$$J_3 + c_1 J_2 + c_2 J_1, \quad J_1^\tau - c_{11} J_1 - c_{21} J_2, \quad J_2^\tau - c_{12} J_1 - c_{22} J_2.$$

Therefore, the invariant equation has the form

$$J_3 = f\left(J_1^\tau - c_{11} J_1 - c_{21} J_2, \ J_2^\tau - c_{12} J_1 - c_{22} J_2\right) - (c_1 J_2 + c_2 J_1).$$

Because of the meaning of the functions $J_1$, $J_1^\tau$, $J_2$, $J_2^\tau$ and $J_3$, we present this equation as

$$y'' = f\left(y_\tau - c_{11} y - c_{21} y', \ y_\tau' - c_{12} y - c_{22} y'\right) - (c_1 y' + c_2 y).$$

**Case** $r = 3$. The Lie algebra is defined by

$$X_1 = \partial_{\bar{x}}, \quad X_2 = \eta_1(\bar{x})\partial_{\bar{y}}, \quad X_3 = \eta_2(\bar{x})\partial_{\bar{y}}, \quad X_3 = \eta_3(\bar{x})\partial_{\bar{y}},$$

where $\eta_1$, $\eta_2$, $\eta_3$ compose a fundamental system of solutions of the equation

$$\eta''' = -(c_1 \eta'' + c_2 \eta' + c_3 \eta). \tag{6.4.29}$$

In Table B.1 these Lie algebras are $L_{26}^4$, $L_{27}^4$, $L_{28}^4$, $L_{31}^4$, $L_{32}^4$, $L_{33}^4$, $L_{34}^4$, $L_{36}^4$ and $L_{37}^4$. Changing the variables (6.4.12), the generators become

$$X_1 = \frac{1}{h_x}\partial_x - \frac{g_x}{h_x g_y}\partial_y, \quad X_2 = \frac{\eta_1(h)}{g_y}\partial_y, \quad X_3 = \frac{\eta_2(h)}{g_y}\partial_y, \quad X_4 = \frac{\eta_3(h)}{g_y}\partial_y.$$

The general solution of the function $\hat{X}_1 J = 0$ is obtained similar to (6.4.21). Applying the generators $X_2^{(3)}$, $X_3^{(3)}$, $X_4^{(3)}$ to the function $J = \Psi(y_1, y_2, y_3, y_4, y_5, y_6, y_7, y_8)$ with

$$y_1 = J_1, \ y_2 = J_1^\tau, \ y_3 = J_2, \ y_4 = J_2^\tau, \ y_5 = J_3, \ y_6 = J_3^\tau, \ y_7 = J_4, \ y_8 = J_4^\tau,$$

$$J_4 = \frac{D(J_3(x, y, y', y''))}{Dh(x, y)}, \quad J_4^\tau = J_4(x - \tau, y_\tau, y_\tau', y_\tau'', y_\tau'''),$$

one obtains the system of differential equations

$$\eta_1 \Psi_{y_1} + \eta_1' \Psi_{y_3} + \eta_1'' \Psi_{y_5} + \eta_1''' \Psi_{y_7} + \eta_1^\tau \Psi_{y_2} + \eta_1'^\tau \Psi_{y_4} + \eta_1''^\tau \Psi_{y_6} + \eta_1'''^\tau \Psi_{y_8} = 0,$$

$$\eta_2 \Psi_{y_1} + \eta_2' \Psi_{y_3} + \eta_2'' \Psi_{y_5} + \eta_2''' \Psi_{y_7} + \eta_2^\tau \Psi_{y_2} + \eta_2'^\tau \Psi_{y_4} + \eta_2''^\tau \Psi_{y_6} + \eta_2'''^\tau \Psi_{y_8} = 0,$$

$$\eta_3 \Psi_{y_1} + \eta_3' \Psi_{y_3} + \eta_3'' \Psi_{y_5} + \eta_3''' \Psi_{y_7} + \eta_3^\tau \Psi_{y_2} + \eta_3'^\tau \Psi_{y_4} + \eta_3''^\tau \Psi_{y_6} + \eta_3'''^\tau \Psi_{y_8} = 0,$$

where $\eta_i^\tau = \eta_i(h(x - \tau))$, $\eta_i'^\tau = \eta_i'(h(x - \tau))$, $\eta_i''^\tau = \eta_i''(h(x - \tau))$, $\eta_i'''^\tau = \eta_i'''(h(x - \tau))$ $(i = 1, 2, 3)$. Here the variables $y_6$, $y_7$ and $y_8$ are introduced for simplicity of the representation of the equations: for second-order DODEs $\Psi_{y_6} = 0$, $\Psi_{y_7} = 0$, $\Psi_{y_8} = 0$. Substituting $\eta_i'''$, and $\eta_i'''^\tau$ found from (6.4.29), the above system of equations has matrix form

$$\Phi \mathbf{z} - \Psi_{y_7} \Phi \mathbf{c} + \Phi^\tau \mathbf{z}^\tau = 0. \tag{6.4.30}$$

Here

$$\Phi = \begin{bmatrix} \eta_1 & \eta_1' & \eta_1'' \\ \eta_2 & \eta_2' & \eta_2'' \\ \eta_3 & \eta_3' & \eta_3'' \end{bmatrix}, \quad \mathbf{c} = \begin{bmatrix} c_3 \\ c_2 \\ c_1 \end{bmatrix}, \quad \mathbf{z} = \begin{bmatrix} \Psi_{y_1} \\ \Psi_{y_3} \\ \Psi_{y_5} \end{bmatrix}, \quad \mathbf{z}^\tau = \begin{bmatrix} \Psi_{y_2} \\ \Psi_{y_4} \\ \Psi_{y_6} \end{bmatrix}, \quad \Phi^\tau = \Phi(h(x - \tau)).$$

Since $\eta_i$ composes a fundamental system of solutions of (6.4.29), $\Phi$ is a fundamental matrix, which has the properties $\Phi(h(x - \tau)) = \Phi(h(x))C$, $\det \Phi \neq 0$ with a nonsingular matrix $C = [c_{ij}]_{3 \times 3}$. Multiplying (6.4.30) by $\Phi^{-1}$, as in the previous case, system (6.4.30) is rewritten as

$$\mathbf{z} - \Psi_{y_7}\mathbf{c} + C\mathbf{z}^\tau = 0,$$

or

$$\Psi_{y_1} - c_3\Psi_{y_7} + c_{11}\Psi_{y_2} + c_{12}\Psi_{y_4} + c_{13}\Psi_{y_6} = 0,$$
$$\Psi_{y_3} - c_2\Psi_{y_7} + c_{21}\Psi_{y_2} + c_{22}\Psi_{y_4} + c_{23}\Psi_{y_6} = 0,$$
$$\Psi_{y_5} - c_1\Psi_{y_7} + c_{31}\Psi_{y_2} + c_{32}\Psi_{y_4} + c_{33}\Psi_{y_6} = 0.$$

Solving these equations and using the conditions $\Psi_{y_6} = \Psi_{y_7} = 0$, the universal invariant of this Lie algebra is found

$$J_1^\tau - c_{11}J_1 - c_{21}J_2 - c_{31}J_3, \quad J_2^\tau - c_{12}J_1 - c_{22}J_2 - c_{32}J_3.$$

Since second-order DODEs are studied, one needs to assume that

$$(c_{31})^2 + (c_{32})^2 \neq 0. \tag{6.4.31}$$

The invariant equation has the form

$$\phi(J_1^\tau - c_{11}J_1 - c_{21}J_2 - c_{31}J_3, \ J_2^\tau - c_{12}J_1 - c_{22}J_2 - c_{32}J_3) = 0,$$

where $\phi(z_1, z_2)$ is an arbitrary function. Because of the meaning of the functions $J_1, J_1^\tau, J_2, J_2^\tau$ and $J_3$, we represent this equation as

$$\phi(y_\tau - c_{11}y - c_{21}y' - c_{31}y'', \ y_\tau' - c_{12}y - c_{22}y' - c_{32}y'') = 0.$$

**Case $r \geq 4$.** Proceeding in the same manner as in the previous case, one obtains the universal invariant of the Lie algebra

$$J_1^\tau - \sum_{i=1}^r c_{i1}J_i, \quad J_2^\tau - \sum_{i=1}^r c_{i2}J_i,$$

where $J_i$ is the representation of $\bar{y}^{(i-1)}$ after the change of variables. The set of equations admitting the generator $L_{50}^5$ is

$$\phi\left(J_1^\tau - \sum_{i=1}^r c_{i1}J_i, \ J_2^\tau - \sum_{i=1}^r c_{i2}J_i\right) = 0,$$

where $\phi(z_1, z_2)$ is an arbitrary function which satisfies the restrictions

$$c_{i1}\phi_{z_1} + c_{i2}\phi_{z_2} = 0, \quad i = 4, \ldots, r.$$

As before, because of the meaning of the functions $J_1, J_1^\tau, J_2, J_2^\tau$ and $J_3$, we represent the invariant equation in the form

$$\phi\left(y_\tau - \sum_{i=1}^r c_{i1}y^{(i-1)}, \ y_\tau' - \sum_{i=1}^r c_{i2}y^{(i-1)}\right) = 0.$$

**Remark 6.4.3** Condition (6.4.31) guarantees that the denominators in the representation of second-order DODE admitting Lie algebras $L_{26}^4, L_{27}^4, L_{28}^4, L_{31}^4, L_{32}^4,$ $L_{33}^4, L_{34}^4, L_{36}^4$ and $L_{37}^4$ are not equal to zero.

**Example 6.4.3** Consider the Lie algebra $L_{21}^3$. This Lie algebra is defined by the generators

$$X_1 = \partial_{\bar{x}}, \quad X_2 = \bar{x}\partial_{\bar{x}}, \quad X_3 = \bar{x}^2\partial_{\bar{x}}$$

which after changing the variables become

$$X_1 = \frac{1}{h_x}\partial_x - \frac{g_x}{h_x g_y}\partial_y, \quad X_2 = \frac{h}{h_x}\partial_x - \frac{hg_x}{h_x g_y}\partial_y, \quad X_3 = \frac{h^2}{h_x}\partial_x - \frac{h^2 g_x}{h_x g_y}\partial_y.$$

Invariants of the first generator are (6.4.21). Applying the second generator $\hat{X}_2$ to the function $\Phi(y_1, y_2, y_3, y_4, y_5)$ with

$$y_1 = J_1, \quad y_2 = J_1^\tau, \quad y_3 = J_2, \quad y_4 = J_2^\tau, \quad y_5 = J_3,$$

one obtains the equation

$$y_3\Phi_{y_3} + y_4\Phi_{y_4} + 2y_5\Phi_{y_5} = 0.$$

The general solution of this equation is $\Phi = \psi(v_1, v_2, v_3, v_4)$, where $\psi$ is an arbitrary function and

$$v_1 = y_1, \quad v_2 = y_2, \quad v_3 = \frac{y_5}{(y_4)^2}, \quad v_4 = \frac{y_5}{(y_3)^2}.$$

Applying the generator $\hat{X}_3$ to the function $\psi(J_1, J_1^\tau, \frac{J_3}{(J_2^\tau)^2}, \frac{J_3}{(J_2)^2})$, one finds

$$v_3\psi_{v_3} + v_4\psi_{v_4} = 0.$$

Thus, the universal invariant of this Lie algebra, $J_1, J_1^\tau, (\frac{J_2^\tau}{J_2})^2$, has no any term with the second-order derivative. Hence, there are no second-order DODEs admitting a Lie algebra equivalent to $L_{21}^3$.

**Example 6.4.4** Consider the Lie Algebra $L_{16}^3$, which is defined by the generators

$$X_1 = \partial_{\bar{x}}, \quad X_2 = \partial_{\bar{y}}, \quad X_3 = (b\bar{x} + \bar{y})\partial_{\bar{x}} + (b\bar{y} - \bar{x})\partial_{\bar{y}}.$$

Changing the variables and using the conditions $\xi_{iy} = 0$ and $\xi_i(x) = \xi_i(x - \tau)$ ($i = 1, 2, 3$), one gets $h_y = g_y = 0$. This is the contradiction to the assumption $\Delta \neq 0$.

### 6.4.6  List of Invariant DODEs

Table B.2 shows the complete group classification of second-order DODEs. The second column in the table is the representation of equivalent of Lie algebra from Table B.1. Generators of this Lie algebra are presented in the third column. The representation of second-order delay ordinary differential equations admitting the Lie algebra is shown in the last column. The representation is given in schematic form: the variables $x, y, y_\tau, y', y'_\tau, y''$ in this table have to be exchanged with their images after a change of the variables.

## 6.5  Equivalence Lie Group for DDE

Most differential equations include arbitrary elements: constants and functions of the independent and dependent variables. A transformation of the independent and dependent variables, and arbitrary elements is called an equivalence transformation of a system of differential equations if it conserves a differential structure of the equations. If a set of equivalence transformations of partial differential equations composes a Lie group of transformations, then the Lie group is called an equivalence group.[14]

Since equivalence transformations allow one to simplify the system of equations in hand, the problem of finding equivalence transformations is one of the milestones in group classification problem.

This section is devoted to a reasonable generalization of the definition of an equivalence Lie group for delay differential equations. The main item in defining an equivalence Lie group for delay differential equations is similar to that used in defining an admitted Lie group. Recall we say that a Lie group is admitted by delay differential equations if the coefficients of the infinitesimal generator of this group satisfy the determining equations. In writing the determining equations of an admitted Lie group of point transformations the Lie–Bäcklund representation of an infinitesimal generator is used. However, even for partial differential equations there does not exist a definition of a Lie–Bäcklund operator of an equivalence Lie group. Hence, this section starts with the presentation of determining equations for the equivalence Lie group of partial differential equations in terms of the Lie–Bäcklund operator. These equations can be obtained by differentiating with respect to the group parameter the transformed system of partial differential equations in which the transformed solution has been substituted.

### 6.5.1  Lie–Bäcklund Representation of Determining Equations for the Equivalence Lie Group of PDEs

Consider a system of partial differential equations with the independent variable $x$, dependent variable $u$, and arbitrary element $\phi$, which transfers a system of differential equations of the given class

---

[14]The infinitesimal approach for finding equivalence group of partial differential equations is given in [17]. The generalization of this approach is considered in [13]. Extension of the last approach to delay differential equations is presented in [14].

$$F^k(x, u, p, \phi) = 0 \quad (k = 1, 2, \dots, s) \qquad (6.5.1)$$

into the system of equations of the same class. Here $(x, u) \in V \subset R^{n+m}$, and $\phi$: $V \to R^t$.

The problem of finding equivalent transformation is formulated as finding a transformation of the space $R^{n+m+t}(x, u, \phi)$ that preserves the equations, while only the change of their representative $\phi = \phi(x, u)$ is accepted. Assume that a one-parameter Lie group of transformations of the space $R^{n+m+t}$ with the group parameter $a$:

$$x' = f^x(x, u, \phi; a), \ u' = f^u(x, u, \phi; a), \ \phi' = f^\phi(x, u, \phi; a) \qquad (6.5.2)$$

satisfies this property. The generator of this Lie group has the form

$$X^e = \xi \partial_x + \eta^u \partial_u + \eta^\phi \partial_\phi,$$

where the coordinates are

$$\xi^i = \xi^i(x, u, \phi), \ \eta^{u^j} = \eta^{u^j}(x, u, \phi), \ \eta^{\phi^k} = \eta^{\phi^k}(x, u, \phi)$$

$$(i = 1, \dots, n; j = 1, \dots, m; k = 1, \dots, t).$$

The equivalent Lie–Bäcklund form of this generator is

$$\hat{X}^e = \zeta^u \partial_u + \zeta^\phi \partial_\phi. \qquad (6.5.3)$$

Here the coordinates are

$$\zeta^{u^j} = \eta^{u^j} - u_x^j \xi, \quad \zeta^{\phi^k} = \eta^{\phi^k} - \xi D_x^e \phi^k,$$

where $D_{x_i}^e = \partial_{x_i} + u_{x_i} \partial_u + (\phi_u u_{x_i} + \phi_{x_i}) \partial_\phi$.

Any solution $u_0(x)$ of system (6.5.1) with the functions $\phi(x, u)$ is transformed by (6.5.2) into the solution $u = u_a(x')$ of system (6.5.1) with the same functions $F^k$, and another (transformed) function $\phi_a(x, u)$. The function $\phi_a(x, u)$ is defined as follows. Solving the relations

$$x' = f^x(x, u, \phi(x, u); a), \quad u' = f^u(x, u, \phi(x, u); a)$$

for $(x, u)$, one obtains

$$x = g^x(x', u'; a), \quad u = g^u(x', u'; a). \qquad (6.5.4)$$

The transformed function is

$$\phi_a(x', u') = f^\phi(x, u, \phi(x, u); a),$$

where one should replace $x$ and $u$ by expressions (6.5.4). In view of the definition of the function $\phi_a(x', u')$, there exists the following identity with respect to $x$ and $u$:

$$\left( \phi_a \circ \left( f^x, f^u \right) \right)(x, u, \phi(x, u); a) = f^\phi(x, u, \phi(x, u); a).$$

The transformed solution $T_a(u) = u_a(x)$ is obtained by solving the relations

$$x' = f^x(x, u_0(x), \phi(x, u_0(x)); a)$$

for $x$ and substituting this solution $x = \psi^x(x'; a)$ into

$$u_a(x') = f^u(x, u_0(x), \phi(x, u_0(x)); a).$$

As for the function $\phi_a$, there is the identity with respect to $x$

$$\left(u_a \circ f^x\right)(x, u_o(x), \phi(x, u_o(x)); a) = f^u(x, u_o(x), \phi(x, u_o(x)); a).$$

(6.5.5)

Formulae for transformations of the partial derivatives $p'_a = f^p(x, u, p, \phi, \ldots, a)$ are obtained by differentiating (6.5.5) with respect to $x'$.

Since the transformed function $u_a(x')$ is a solution of system (6.5.1) with the transformed arbitrary element $\phi_a(x', u')$, the equations

$$F^k(x', u_a(x'), p'_a(x'), \phi_a(x', u_a(x'))) = 0 \quad (k = 1, 2, \ldots, s) \qquad (6.5.6)$$

are satisfied for an arbitrary $x'$. Because of a one-to-one correspondence between $x$ and $x'$ one has

$$F^k(f^x(z(x), a), f^u(z(x), a), f^p(z_p(x), a), f^\phi(z(x))) = 0 \quad (k = 1, 2, \ldots, s)$$

(6.5.7)

where $z(x) = (x, u_o(x), \phi(x, u_o(x)))$, $z_p(x) = (x, u_o(x), \phi(x, u_o(x)), p_o(x), \ldots)$.

Differentiating (6.5.6) with respect to the group parameter $a$, and setting $a = 0$, one obtains the determining equations in Lie–Bäcklund form:[15]

$$\tilde{X}^e F^k(x, u, p, \phi)_{|(S)} = 0 \quad (k = 1, 2, \ldots, s). \qquad (6.5.8)$$

The prolonged operator for the equivalence Lie group

$$\tilde{X}^e = \hat{X}^e + \zeta^{u_x} \partial_{u_x} + \cdots$$

has the following coordinates related to the dependent functions

$$\zeta^{u_\lambda} = D_\lambda^e \zeta^u, \qquad D_\lambda^e = \partial_\lambda + u_\lambda \partial_u + (\phi_u u_\lambda + \phi_\lambda) \partial_\phi,$$

where $\lambda$ stands for $x_i$ $(i = 1, 2, \ldots, n)$. The sign $|(S)$ means that the equations $\tilde{X}^e F^k(x, u, p, \phi)$ are considered on any solution $u_o(x)$ of (6.5.1).

The set of transformations, which is generated by one-parameter Lie groups corresponding to the generators $X^e$, is called an equivalence Lie group. This group is denoted by $GS^e$.

The determining equations (6.5.8) were obtained by using the existence of the solution of (6.5.1). In constructing (6.5.8) one can use a geometrical approach in which the equivalence group is defined by (6.5.8) without the requirement of the existence of a solution of (6.5.1). In this case the sign $|(S)$ means that the equations $\tilde{X}^e F^k(x, u, p, \phi)$ are considered on the manifold defined by (6.5.1). The difference between these two approaches consists in defining the sign $|(S)$. Note that the same difference between the geometrical approach and the others lies in the definitions for obtaining an admitted Lie group.

---

[15]In contrast, differentiating (6.5.7) with respect to the group parameter $a$, and setting $a = 0$, one obtains the determining equations of an admitted Lie group in the classical form [13].

### 6.5.2 Potential Equivalence Lie Group of Delay Differential Equations

Let us consider delay differential equations which include arbitrary elements. The arbitrary elements are functions and constants which are not specified in the equations. For constructing an equivalence Lie group for these delay differential equations one can apply similar procedure as described above.

Firstly determining equations of an equivalence Lie group are constructed. These equations are obtained on the basis that a Lie group of transformations of the independent, dependent variables, and arbitrary elements transforms a solution of the original system of equations into the solution of a system of equations which differs from the original system only by arbitrary elements. Differentiating with respect to the group parameter and assigning it to zero, one obtains the determining equations.

A solution of the determining equations gives the generator of a Lie group. This Lie group of transformations is called a potential equivalence Lie group. Notice that for partial differential equations[16] by virtue of the inverse function theorem a potential equivalence Lie group simply becomes an equivalence Lie group.

An example of obtaining a potential equivalence Lie group is given in the next section, where the reaction diffusion equation with a delay is studied.

## 6.6  The Reaction–Diffusion Equation with a Delay

The reaction–diffusion delay differential equation

$$u_t(t, x) = u_{xx}(t, x) + g(x, u(t, x), u(t - \tau, x)) \quad (t > t_o). \qquad (6.6.1)$$

arises in many fields of application. The theory and applications of this equation can be found in [23]. The complete group classification of the reaction diffusion equation with delay is presented in this section.[17] For differential equations, an admitted Lie group allows one to find invariant solutions. Although for delay differential equations such theorems do not exist, there is the assumption that an admitted Lie group of delay differential equations can be also applied for obtaining invariant solutions. This section confirms this assumption. In this section the found admitted Lie groups are applied for the construction of invariant solutions of the reaction diffusion equation with a delay.

**Remark 6.6.1** Symmetry-preserving discrete schemes for some heat transfer equations were studied in [1].

---

[16]For partial differential equations, the equivalence group is found by solving the determining equations, and conversely: any solution of the determining equations composes a Lie group of equivalence transformations of partial differential equations.

[17]Application of the group analysis method for studying this equation is given in [14].

### 6.6.1 The Cauchy Problem

The theory of existence of solutions of (6.6.1) can be found in [23]. For example, the initial conditions for the Cauchy problem are

$$u(s, x) = \varphi(s, x) \quad (t_o - \tau \leq s \leq t_o),$$

where $\varphi(s, x)$ is an arbitrary function. Due to the arbitrariness of $\varphi(s, x)$ one can conclude that the values $u(t_o, x_o)$, $u(t_o - \tau, x_o)$, $u_x(t_o, x_o)$, $u_x(t_o - \tau, x_o)$, $u_{xx}(t_o, x_o)$, $u_{xx}(t_o - \tau, x_o)$ and other derivatives are arbitrary. In constructing Lie groups this property allows one to split determining equations.

### 6.6.2 The Equivalence Lie Group

For the sake of simplicity the new dependent variable is introduced $v$, which is related to $u$ by the formula

$$v(t, x) = u(t - \tau, x). \tag{6.6.2}$$

In view of (6.6.2) the equation (6.6.1) becomes the partial differential equations with two dependent variables

$$S \equiv u_t - (u_{xx} + g) = 0, \tag{6.6.3}$$

with the arbitrary element $g = g(x, u, v)$. The generator of the Lie group of equivalent transformations takes the form

$$X^e = \xi \partial_x + \eta \partial_t + \zeta \partial_u + \zeta^v \partial_v + \zeta^g \partial_g,$$

where $\xi = \xi(t, x, u, v, g)$, $\eta = \eta(t, x, u, v, g)$, $\zeta = \zeta(t, x, u, v, g)$, $\zeta^v = \zeta^v(t, x, u, v, g)$, $\zeta^g = \zeta^g(t, x, u, v, g)$.

Applying the algorithm described earlier to (6.6.3), one obtains the determining equation

$$\left( \zeta^{u_t} - \zeta^{u_{xx}} - \zeta^g + \xi D_x g + \eta D_t g \right)_{|(S)} = 0, \tag{6.6.4}$$

where

$$\zeta^{u_t} = D_t(\zeta - \xi u_c - \eta u_t), \quad \zeta^{u_{xx}} = D_x^2(\zeta - \xi u_x - \eta u_t). \tag{6.6.5}$$

The determining equation of an admitted Lie group related to (6.6.2) is

$$\left\{ \zeta^v(z(t, x)) - \zeta(z(t - \tau, x)) - v_t(t, x) \big( \xi(z(t, x)) - \xi(z(t - \tau, x)) \big) \right.$$
$$\left. - v_x(t, x) \big( \eta(z(t, x)) - \eta(z(t - \tau, x)) \big) \right\}_{|(S)} = 0, \tag{6.6.6}$$

where

$$z(t, x) = (t, x, u(t, x), v(t, x), g(x, u(t, x), v(t, x))).$$

Substituting the coefficients (6.6.5) into (6.6.4) and replacing the derivatives

$$u_{tt} = u_t g_u + g_v v_t + u_{xxt}, \quad u_{xt} = v_x g_v + g_u u_x + u_{xxx} + g_x,$$

$$u_t = u_{xx} + g, \quad v_t = v_{xx} + \bar{g},$$

found from (6.6.3), the determining equation (6.6.4) becomes

$$-v_x^2 u_x \xi_{vv} - v_x^2 u_{xx} \eta_{vv} + v_x^2(-\eta_{vv}g - 2\eta_v g_v + \zeta_{vv}) - 2v_x u_x^2 \xi_{uv} - 2v_x u_x u_{xx} \eta_{uv}$$

$$+ 2v_x u_x(-\eta_{uv}g - \eta_u g_v - \eta_v g_u - \xi_{xv} + \zeta_{uv}) - 2v_x u_{xx}(\eta_{xv} + \xi_v) - u_x^3 \xi_{uu}$$

$$+ 2v_x(-\eta_{xv}g - \eta_v g_x - \eta_x g_v + \zeta_{xv}) - u_x^2 u_{xx} \eta_{uu} - 2u_x u_{xx}(\eta_{xu} + \xi_u)$$

$$+ u_x^2(-\eta_{uu}g - 2\eta_u g_u - 2\xi_{xu} + \zeta_{uu}) - 2v_x u_{xxx} \eta_v - 2u_x u_{xxx} \eta_u - 2u_{xxx} \eta_x$$

$$+ u_x(\bar{g}\xi_v - 2\eta_{xu}g - 2\eta_u g_x - 2\eta_x g_u + \xi_t + \xi_u g - \xi_{xx} + 2\zeta_{xu})$$

$$+ u_{xx}(\bar{g}\eta_v + \eta_t + \eta_u g - \eta_{xx} - 2\xi_x)$$

$$+ \bar{g}(\eta_v g - \zeta_v) + \eta_t g + \eta_u g^2 - \eta_{xx}g - 2\eta_x g_x - \zeta_t - \zeta_u g + \zeta_{xx} + \zeta^g = 0.$$

Splitting this equation with respect to $u_x, u_{xx}, v_x, u_{xxx}$ and $\bar{g}$ one obtains[18]

$$\eta_t g + \eta_u g^2 - \eta_{xx}g - 2\eta_x g_x - \zeta_t - \zeta_u g + \zeta_{xx} + \zeta^g = 0, \qquad (6.6.7)$$

$$-2\eta_{xu}g - 2\eta_u g_x - 2\eta_x g_u + \xi_t + \xi_u g - \xi_{xx} + 2\zeta_{xu} = 0, \qquad (6.6.8)$$

$$\eta_t + \eta_u g - \eta_{xx} - 2\xi_x = 0, \qquad (6.6.9)$$

$$\eta_x = 0, \quad \eta_u = 0, \quad \eta_v = 0, \quad \eta_{xu} + \xi_u = 0,$$

$$\eta_{xv} + \xi_v = 0, \quad \xi_v = 0, \quad \eta_v g - \zeta_v = 0, \qquad (6.6.10)$$

$$\xi_{uu} = 0, \quad \eta_{uu} = 0, \quad \eta_{vv} = 0, \quad \eta_{uv} = 0, \quad \xi_{vv} = 0, \quad \xi_{uv} = 0,$$

$$-\eta_{xv}g - \eta_v g_x - \eta_x g_v + \zeta_{xv} = 0, \quad -\eta_{vv}g - 2\eta_v g_v + \zeta_{vv} = 0,$$

$$-\eta_{uv}g - \eta_u g_v - \eta_v g_u - \xi_{xv} + \zeta_{uv} = 0, \qquad (6.6.11)$$

$$-\eta_{uu}g - 2\eta_u g_u - 2\xi_{xu} + \zeta_{uu} = 0.$$

From (6.6.10) one obtains

$$\eta_x = 0, \quad \eta_u = 0, \quad \eta_v = 0, \quad \xi_u = 0, \quad \xi_v = 0, \quad \zeta_v = 0.$$

Differentiating (6.6.9) with respect to $x$, one gets $\xi_{xx} = 0$. Hence $\xi = \xi_1 x + \xi_0$, where $\xi_0 = \xi_0(t)$ and $\xi_1 = \xi_1(t)$, and then $\xi_1 = \eta_t/2$. The general solution of (6.6.11) is $\zeta = \zeta_1 u + \zeta_0$, where $\zeta_1 = \zeta_1(t, x)$, $\zeta_0 = \zeta_0(t, x)$. Solving (6.6.8), one obtains $\zeta_1 = -\xi_0' x/2 - \eta'' x^2/8 + \zeta_{10}$, where $\zeta_{10} = \zeta_{10}(t)$.

For the sake of simplicity we study the case $g_x = 0$. The assumption that the function $g$ does not depend on $t$ and $x$ gives the conditions

$$\zeta_t = 0, \quad \zeta_x = 0, \quad \zeta_t^g = 0, \quad \zeta_x^g = 0.$$

These equations give $\eta = 2k_3 t + k_4$, $\xi_0 = k_7$, $\zeta_{0x} = 0$. From (6.6.7) one finds

$$\zeta^g = \zeta_{0t} - 2k_3 g + g\zeta_{10}.$$

---

[18]One could also split with respect to $g_u, g_v$.

From the equation $\zeta_t^g = 0$ one obtains $\zeta_{0tt} = 0$, or

$$\zeta_0 = k_2 t + k_1.$$

Thus,

$$\xi = 2k_3 t + k_4, \ \eta = k_3 x + k_6, \ \zeta = k_1 + k_2 t + k_5 u$$

or

$$X^e = k_1 X_1^e + k_2 X_2^e + k_3 X_3^e + k_4 X_4^e + \zeta_{10} X_5^e + \xi_0 X_6^e + \zeta^v \partial_v,$$

where

$$X_1^e = \partial_u, \ X_2^e = \partial_g + t \partial_u, \ X_3^e = -2g \partial_g + 2t \partial_t + x \partial_x,$$
$$X_4^e = \partial_t, \ X_5^e = g \partial_g + u \partial_u, \ X_6^e = \partial_x.$$

Equation (6.6.6) becomes

$$\zeta^v(z(t,x)) = k_1 + k_2(t - \tau) + k_5 v(t, x).$$

This gives

$$\zeta^v(t, x, u, v) = k_1 + k_2(t - \tau) + k_5 v.$$

## 6.6.3 Admitted Lie Group of Equation

The generator of the Lie group admitted by (6.6.1) is

$$X = \xi \partial_x + \eta \partial_t + \zeta \partial_u,$$

where $\xi$, $\eta$ and $\zeta$ are functions of $x, t$ and $u$.

According to the algorithm for constructing determining equations of an admitted Lie group, one obtains

$$-\zeta^{u_t} + \zeta^{u_{xx}} + g_u \zeta^u + g_{\bar{u}} \zeta^{\bar{u}} = 0, \tag{6.6.12}$$

where

$$\zeta^u = \zeta - u_x \xi - u_t \eta, \ \zeta^{\bar{u}} = \bar{\zeta} - \bar{u}_x \bar{\xi} - \bar{u}_t \bar{\eta}, \ \zeta^{u_x} = D_x \zeta^u,$$
$$\zeta^{u_{xx}} = D_x \zeta^{u_x}, \ \zeta^{u_t} = D_t \zeta^u.$$

Here the bar over a function $f(x, t)$ means $\bar{f}(x, t) = f(x, t - \tau)$. The determining equation has to be satisfied for any solution $u(x, t)$ of (6.6.1). Since the determining equation is considered on a solution of (6.6.1), the value $\bar{f}$ for a function $f(x, t, u)$ is defined as $\bar{f}(x, t) = f(x, t - \tau, u(x, t - \tau))$.

Substituting into the determining equation (6.6.12) the derivatives $u_t, u_{xt}, u_{tt}, \bar{u}_t$ found from (6.6.1) and its prolongations, one obtains

$$\bar{g}g_{\bar{u}}(\eta - \bar{\eta}) - 2\bar{u}_x u_x \eta_u g_{\bar{u}} + \bar{u}_x g_{\bar{u}}(-2\eta_x + \xi - \bar{\xi}) + \bar{u}_{xx}g_{\bar{u}}(\eta - \bar{\eta})$$
$$- u_x^3 \xi_{uu} - u_x^2 u_{xx}\eta_{uu} + u_x^2(-\eta_{uu}g - 2\eta_u g_u - 2\xi_{xu} + \zeta_{uu})$$
$$- 2u_x u_{xx}(\eta_{xu} + \xi_u) - 2u_x u_{xxx}\eta_u + u_x(-2\eta_{xu}g - 2\eta_u g_x - 2\eta_x g_u$$
$$+ \xi_t + \xi_u g - \xi_{xx} + 2\zeta_{xu}) + u_{xx}(\eta_t + \eta_u g - \eta_{xx} - 2\xi_x) - 2u_{xxx}\eta_x$$
$$+ \eta_t g + \eta_u g^2 - \eta_{xx}g - 2\eta_x g_x + g_u \zeta + g_{\bar{u}}\zeta + g_x \xi - \zeta_t - \zeta_u g + \zeta_{xx} = 0,$$

$$(6.6.13)$$

where $\bar{g} = g(u(x, t - \tau), u(x, t - 2\tau))$. Splitting this equation with respect to the derivatives $u_x, u_{xx}, u_{xxx}, \bar{u}_x, \bar{u}_{xx}$ and using the property $g_{\bar{u}} \neq 0$, one comes to the equations

$$\eta_t g + \eta_u g^2 - \eta_{xx}g - 2\eta_x g_x + g_u \zeta + g_{\bar{u}}\bar{\zeta} + g_x \xi$$
$$- \zeta_t - \zeta_u g + \zeta_{xx} = 0, \qquad\qquad (6.6.14)$$

$$-2\eta_{xu}g - 2\eta_u g_x - 2\eta_x g_u + \xi_t + \xi_u g - \xi_{xx} + 2\zeta_{xu} = 0, \qquad (6.6.15)$$

$$-\eta_{uu}g - 2\eta_u g_u - 2\xi_{xu} + \zeta_{uu} = 0, \qquad\qquad (6.6.16)$$

$$\eta_t + \eta_u g - \eta_{xx} - 2\xi_x = 0, \qquad\qquad (6.6.17)$$

$$\eta_{xu} + \xi_u = 0, \qquad\qquad (6.6.18)$$

$$\xi_{uu} = 0, \ \eta_{uu} = 0, \ \eta_x = 0, \ \eta_u = 0, \qquad\qquad (6.6.19)$$

$$-2\eta_x + \xi - \bar{\xi} = 0, \qquad\qquad (6.6.20)$$

$$\eta_u = 0, \qquad\qquad (6.6.21)$$

$$\eta - \bar{\eta} = 0. \qquad\qquad (6.6.22)$$

Simplification of (6.6.18)–(6.6.22) yields

$$\eta_u = 0, \ \eta_x = 0, \ \xi_u = 0,$$

and

$$\xi(x, t - \tau) = \xi(x, t), \quad \eta(t - \tau) = \eta(t).$$

From (6.6.16) and (6.6.17) one gets $\xi = x\eta_t/2 + \xi_0$ and $\zeta = u\zeta_1 + \zeta_0$, where $\xi_0 = \xi_0(t)$, $\zeta_1 = \zeta_1(x, t)$ and $\zeta_0 = \zeta_0(x, t)$. Substituting this into (6.6.15) one obtains

$$\eta_{tt}x + 2\xi_{0t} + 4\zeta_{1x} = 0,$$

and integrating the equation with respect to $x$ one gets

$$\zeta_1 = -\eta_{tt}x^2/8 - x\xi_{0t}/2 + \zeta_{10},$$

where $\zeta_{10} = \zeta_{10}(t)$. Thus, (6.6.14) is the only unsolved equation of the set (6.6.14)–(6.6.22). This equation becomes

$$8g_u\zeta_0 + 8g_{\bar{u}}\bar{\zeta}_0 + g_u u(-\eta_{tt}x^2 - 4\xi_{0t}x + 8\zeta_{10})$$
$$+ g_{\bar{u}}\bar{u}(-\eta_{tt}x^2 - 4\xi_{0t}x + 8\bar{\zeta}_{10}) + 4g_x(\eta_t x + 2\xi_0)$$
$$+ g(\eta_{tt}x^2 + 8\eta_t + 4\xi_{0t}x - 8\zeta_{10}) + \eta_{ttt}ux^2$$
$$- 2\eta_{tt}u + 4\xi_{0tt}ux - 8\zeta_{0t} + 8\zeta_{0xx} - 8\zeta_{10t}u = 0. \qquad (6.6.23)$$

Differentiating (6.6.23) with respect to $u$ and $\bar{u}$, one obtains

$$8\zeta_0 g_{uu} + 8\bar{\zeta}_0 g_{u\bar{u}} + 8g_u \eta_t + \eta_{ttt} x^2 - \eta_{tt} g_{u\bar{u}} \bar{u} x^2 - \eta_{tt} g_{uu} u x^2$$
$$- 2\eta_{tt} + 4\eta_t g_{xu} x - 4g_{u\bar{u}} \xi_{0t} \bar{u} x$$
$$+ 8g_{u\bar{u}} \bar{u} \zeta_{10} + 8g_{xu} \xi_0 - 4g_{uu} \xi_{0t} u x$$
$$+ 8g_{uu} u \zeta_{10} + 4\xi_{0tt} x - 8\zeta_{10t} = 0, \tag{6.6.24}$$

$$8\zeta_0 g_{u\bar{u}} + 8\bar{\zeta}_0 g_{\bar{u}\bar{u}} + 8g_{\bar{u}} (\eta_t - \zeta_{10} + \bar{\zeta}_{10})$$
$$- \eta_{tt} g_{u\bar{u}} u x^2 - \eta_{tt} g_{\bar{u}\bar{u}} \bar{u} x^2 + 4\eta_t g_{x\bar{u}} x$$
$$- 4g_{u\bar{u}} \xi_{0t} u x + 8g_{u\bar{u}} u \zeta_{10} + 8g_{x\bar{u}} \xi_0$$
$$- 4g_{\bar{u}\bar{u}} \xi_{0t} \bar{u} x + 8g_{\bar{u}\bar{u}} \bar{u} \bar{\zeta}_{10} = 0. \tag{6.6.25}$$

Equations (6.6.24) and (6.6.25) are linear algebraic equations with respect to $\zeta_0$ and $\bar{\zeta}_0$. The determinant of the matrix of this linear system of equations is equal to

$$\Delta = g_{u\bar{u}}^2 - g_{uu} g_{\bar{u}\bar{u}}.$$

## 6.6.4 Case $\Delta \neq 0$

For $\Delta \neq 0$, $\zeta_0$ and $\bar{\zeta}_0$ are given by

$$\zeta_0 = (4\eta_t (2(g_u g_{\bar{u}\bar{u}} - g_{\bar{u}} g_{u\bar{u}}) + x(g_{xu} g_{\bar{u}\bar{u}} - g_{u\bar{u}} g_{x\bar{u}})) + \eta_{ttt} g_{\bar{u}\bar{u}} x^2$$
$$+ \eta_{tt} (u x^2 \Delta - 2g_{\bar{u}\bar{u}}) + 4\xi_{0t} x u \Delta + 4\xi_{0tt} x g_{\bar{u}\bar{u}} + 8\xi_0 (g_{xu} g_{\bar{u}\bar{u}} - g_{u\bar{u}} g_{x\bar{u}})$$
$$+ 8g_{\bar{u}} g_{u\bar{u}} (\zeta_{10} - \bar{\zeta}_{10}) - 8u\Delta \bar{\zeta}_{10} - 8g_{\bar{u}\bar{u}} \zeta_{10t})/(8\Delta),$$

$$\bar{\zeta}_0 = (4\eta_t (2(g_{\bar{u}} g_{uu} - g_u g_{u\bar{u}}) + x(g_{uu} g_{x\bar{u}} - g_{u\bar{u}} g_{xu})) - \eta_{ttt} g_{u\bar{u}} x^2$$
$$+ \eta_{tt} (\bar{u} x^2 \Delta + 2g_{u\bar{u}}) + 4\xi_{0t} \bar{u} x \Delta - 4g_{u\bar{u}} \xi_{0tt} x + 8\xi_0 (g_{uu} g_{x\bar{u}} - g_{u\bar{u}} g_{xu})$$
$$- 8g_{\bar{u}} g_{uu} (\zeta_{10} - \bar{\zeta}_{10}) - 8g_{u\bar{u}}^2 \bar{u} \bar{\zeta}_{10} + 8g_{u\bar{u}} \zeta_{10t} + 8g_{uu} g_{\bar{u}\bar{u}} \bar{u} \bar{\zeta}_{10})/(8\Delta).$$

It is assumed here that[19]

$$g_x = 0.$$

Notice that in this case the kernel of admitted Lie algebras contains translations in $x$ and $t$:

$$X_1 = \partial_t, \qquad X_2 = \partial_x.$$

Splitting (6.6.23) with respect to $x$, one has

---

[19] The general case where $g_x \neq 0$ is too complicated for solving.

$$\eta_{tttt}\alpha_4 + \eta_{ttt}\alpha_3 + \alpha_2\eta_{tt} = 0, \quad \xi_{0ttt}\beta_3 + \xi_{0tt}\beta_2 + \beta_1\xi_{0t} = 0, \quad (6.6.26)$$

$$2\eta_{ttt}g_{\bar{u}\bar{u}} + 4\eta_t(g_u^2 g_{\bar{u}\bar{u}} - 2g_u g_{\bar{u}} g_{u\bar{u}} + g_{\bar{u}}^2 g_{uu} + g\Delta) - 5\eta_{tt}\alpha_3$$
$$+ 4g_{\bar{u}}(g_u g_{u\bar{u}} - g_{\bar{u}} g_{uu})(\zeta_{10} - \bar{\zeta}_{10}) - 4g\Delta\zeta_{10}$$
$$- 4g_u g_{\bar{u}\bar{u}}\zeta_{10t} + 4g_{\bar{u}} g_{u\bar{u}} 4\bar{\zeta}_{10t} + 4g_{\bar{u}\bar{u}}\zeta_{10tt} = 0, \quad (6.6.27)$$

where

$$\beta_3 = \alpha_4 = -g_{\bar{u}\bar{u}}, \quad \beta_2 = \alpha_3 = -(g_{u\bar{u}}g_{\bar{u}} - g_u g_{\bar{u}\bar{u}}), \quad \beta_1 = \alpha_2 = g\Delta.$$

The assumption $\eta_{tt}^2 + \xi_{0t}^2 \neq 0$ is in a contradiction with the condition $\Delta \neq 0$. Since $\eta(t - \tau) = \eta(t)$ and $\xi(t - \tau) = \xi(t)$, one finds that $\eta(t)$ and $\xi(t)$ are constant. Hence, $\zeta_0$, $\bar{\zeta}_0$ and (6.6.27) become

$$\zeta_0 = \alpha_1\zeta_{10t} + \beta_1\zeta_{10} + \gamma_1\bar{\zeta}_{10}, \quad \bar{\zeta}_0 = \alpha_2\zeta_{10t} + \beta_2\zeta_{10} + \gamma_2\bar{\zeta}_{10},$$

$$\zeta_{10}(g_{\bar{u}}(g_{u\bar{u}}g_u - g_{uu}g_{\bar{u}}) - g\Delta) - \bar{\zeta}_{10}g_{\bar{u}}(g_{u\bar{u}}g_u - g_{uu}g_{\bar{u}})$$
$$+ g_{\bar{u}}g_{u\bar{u}}\bar{\zeta}_{10t} - g_u g_{\bar{u}\bar{u}}\zeta_{10t} + g_{\bar{u}\bar{u}}\zeta_{10tt} = 0, \quad (6.6.28)$$

where

$$\alpha_1 = -g_{\bar{u}\bar{u}}/\Delta, \quad \beta_1 = -u + g_{u\bar{u}}g_{\bar{u}}/\Delta, \quad \gamma_1 = -g_{\bar{u}}g_{u\bar{u}}/\Delta,$$
$$\alpha_2 = g_{u\bar{u}}/\Delta, \quad \beta_2 = -g_{\bar{u}}g_{uu}/\Delta, \quad \gamma_2 = -\bar{u} + g_{\bar{u}}g_{uu}/\Delta.$$

The case $\zeta_{10} = 0$ is trivial and corresponds to the kernel of admitted Lie algebras. Extension of the kernel is possible if $\zeta_{10} \neq 0$. Two cases are necessary to consider here. In the first case we have

$$\zeta_{10t} = -k_0\zeta_{10}, \quad \bar{\zeta}_{10} = -k_1\zeta_{10} \quad (k_1 \neq 0), \quad (6.6.29)$$

and in the second case one obtains

$$\zeta_{10t} = -h_2\zeta_{10} - h_1\bar{\zeta}_{10},$$
$$\beta_1 - h_2\alpha_1 = p_1, \quad \gamma_1 - h_1\alpha_1 = p_2, \quad \beta_2 - h_2\alpha_2 = p_3, \quad \gamma_2 - h_1\alpha_2 = p_4. \quad (6.6.30)$$

In (6.6.29) and (6.6.30) $k_i$, $h_i$ $(i = 1, 2)$, and $p_j$ $(j = 1, 2, 3, 4)$ are constant. Notice that the case where $h_1 = 0$ is reduced to (6.6.29). Hence, it is assumed that $h_1 \neq 0$.

### 6.6.4.1 Case (6.6.29)

In this case

$$\zeta_0 = (\beta_1 - \alpha_1 k_0 - \gamma_1 k_1)\zeta_{10}, \quad \bar{\zeta}_0 = (\beta_2 - \alpha_2 k_0 - \gamma_2 k_1)\zeta_{10}.$$

Since $\zeta_0$ and $\bar{\zeta}_0$ do not depend on $u$ and $\bar{u}$, one obtains

$$\beta_1 - k_0\alpha_1 - k_1\gamma_1 = C_o, \quad \beta_2 - k_0\alpha_2 - k_1\gamma_2 = -C_o k_1.$$

Because $\Delta \neq 0$, the previous system of equations can be solved with respect to $g_{u\bar{u}}$ and $g_{uu}$:

$$g_{u\bar{u}} = (g_{\bar{u}\bar{u}}k_1(C_o + \bar{u}) + g_{\bar{u}}(k_1 + 1))/(C_o + u),$$
$$g_{uu} = (k_1(\bar{u} + C_o)g_{u\bar{u}} - k_0)/(u + C_o). \quad (6.6.31)$$

Substituting these relations into (6.6.28), it becomes

$$-g_{\bar{u}}k_1(C_o + \bar{u}) + g_u(C_o + u) - g + k_0(u + C_o) = 0. \tag{6.6.32}$$

Any function $g(u, \bar{u})$ satisfying (6.6.32) also satisfies (6.6.31). By virtue of the equivalence transformation corresponding to the generator $X_1^e$, the constant $C_o$ is unessential since by shifting the dependent variable $u$ it can be reduced to zero. The general solution of (6.6.32) is

$$g(u, \bar{u}) = u(-k_0 \ln(u) + \psi(\bar{u}u^{k_1})), \tag{6.6.33}$$

where $\psi$ is an arbitrary function. Notice also that the general solution of (6.6.29) is $\zeta_{10} = Ce^{-k_0 t}$, where $k_1 = -e^{k_0\tau}$. Thus, the extension of the kernel of admitted generators is

$$X = e^{-k_0 t}u\partial_u. \tag{6.6.34}$$

### 6.6.4.2 Case (6.6.30)

In this case

$$\zeta_0 = p_1\zeta_{10} + p_2\bar{\zeta}_{10}, \quad \bar{\zeta}_0 = p_3\zeta_{10} + p_4\bar{\zeta}_{10}.$$

Since $h_1 \neq 0$ we have

$$\gamma_1 - h_1\alpha_1 - p_2 = 0, \quad \beta_1 - h_2\alpha_1 - p_1 = 0,$$
$$\gamma_2 - h_1\alpha_2 - p_4 = 0, \quad \beta_2 - h_2\alpha_2 - p_3 = 0.$$

These equations can be solved with respect to $g_{uu}$, $g_{u\bar{u}}$, $g_{\bar{u}\bar{u}}$ and $g_{\bar{u}}$ to give

$$g_{uu} = \frac{(h_1 p_3 - h_2 p_4 - h_2\bar{u})}{p_1 p_4 + p_1\bar{u} - p_2 p_3 + p_4 u + u\bar{u}}, \quad g_{u\bar{u}} = \frac{g_{\bar{u}}(p_3 + p_4 + \bar{u})}{p_1 p_4 + p_1\bar{u} - p_2 p_3 + p_4 u + u\bar{u}},$$

$$g_{\bar{u}\bar{u}} = \frac{g_{\bar{u}}^2(p_1 + p_2 + u)((p_1 + u)h_1 - h_2 p_2)(p_4 + \bar{u} + p_3)}{(h_1 p_1 + h_1 u - h_2 p_2)(p_1 p_4 + p_1\bar{u} - p_2 p_3 + p_4 u + u\bar{u})}, \quad g_{\bar{u}} = \frac{-h_1 p_1 - h_1 u + h_2 p_2}{p_3 + p_4 + \bar{u}}.$$

Considering $(g_{\bar{u}})_u - g_{u\bar{u}} = 0$ and $(g_{\bar{u}})_{\bar{u}} - g_{\bar{u}\bar{u}} = 0$ one obtains

$$h_1(p_1 + p_2 + u)p_3 - h_2(p_3 - p_4 - \bar{u})p_2 = 0,$$
$$(h_1(u + p_1) - h_2 p_2)((u + p_1 + p_2)p_3 + p_2(\bar{u} + p_3 + p_4)) = 0.$$

Since $h_1 \neq 0$, one finds $p_3 = 0$ and $p_2 = 0$. This simplifies $\bar{\zeta}_0 = \zeta_{10}p_4$ and $\zeta_0 = \zeta_{10}p_1$, giving $p_4 = p_1$. Hence, $g_{\bar{u}} + h_1(p_1 + u)/(p_1 + \bar{u}) = 0$, and (6.6.28) becomes

$$g_u = \frac{g}{(p_1 + u)} - h_2.$$

The general solution for the function $g$ is

$$g = -(p_1 + u)(h_2\ln(p_1 + u) + h_1\ln(p_1 + \bar{u}) + k_3).$$

As before the constant $p_1$ is unessential and can be reduced to zero, that is,

$$g = -u(h_2\ln(u) + h_1\ln(\bar{u}) + k_3). \tag{6.6.35}$$

In this case an extension of the kernel of admitted generators is given by

$$X = q(t)u\partial_u, \tag{6.6.36}$$

where the function $q(t)$ is a solution of the delay differential equation

$$q'(t) + h_2 q(t) + h_1 q(t - \tau) = 0. \tag{6.6.37}$$

### 6.6.4.3 Case $\alpha_{1u} \neq 0$

Differentiating $\zeta_0$ with respect to $u$ and $\bar{u}$, one obtains

$$\zeta_{10t} + \frac{\beta_{1u}}{\alpha_{1u}}\zeta_{10} + \frac{\gamma_{1u}}{\alpha_{1u}}\bar{\zeta}_{10} = 0. \tag{6.6.38}$$

Differentiating the previous equation with respect to $u$ and $\bar{u}$, it gives

$$\left(\frac{\beta_{1u}}{\alpha_{1u}}\right)_u \zeta_{10} + \left(\frac{\gamma_{1u}}{\alpha_{1u}}\right)_u \bar{\zeta}_{10} = 0,$$

$$\left(\frac{\beta_{1u}}{\alpha_{1u}}\right)_{\bar{u}} \zeta_{10} + \left(\frac{\gamma_{1u}}{\alpha_{1u}}\right)_{\bar{u}} \bar{\zeta}_{10} = 0.$$

For non constant either $\beta_{1u}/\alpha_{1u}$ or $\gamma_{1u}/\alpha_{1u}$ the above system of equations belongs to Case (6.6.29), and hence we have the only case of interest left, which corresponds to

$$\frac{\beta_{1u}}{\alpha_{1u}} = h_2, \quad \frac{\gamma_{1u}}{\alpha_{1u}} = h_1,$$

where $h_1$ and $h_2$ are constant.

Substituting the derivative $\zeta_{10t}$ into the equations

$$\zeta_{0\bar{u}} = 0, \quad \bar{\zeta}_{0u} = 0, \quad \bar{\zeta}_{0\bar{u}} = 0,$$

one has

$$(\beta_{1\bar{u}} - h_2\alpha_{1\bar{u}})\zeta_{10} + (\gamma_{1\bar{u}} - h_1\alpha_{1\bar{u}})\bar{\zeta}_{10} = 0,$$

$$(\beta_{2u} - h_2\alpha_{2u})\zeta_{10} + (\gamma_{2u} - h_1\alpha_{2u})\bar{\zeta}_{10} = 0,$$

$$(\beta_{2\bar{u}} - h_2\alpha_{2\bar{u}})\zeta_{10} + (\gamma_{2\bar{u}} - h_1\alpha_{2\bar{u}})\bar{\zeta}_{10} = 0.$$

If one of the coefficients in this linear system of equations for $\zeta_{10}$, $\bar{\zeta}_{10}$ is not equal to zero, then one comes to Case (6.6.29). When all coefficients vanish, one gets Case (6.6.30).

A similar result is obtained for $(\alpha_{1\bar{u}})^2 + (\alpha_{2u})^2 + (\alpha_{2\bar{u}})^2 \neq 0$. Hence, to proceed one needs to study the only case $\alpha_1 = \text{const}$, $\alpha_2 = \text{const}$.

### 6.6.4.4 Case $\alpha_1 = \text{const}$, $\alpha_2 = \text{const}$

Assume that $\alpha_1 = p_1$, $\alpha_2 = p_2$ where $p_1$ and $p_2$ are constant. Notice that because of $\Delta \neq 0$, one has $p_1^2 + p_2^2 \neq 0$.

First it is considered the case for which $p_1 \neq 0$. From the equations $\alpha_1 = p_1$ and $\alpha_2 = p_2$ one finds the derivatives $g_{u\bar{u}}$ and $g_{uu}$:

$$g_{u\bar{u}} = -g_{\bar{u}\bar{u}} p_2 / p_1, \quad g_{uu} = (g_{\bar{u}\bar{u}} p_2^2 + p_1)/p_1^2. \tag{6.6.39}$$

The first equation can be integrated to give

$$g_u = -p_2 g_{\bar{u}} / p_1 + \beta,$$

where $\beta = \beta(u)$ is an arbitrary function of the integration. Substituting this into the second equation of (6.6.39) and integrating it, one obtains

$$\beta = u/p_1 + C_o,$$

where $C_o$ is constant. Notice that in this case $\Delta = g_{\bar{u}\bar{u}}/p_1 \neq 0$. Differentiating $\zeta_0$ with respect to $u$ and $\bar{u}$, one has

$$g_{\bar{u}\bar{u}} p_2^2 (\zeta_{10} - \bar{\zeta}_{10}) + p_1 \zeta_{10} = 0, \quad (\zeta_{10} - \bar{\zeta}_{10}) g_{\bar{u}\bar{u}} p_2 = 0.$$

This gives a contradiction to $\zeta_{10} \neq 0$.

If $p_1 = 0$, then $p_2 \neq 0$ and from the equations $\alpha_1 = p_1$ and $\alpha_2 = p_2$ one finds

$$g_{\bar{u}\bar{u}} = 0, \quad g_{u\bar{u}} = 1/p_2.$$

The equation $\zeta_{0u} = 0$ also gives $\zeta_{10} = 0$, which is a contradiction.

## 6.6.5 Case $\Delta = 0$

### 6.6.5.1 Case $g_{\bar{u}\bar{u}} \neq 0$

Let $g_{\bar{u}\bar{u}} \neq 0$. In this case the general solution of the equation $\Delta = 0$ is

$$g_u = \phi(g_{\bar{u}}),$$

where $\phi$ is an arbitrary function.

Excluding $\zeta_0$ and $\bar{\zeta}_0$ from (6.6.24) and (6.6.25), one finds

$$8g_{\bar{u}}\phi'(-\eta_t + \zeta_{10} - \bar{\zeta}_{10}) + \eta_{ttt} x^2 - 2\eta_{tt} + 8\eta_t \phi + 4\xi_{0tt} x - 8\zeta_{10t} = 0. \tag{6.6.40}$$

Splitting this equation with respect to $x$, one has

$$\eta_{ttt} = 0, \quad \xi_{0tt} = 0.$$

Hence,

$$\eta = a_2 t^2 + a_1 t + a_0, \quad \xi_0 = b_1 t + b_0,$$

where $a_0, a_1, a_2, b_1, b_0$ are constants. Since $\xi_0(t) = \xi_0(t - \tau)$ and $\eta(t) = \eta(t - \tau)$, one gets $\xi_0 = b_0$ and $a_2 = a_1 = 0$. Notice that a nontrivial extension of the kernel of admitted Lie groups exists provided that $\zeta_0^2 + \zeta_{10}^2 \neq 0$.

Equation (6.6.40) becomes

$$g_{\bar{u}}\phi'(\zeta_{10} - \bar{\zeta}_{10}) - \zeta_{10t} = 0. \tag{6.6.41}$$

If $\phi' = 0$, then $\bar{\zeta}_{10} = \zeta_{10}$ are constant. In this case (6.6.25) is

$$\bar{\zeta}_0 + \bar{u}\bar{\zeta}_{10} = 0.$$

This leads to $\zeta_0 = 0$ and $\zeta_{10} = 0$, which means that there is no any extension of the kernel of admitted Lie groups. Hence, for extension of the kernel one needs to study $\phi' \neq 0$.

Differentiating (6.6.40) with respect to $\bar{u}$, one obtains

$$(g_{\bar{u}}\phi')_{\bar{u}}(\zeta_{10} - \bar{\zeta}_{10}) = 0. \tag{6.6.42}$$

Let $\bar{\zeta}_{10} - \zeta_{10} = 0$. Equation (6.6.41) gives that $\zeta_{10}$ is constant. Equations (6.6.24) and (6.6.25) are reduced to the equation

$$\phi' = -\frac{\bar{\zeta}_0 + \bar{u}\zeta_{10}}{\zeta_0 + u\zeta_{10}}. \tag{6.6.43}$$

Differentiating this equation with respect to $t$ and $x$, one gets

$$\phi'\zeta_{0t} + \bar{\zeta}_{0t} = 0, \quad \phi'\zeta_{0x} + \bar{\zeta}_{0x}. \tag{6.6.44}$$

Assuming $\phi'' \neq 0$ one obtains from (6.6.44) that $\zeta_0$ is also constant. By virtue of the inverse function theorem one has from (6.6.43)

$$g_{\bar{u}} = h\left(\frac{\bar{\zeta}_0 + \bar{u}\zeta_{10}}{\zeta_0 + u\zeta_{10}}\right),$$

where $h$ is the inverse function of $\phi'$. Because $g_{\bar{u}\bar{u}} \neq 0$, the constant $\zeta_{10} \neq 0$. Using the equivalence transformation corresponding to the generators $X_1^e$ and $X_5^e$, one can set $\zeta_0 = 0$, $\zeta_{10} = 1$. Since $g_u = \phi(g_{\bar{u}})$, integrating these equations, one finds

$$g(u, \bar{u}) = uh\left(\frac{\bar{u}}{u}\right) + k_1 u + k_0, \tag{6.6.45}$$

where $k_0$ and $k_1$ are integration constants.

Equation (6.6.23) becomes

$$g_u u + g_{\bar{u}}\bar{u} - g = 0. \tag{6.6.46}$$

Substituting in this equation the function $g$, one finds that $k_0 = 0$. The extension of the kernel of admitted Lie algebras is given by the generator

$$X = u\partial_u.$$

Let $\phi'' = 0$ or

$$g_u = k_1 g_{\bar{u}} - k_0, \tag{6.6.47}$$

where $k_0$ and $k_1 \neq 0$ are constant. Equation (6.6.43) gives $\zeta_{10} = 0$ and

$$\bar{\zeta}_0 = -k_1\zeta_0. \tag{6.6.48}$$

Equation (6.6.23) becomes

$$\zeta_{0t} = \zeta_{0xx} - k_0\zeta_0.$$

If there exists a solution $q(t,x)$ of the partial differential equation

$$q_t = q_{xx} - k_0 q, \tag{6.6.49}$$

satisfying the condition

$$q(t-\tau,x) = -k_1 q(t,x), \tag{6.6.50}$$

then the extension of the kernel is given by the generator

$$X = q(t,x)\partial_u.$$

Let $\bar{\zeta}_{10} - \zeta_{10} \neq 0$. Equation (6.6.42) leads to $(g_{\bar{u}}\phi')_{\bar{u}} = 0$ or $\phi = k_1\ln(g_{\bar{u}}) + k_0$, where $k_0$ and $k_1 \neq 0$ are integration constants. Equations (6.6.40) and (6.6.25) give

$$\bar{\zeta}_{10} = \zeta_{10} - \zeta_{10t}/k_1, \tag{6.6.51}$$

$$g_{\bar{u}\bar{u}} = \frac{g_{\bar{u}}^2(\zeta_{10} - \bar{\zeta}_{10})}{k_1(\zeta_0 + u\zeta_{10}) + g_{\bar{u}}(\bar{\zeta}_0 + \bar{u}\bar{\zeta}_{10})}. \tag{6.6.52}$$

Differentiating (6.6.52) with respect to $x$ and $t$, one obtains

$$g_{\bar{u}}\bar{\zeta}_{0x} + k_1\zeta_{0x} = 0, \tag{6.6.53}$$

and

$$g_{\bar{u}}\left(\bar{u}(\bar{\zeta}_{10}\zeta_{10t} - \bar{\zeta}_{10t}\zeta_{10}) + \bar{\zeta}_{0t}(\bar{\zeta}_{10} - \zeta_{10}) - \bar{\zeta}_0(\bar{\zeta}_{10t} - \zeta_{10t})\right)$$
$$+ k_1\left(u(\bar{\zeta}_{10}\zeta_{10t} - \bar{\zeta}_{10t}\zeta_{10}) + \zeta_{0t}(\bar{\zeta}_{10} - \zeta_{10}) - \zeta_0(\bar{\zeta}_{10t} - \zeta_{10t})\right) = 0. \tag{6.6.54}$$

Since $g_{\bar{u}\bar{u}} \neq 0$, the equation (6.6.53) gives $\zeta_0 = \zeta_0(t)$.

Notice that

$$\bar{\zeta}_{10}\zeta_{10t} - \bar{\zeta}_{10t}\zeta_{10} = \frac{1}{k_1}(\zeta_{10}\zeta_{10tt} - \zeta_{10t}^2).$$

Assuming $\bar{\zeta}_{10}\zeta_{10t} - \bar{\zeta}_{10t}\zeta_{10} \neq 0$, the equation (6.6.53) can be solved with respect to $g_{\bar{u}}$:

$$g_{\bar{u}} = -k_1\frac{u+b}{\bar{u}+c}, \tag{6.6.55}$$

where

$$b = \frac{\bar{\zeta}_{0t}(\bar{\zeta}_{10} - \zeta_{10}) - \bar{\zeta}_0(\bar{\zeta}_{10t} - \zeta_{10t})}{(\bar{\zeta}_{10}\zeta_{10t} - \bar{\zeta}_{10t}\zeta_{10})}, \qquad c = \frac{\zeta_{0t}(\bar{\zeta}_{10} - \zeta_{10}) - \zeta_0(\bar{\zeta}_{10t} - \zeta_{10t})}{(\bar{\zeta}_{10}\zeta_{10t} - \bar{\zeta}_{10t}\zeta_{10})}.$$

Differentiating (6.6.55) with respect to $t$, we find that $b$ and $s$ are constants. Substituting the derivatives $g_{\bar{u}}$ and $g_{\bar{u}\bar{u}}$, found from (6.6.55), into (6.6.52), one obtains

$$\bar{u}(\zeta_0 - b\zeta_{10}) - u(\bar{\zeta}_0 - c\bar{\zeta}_{10}) + c\zeta_0 - b\bar{\zeta}_0 + cb(\bar{\zeta}_{10} - \zeta_{10}) = 0.$$

This leads to $\zeta_0 = b\zeta_{10}$ and $\bar{\zeta}_0 = c\bar{\zeta}_{10}$. Because $\bar{\zeta}_{10} = \zeta_{10}(t-\tau)$, $\bar{\zeta}_0 = \zeta_0(t-\tau)$, and $\bar{\zeta}_{10} - \zeta_{10} \neq 0$, one gets $c = b$. By virtue of the equivalence transformations corresponding to the generator $X_1^e$, one can assume that $b = 0$. Integrating the obtained derivatives $g_u$ and $g_{\bar{u}}$, one gets

$$g(u, \bar{u}) = \lambda u - k_1 \ln(\bar{u}/u) + \gamma, \tag{6.6.56}$$

where $\lambda$ and $\gamma$ are constant. Substituting (6.6.56) into (6.6.23), one obtains $\gamma = 0$. The extension of the kernel of admitted Lie algebras is given by the generator

$$X = q(t)u\partial_u, \tag{6.6.57}$$

where $q(t)$ is a solution of the delay differential equation (6.6.51):

$$q'(t) = k_1(q(t) - q(t-\tau)). \tag{6.6.58}$$

Assume that $\bar{\zeta}_{10}\zeta_{10t} - \bar{\zeta}_{10t}\zeta_{10} = 0$. As was noticed the function $\zeta_{10}(t)$ has to satisfy the equation

$$\zeta_{10}\zeta_{10tt} - \zeta_{10t}^2 = 0.$$

Hence, $\zeta_{10}(t) = Ce^{\lambda t}$, where $C$ and $\lambda$ are constant such that $C\lambda \neq 0$. In this case one has

$$\bar{\zeta}_{10} = k\zeta_{10},$$

where $k = e^{-\lambda\tau}$. Hence, (6.6.54) becomes

$$g_{\bar{u}}\left(\bar{\zeta}_{0t}\zeta_{10} - \bar{\zeta}_0\zeta_{10t}\right) + k_1\left(\zeta_{0t}\zeta_{10} - \zeta_0\zeta_{10t}\right) = 0.$$

Since $g_{\bar{u}\bar{u}} \neq 0$, one obtains

$$\zeta_0 = \alpha\zeta_{10},$$

with constant $\alpha$. Without loss of generality one can set $\alpha = 0$. Then (6.6.23) becomes

$$ug_u + k\bar{u}g_{\bar{u}} = g + \lambda u.$$

The general solution of this equation is

$$g(u, \bar{u}) = \lambda u \ln(u) + u\psi(\bar{u}u^{-k}). \tag{6.6.59}$$

By virtue of the relation $g_u = k_1 g_{\bar{u}} + k_0$, the function $\psi(z)$ has to satisfy the ordinary differential equation

$$k_1 \ln(\psi') + kz\psi' = \psi + \lambda - k_0. \tag{6.6.60}$$

The extension of the kernel of admitted Lie algebras is given by the generator

$$X = e^{\lambda t}u\partial_u. \tag{6.6.61}$$

### 6.6.5.2 Case $g_{\bar{u}\bar{u}} = 0$

Assuming that $g_{\bar{u}\bar{u}} = 0$, one has

$$g(u, \bar{u}) = k_1 \bar{u} + h(u),$$

where $k_1 \neq 0$ is a constant. Hence (6.6.25) gives

$$\bar{\zeta}_{10} = \zeta_{10} - \eta_t.$$

Furthermore, if we let $g_{uu} = h'' \neq 0$, we can define from (6.6.24)

$$\zeta_0 = u(x^2 \eta_{tt} + 4x\xi_{0t} - 8\zeta_{10})/8 + (2\eta_{tt} - 8\eta_t h' + 8\zeta_{10t} - \eta_{ttt} x^2 - 4\xi_{0tt} x)/(8h'').$$

Since $\zeta_{0u} = 0$, then

$$(\eta_{ttt} h''' + \eta_{tt} h''^2)x^2 + 4(h''' \xi_{0tt} + h''^2 \xi_{0t})x - 2\eta_{tt} h'''$$
$$+ 8(\eta_t h''' h' - \eta_t h''^2 - h''' \zeta_{10t} - h''^2 \zeta_{10}) = 0.$$

The last equation can be split with respect to $x$ so that

$$\eta_{ttt}(h'''/h''^2) + \eta_{tt} = 0,$$
$$\xi_{0tt}(h'''/h''^2) + \xi_{0t} = 0,$$

and

$$(-\eta_{tt} h''' + 4\eta_t h''' h' - 4\eta_t h''^2)/(4h''^2) - (\zeta_{10t}(h'''/h''^2) + \zeta_{10}) = 0.$$

$$(6.6.62)$$

Differentiating the first and the second equations with respect to $u$, one obtains

$$\eta_{ttt}(h'''/h''^2)' = 0, \quad \xi_{0tt}(h'''/h''^2)' = 0.$$

Notice that if $(h'''/h''^2)' \neq 0$, then $\eta_{tt} = 0$, $\xi_{0t} = 0$, and because of $\eta(t - \tau) = \eta(t)$, $\xi(t - \tau) = \xi(t)$, one obtains that $\eta = \text{const}$ and $\xi = \text{const}$. In this case (6.6.62) gives

$$\zeta_{10} = 0,$$

which corresponds to the kernel of admitted Lie algebras. Thus, one needs to study the case $(h'''/h''^2)' = 0$ or $h''' = K h''^2$ with some constant $K$. This case also leads to the same result, that is, there is no extension of the kernel. In fact, differentiating (6.6.62) with respect to $u$, one obtains $\eta_t K = 0$. If $K = 0$, then $\eta_{tt} = 0$, which also gives that $\eta = \text{const}$. This leads to $\bar{\zeta}_{10} = \zeta_{10}$. Similar analysis of the equations $\xi_{0tt} K + \xi_{0t} = 0$ and $\xi(t - \tau) = \xi(t)$ gives that $\xi_0 = \text{const}$. The function $\zeta_{10}(t)$ has to satisfy the equations

$$\zeta_{10t} K + \zeta_{10} = 0, \quad \bar{\zeta}_{10} = \zeta_{10}.$$

The general solution of these equations is $\zeta_{10} = 0$. Thus, the case $g_{uu} \neq 0$ does not give extensions of the kernel.

### 6.6.5.3 Case $g_{\bar{u}\bar{u}} = 0$ and $g_{uu} = 0$

We extend our study to the case of a linear function

$$g(u, \bar{u}) = k_1\bar{u} + k_2u + k, \tag{6.6.63}$$

where $k_1 \neq 0$. In this case (6.6.24) becomes

$$\eta_{ttt}x^2 + 4\xi_{0tt}x - 2(\eta_{tt} - 4\eta_tk_2 + 4\zeta_{10t}) = 0.$$

Splitting this equation with respect to $x$, one finds

$$\eta_{ttt} = 0, \quad \xi_{0tt} = 0, \quad \eta_{tt} - 4\eta_tk_2 + 4\zeta_{10t} = 0.$$

By virtue of $\eta(t - \tau) = \eta(t)$ and $\xi(t - \tau) = \xi(t)$ the values $\eta$, $\xi$ and $\zeta_{10}$ are constant. Equation (6.6.23) becomes

$$\zeta_{0t} = \zeta_{0xx} + k_2\zeta_0 + k_1\bar{\zeta}_0 - \zeta_{10}k.$$

If $k_2 + k_1 \neq 0$, then by using the equivalence transformation related to the generator $X_1^e$ the constant $k_0$ can be reduced to zero. In this case the extension of the kernel is given by the generators $X = u\partial_u$ and $X_q = q(t, x)\partial_u$, where the function $q(t, x)$ satisfies the delay partial differential equation

$$q_t(t, x) = q_{xx}(t, x) + k_2q(t, x) + k_1q(t - \tau, x). \tag{6.6.64}$$

If $k_2 + k_1 = 0$, then introducing $q = \zeta_0 - \zeta_{10}kx^2/2$, one gets that the extension of the kernel is given by the generators $X_q$ and

$$X = (2u + kx^2)\partial_u. \tag{6.6.65}$$

## 6.6.6  Summary of the Group Classification

**Case 1.** Combining (6.6.33) and (6.6.59),

$$g(u, \bar{u}) = u(-k_0\ln(u) + \psi(\bar{u}u^{k_1})), \tag{6.6.66}$$

where $\psi$ is an arbitrary function, $k_1 = -e^{k_0\tau}$, and

$$X = e^{-k_0t}u\partial_u. \tag{6.6.67}$$

**Case 2.** Combining (6.6.35) and (6.6.56),

$$g = -u(h_2\ln(u) + h_1\ln(\bar{u}) + k_3), \tag{6.6.68}$$

and

$$X = q(t)u\partial_u, \tag{6.6.69}$$

where the function $q(t)$ is a solution of the delay differential equation

$$q'(t) + h_2q(t) + h_1q(t - \tau) = 0. \tag{6.6.70}$$

A particular solution of (6.6.70) is $q = e^{-k_0t}$, where $k_0 = h_2 + h_1e^{k_0\tau}$.

**Case 3.** The general solution of (6.6.47) is

$$g(u, \bar{u}) = -k_0 u + \psi(\bar{u} + k_1 u) \quad (k_1 \neq 0), \tag{6.6.71}$$

where $\psi$ is an arbitrary function. The extension of the kernel is given by the generator

$$X = q(t, x)\partial_u, \tag{6.6.72}$$

where $q(t, x)$ is a solution of equation

$$q_t = q_{xx} - k_0 q, \tag{6.6.73}$$

satisfying the condition

$$q(t - \tau, x) = -k_1 q(t, x). \tag{6.6.74}$$

For particular cases of $k_0$ and $k_1$ the problem (6.6.73), (6.6.74) has a solution. For example, let $k_1 = -1$, then a solution can be sought in the form $q = q(x)$, where

$$q''(x) - k_0 q(x) = 0.$$

For $\tau$, $k_0$ and $k_1$ which obey the relation $k_1 = -e^{k_0\tau}$ there exists the particular solution of the problem (6.6.73) and (6.6.74), $q = e^{-k_0 t}$.

**Case 4.** If $g(u, \bar{u})$ is a linear function

$$g(u, \bar{u}) = k_1 \bar{u} + k_2 u + k \quad (k_1 \neq 0), \tag{6.6.75}$$

then the extension of the kernel of admitted generators consists of the generators

$$X_q = q(t, x)\partial_u, \tag{6.6.76}$$

where the function $q(t, x)$ satisfies the reaction–diffusion equation with a delay:

$$q_t(t, x) = q_{xx}(t, x) + k_2 q(t, x) + k_1 q(t - \tau, x), \tag{6.6.77}$$

and one more generator, which depends on the value of the constants $k_1$ and $k_2$. For $k_2 + k_1 \neq 0$ one can take $k = 0$, and gets the additional generator in the form

$$X = u\partial_u. \tag{6.6.78}$$

In the case of $k_2 + k_1 = 0$, the additional generator is

$$X = (2u + kx^2)\partial_u. \tag{6.6.79}$$

A particular solution of (6.6.77) is $q = e^{-k_0 t}$, where $k_0 = -(k_1 e^{k_0\tau} + k_2)$.

### 6.6.7 Invariant Solutions

Invariant solutions can be sought for a subalgebra of an admitted Lie algebra. Substantially different invariant solutions are obtained on the base of an optimal system of admitted subalgebras. The set of all generators nonequivalent with respect to automorphisms composes an optimal system of one dimensional subalgebras [17]. This set is used for constructing nonequivalent invariant solutions. Equivalence of invariant solutions is considered with respect to an admitted Lie algebra.

Apart from automorphisms for constructing the optimal system of subalgebras one has to use involutions. Equations (6.6.1) possess the involution $E$ corresponding to the change $x \to -x$.

### 6.6.7.1 Optimal System of Subalgebras

Let us consider the algebra $L_3 = \{X_1, X_2, X_3\}$, with the commutator table

|       | $X_1$    | $X_2$ | $X_3$     |
|-------|----------|-------|-----------|
| $X_1$ | 0        | 0     | $-k_0 X_3$ |
| $X_2$ | 0        | 0     | 0         |
| $X_3$ | $k_0 X_3$ | 0     | 0.        |

Such algebras are admitted by (6.6.1) with the function $g(u, \bar{u})$ in (6.6.66), (6.6.68), (6.6.71) and (6.6.75). The generator $X_2$ composes a center of the algebra $L_3$.

The coordinates $(x_1, x_2, x_3)$ of the generator

$$X = x_1 X_1 + x_2 X_2 + x_3 X_3$$

are simplified [17] by the automorphisms $A_1$ and $A_3$, which are defined by the table of commutators

$$A_1 : x_3' = x_3 e^{-k_0 a_1}, \qquad A_3 : x_3' = x_3 + k_0 x_1 a_2.$$

Here only transformed coordinates are presented.

The optimal system of subalgebras of the algebra $L_3$ with $k_0 \neq 0$ consists of the subalgebras

$$H_1 = X_3 + \alpha X_2, \quad H_2 = X_1 + \alpha X_2, \quad H_3 = X_2,$$

where $\alpha$ is an arbitrary constant.

Representations of the invariant solutions corresponding to the subalgebras $H_2$ and $H_3$ are

$$u = \varphi(x - \alpha t), \quad u = \varphi(t),$$

respectively. It is obvious that these representations reduce the number of the independent variables.

### 6.6.7.2 Invariant Solutions with Respect to $H_1$

**Case 1.** For the function (6.6.66) the generator $X_3 = q(t)\partial_u$ and the representation of an invariant solution is

$$u = e^{\beta x q(t)} \varphi(t),$$

where $\beta = 1/\alpha$ and $q(t) = e^{-k_0 t}$. The reduced equation is

$$\varphi'(t) = \varphi(t)\big(\beta^2 q^2 - k_0 \ln(\varphi(t)) + \psi(\varphi(t - \tau)\varphi^{k_1}(t))\big). \tag{6.6.80}$$

**Case 2.** For the function (6.6.68) the generator $X_3$ and the representation of an invariant solution is the same as in the previous case. The reduced equation is

$$\varphi'(t) = \varphi(t)\big(\beta^2 q^2 - h_2 \ln(\varphi(t)) - h_1 \ln(\varphi(t - \tau)) + k_3\big). \tag{6.6.81}$$

**Case 3.** For the function (6.6.71) in the case, where $k_1 = -e^{k_0\tau}$, the generator $X_3 = e^{-k_0 t}\partial_u$, and the representation of an invariant solution is $u = \beta x e^{-k_0 t} + \varphi(t)$. The reduced equation is

$$\varphi'(t) = -k_0\varphi(t) + \psi\left(\varphi(t - \tau) + k_1\varphi(t)\right). \tag{6.6.82}$$

**Case 4.** If the function $g(u, \bar{u})$ is as given in (6.6.75) and the generator $X_3 = e^{-k_0 t}\partial_u$, then the invariant solution is $u = \beta x e^{-k_0 t} + \varphi(t)$, where the function $\varphi(t)$ satisfies the reduced equation

$$\varphi'(t) = k_1\varphi(t - \tau) + k_2\varphi(t) + k. \tag{6.6.83}$$

# References

1. Bakirova, M.I., Dorodnitsyn, V.A., Kozlov, R.V.: Symmetry-preserving discrete schemes for some heat transfer equations. http://arxiv.org/abs/math/0402367v1, pp. 1–21 (2004)
2. Bellman, R., Cooke, K.L.: Differential-Difference Equations. Academic Press, New York (1963)
3. Dorodnitsyn, V.A.: Group Properties of Difference Equations. Fizmatlit, Moscow (2001) (in Russian)
4. Dorodnitsyn, V., Kozlov, R., Winternitz, P.: Lie group classification of second order difference equations. J. Nonlinear Math. Phys. **41**(1), 480–504 (2000)
5. Dorodnitsyn, V., Kozlov, R., Winternitz, P.: Integration of second order ordinary difference equations. J. Nonlinear Math. Phys. **10**(2), 41–56 (2003)
6. Driver, R.D.: Ordinary and Delay Differential Equations. Springer, New York (1977)
7. Gonzalez-Lopez, A., Kamran, N., Olver, P.J.: Lie algebras of differential operators in two complex variables. Am. J. Math. **114**, 1163–1185 (1992)
8. Gonzalez-Lopez, A., Kamran, N., Olver, P.J.: Lie algebras of vector fields in the real plane. Proc. Lond. Math. Soc. **64**, 339–368 (1992)
9. Hale, J.: Functional Differential Equations. Springer, New York (1977)
10. Hearn, A.C.: REDUCE Users Manual, ver. 3.3. The Rand Corporation CP 78, Santa Monica (1987)
11. Ibragimov, N.H. (ed.): CRC Handbook of Lie Group Analysis of Differential Equations, vols. 1, 2, 3. CRC Press, Boca Raton (1994,1995,1996)
12. Kolmanovskii, V., Myshkis, A.: Applied Theory of Functional Differential Equations. Kluwer Academic, Dordrecht (1992)
13. Meleshko, S.V.: Methods for Constructing Exact Solutions of Partial Differential Equations. Springer, New York (2005)
14. Meleshko, S.V., Moyo, S.: On the complete group classification of the reaction–diffusion equation with a delay. J. Math. Anal. Appl. **338**, 448–466 (2008)
15. Myshkis, A.D.: Linear Differential Equations with Retarded Argument. Nauka, Moscow (1972)
16. Nesterenko, M.: Transformation groups on real plane and their differential invariants. http://arxiv.org/abs/math-ph/0512038, pp. 1–15 (2006)
17. Ovsiannikov, L.V.: Group Analysis of Differential Equations. Nauka, Moscow (1978). English translation by Ames, W.F. (ed.), published by Academic Press, New York (1982)
18. Pontriagin, L.S.: Ordinary Differential Equations. Nauka, Moscow (1974)
19. Pue-on, P.: Group classification of second-order delay ordinary differential equations. Ph.D. thesis, School of Mathematics, Suranaree University of Technology, Nakhon Ratchasima, Thailand (2008)
20. Pue-on, P., Meleshko, S.V.: Group classification of second-order delay ordinary differential equation. Commun. Nonlinear Sci. Numer. Simul. (2010). doi:10.1016/j.cnsns.2009.06.013

21. Tanthanuch, J.: Application of group analysis to delay differential equations. Ph.D. thesis, School of Mathematics, Suranaree University of Technology, Nakhon Ratchasima, Thailand (2003)
22. Tanthanuch, J., Meleshko, S.V.: On definition of an admitted lie group for functional differential equations. Commun. Nonlinear Sci. Numer. Simul. **9**(1), 117–125 (2004)
23. Wu, J.: Theory and Applications of Partial Functional Differential Equations. Springer, New York (1996)

# Appendix A

## A.1 Optimal Systems of Subalgebras

The generators admitted by the gas dynamics equations are

$$Y_1 = \partial_x, \ Y_2 = \partial_y, \ Y_3 = \partial_z, \ Y_4 = t\partial_x + \partial_u,$$
$$Y_5 = t\partial_y + \partial_v, \ Y_6 = t\partial_z + \partial_w,$$
$$Y_7 = y\partial_z - z\partial_y + v\partial_w - w\partial_v, \ Y_8 = z\partial_x - x\partial_z + w\partial_u - u\partial_w,$$
$$Y_9 = x\partial_y - y\partial_x + u\partial_v - v\partial_u, \ Y_{10} = \partial_t,$$
$$Y_{11} = t\partial_t + x\partial_x + y\partial_y + z\partial_z - f\partial_f,$$
$$Y_{12} = t\partial_t - u\partial_u - v\partial_v - w\partial_w + (\gamma + 2)f\partial_f.$$

The table of commutators can be written in a symbolical form as follows.

|     | 1  | 2  | 3  | 4 | 5 | 6 | 7  | 8  | 9  | 10 | 11  | 12  |
|-----|----|----|----|---|---|---|----|----|----|----|-----|-----|
| 1   | 0  | 0  | 0  | 0 | 0 | 0 | 0  | -3 | 2  | 0  | 1   | 0   |
| 2   | 0  | 0  | 0  | 0 | 0 | 0 | 3  | 0  | -1 | 0  | 2   | 0   |
| 3   | 0  | 0  | 0  | 0 | 0 | 0 | -2 | 1  | 0  | 0  | 3   | 0   |
| 4   | 0  | 0  | 0  | 0 | 0 | 0 | 0  | -6 | 5  | -1 | 0   | -4  |
| 5   | 0  | 0  | 0  | 0 | 0 | 0 | 6  | 0  | -4 | -2 | 0   | -5  |
| 6   | 0  | 0  | 0  | 0 | 0 | 0 | -5 | 4  | 0  | -3 | 0   | -6  |
| 7   | 0  | -3 | 2  | 0 | -6| 5 | 0  | -9 | 8  | 0  | 0   | 0   |
| 8   | 3  | 0  | -1 | 6 | 0 | -4| 9  | 0  | -7 | 0  | 0   | 0   |
| 9   | -2 | 1  | 0  | -5| 4 | 0 | -8 | 7  | 0  | 0  | 0   | 0   |
| 10  | 0  | 0  | 0  | 1 | 2 | 3 | 0  | 0  | 0  | 0  | 10  | 10  |
| 11  | -1 | -2 | -3 | 0 | 0 | 0 | 0  | 0  | 0  | -10| 0   | 0   |
| 12  | 0  | 0  | 0  | 4 | 5 | 6 | 0  | 0  | 0  | -10| 0   | 0   |

Here numbers of corresponding basis generators are only presented. Notice that the table of commutators of the generators $X_i$ $(i = 1, 2, \ldots)$ admitted by the full Boltzmann equation is written in the same symbolical form.

Y.N. Grigoriev et al., *Symmetries of Integro-Differential Equations*,
Lecture Notes in Physics 806,
DOI 10.1007/978-90-481-3797-8, © Springer Science+Business Media B.V. 2010

## A.1.1  Six and Seven-Dimensional Subalgebras of the Lie Algebra $L_{11}(Y)$

Here a part of the normalized optimal system of subalgebras of the Lie algebra $L_{11}(Y)$ [36] (Chap. 3) is presented. Subalgebra-representatives are notated by a pair of numbers $r, i$, where $r$ is the dimension and $i$ is the number of the subalgebra of the dimension $r$ which is given in the first column. In the second column a basis of the subalgebra is presented in a symbolical form as in the table of commutators. The restrictions for parameters are also given in this column. Notice that $\mu \neq 0$ in all subalgebras and the absence of restrictions for the parameters means that they are arbitrary real numbers. In the third column the normalizer of the subalgebra is presented. The sign $=$ means that the subalgebra is self-normalized.

| $r = 6$ | | | $r = 7$ | | |
|---|---|---|---|---|---|
| 1 | 1,2,3; 7; 10; 11; | 7,4 | 1 | 1,2,3; 7,8,9; 11; | $=7,1$ |
| 2 | 2,3;5,6; 10;7+$\mu$11 | 7,3 | 2 | 4,5,6;7,8,9;11 | $=7,2$ |
| 3 | 1,2,3;4; 10;7+$\alpha$11 | 7,4 | 3 | 2,3;5,6; 7;10;11 | $=7,3$ |
| 4 | 1;4,5,6;7; 11 | $=6,4$ | 4 | 12,3;4;7; 10;11 | $=7,4$ |
| 5 | 1,2,3;4;7; 11 | $=6,5$ | 5 | 1,2,3;5,6; 10; $\beta$4+7+$\alpha$11 | 9,1 |
| 6 | 2,3;5,6;$\alpha$4+7;$\beta$4+11 | 7,6 | 6 | 2,3;4,5,6;7;11 | $=7,6$ |
| 7 | 2,3; 4,5,6; 7+$\mu$11 | 7,6 | 7 | 1,2,3; 5,6;$\alpha$4+7;$\beta$4+11 | 8,4 |
| 8 | 1,2,3;5,6;$\beta$4+7+$\alpha$11 | 8,4 | 8 | 1,2,3;4,5,6;7+$\mu$11 | 8,4 |
| 9 | 4,5,6;7,8,9 | 7,2 | 9 | 1,2,3;7,8,9;10 | 8,1 |
| 10 | 1,2,3;7,8,9 | 8,1 | 10 | 1,2,3;5,6;$\alpha$4+7;4+10 | $8,3^0$ |
| 11 | 2,3;5,6; 1+7; 10 | $7,5^{00}$ | 11 | 1,2,3;4,5,6;7+10 | $8,3^0$ |
| 12 | 2,3;5,6;$\alpha$1+7;4+10 | $7,10^0$ | 12 | 1,2,3;5,6;$\alpha$4+11 | 9,1 |
| 13 | 2,3;5,6;7; 10 | $8,2^{00}$ | 13 | 1,2,3;4,5,6; 11 | 10,1 |
| 14 | 2,3;4,5,6;1+7 | $7,8^0$ | 14 | 1,2,3;4,5,6;10 | 11,1 |
| 15 | 2,3;4,5,6;7 | 8,4 | | | |
| 16 | 1,2,3; 5,6; 7+10 | $7,5^{00}$ | | | |
| 17 | 2,3; 5,6; 10,11 | 7,3 | | | |
| 18 | 1,2,3;4;10;$\mu$6+11 | 8,5 | | | |
| 19 | 1,2,3;4; 10; 11 | 9,1 | | | |
| 20 | 1,2,3;5,6;$\alpha$4+11 | 8,4 | | | |
| 21 | 2,3;4,5, 6; 11 | 7,6 | | | |
| 22 | 1,2,3;5,6;10 | 9,1 | | | |
| 23 | 1,2,3;5,6;4+10 | $8,3^0$ | | | |
| 24 | 1,2,3;4,5,6 | 11,1 | | | |
| 25 | 1,2,3;5,6;$\beta$4+7 | 9,1 | | | |

## A.1.2 Six-Dimensional Subalgebras of the Lie Algebra $L_{12}(Y)$

Here a part of the optimal system of subalgebras [16] (Chap. 3) is presented.

| $r=6$ | | | $r=6$ | | |
|---|---|---|---|---|---|
| 1 | 1,2,3,5,6,7 | 6,8 | 75 | $2,3,5,6,10,11+\alpha12$ | 6,17 |
| 2 | 4,5,6,7,8,9 | 6,9 | 76 | 2,3,5,6,10,12 | 5,33 |
| 3 | 1,2,3,7,8,9 | 6,10 | 77 | $2,3,5,6,10,1\pm12$ | 5,33 |
| 4 | 1,2,3,4,10,6+11 | 6,18 | 78 | $2,3,5,6,10,1+7+\alpha12$ | 6,11 |
| 5 | 1,2,3,5,6,4+11 | 6,20 | 79 | $1,2,3,6,4+10,2\cdot11-12$ | 5,34 |
| 6 | 1,2,3,5,6,10 | 6,22 | 82 | $2,3,4,5,6,7+\mu11+\beta12$ | 6,7 |
| 7 | 1,2,3,5,6,4+10 | 6,23 | 83 | $2,3,4,5,6,7+\alpha12$ | 6,15 |
| 8 | 1,2,3,4,5,6 | 6,24 | 84 | $2,3,4,5,6,11+\alpha12$ | 6,21 |
| 9 | 7,8,9,10,11,12 | 5,1 | 85 | 2,3,4,5,6,12 | 5,35 |
| 10 | 1,4,7,10,11,12 | 5,2 | 86 | $2,3,4,5,6,1+7+\alpha12$ | 6,14 |
| 35 | $1,2,3,10,\alpha4+11,\beta4+7$ | 5,22 | 88 | $2,3,4,5,1+6,11-12$ | 5,36 |
|  | $\alpha^2+\beta^2=1$ | 5,23 | 91 | $1,2,3,5,6,7+\mu11+\beta12$ | 5,37 |
| 38 | $1,2,3,4,10,7+\alpha11+\beta12$ | 6,3 |  | $\beta(\mu+\beta)\neq0$ |  |
| 39 | $1,2,3,4,10,\alpha11+12$ | 5,22 | 92 | $1,2,3,5,6,7+\mu11$ | 6,8 |
| 40 | 1,2,3,4,10,11 | 6,19 | 93 | $1,2,3,5,6,7+\alpha(11-12)$ | 5,37 |
| 42 | $1,2,3,10,11,7+\mu12$ | 5,23 | 94 | 1,2,3,5,6,7 | 6,25 |
| 43 | 1,2,3,7,10,11 | 6,1 | 95 | $1,2,3,5,6,\alpha11+12\ (\alpha\neq-1)$ | 5,37 |
| 44 | 1,2,3,10,11,12 | 5,23 | 96 | 1,2,3,5,6,11 | 6,20 |
| 46 | $1,4,5,6,11,7+\alpha12$ | 6,4 | 97 | $1,2,3,5,6,11-12$ | 5,37 |
| 47 | 1,4,5,6,11,12 | 5,24 | 99 | $1,2,3,5,6,10+11-12$ | 5,37 |
| 49 | $2,3,4+\alpha5,6,11,12$ | 5,25 | 101 | $1,2,3,5,6,7+10+\mu(-11+12)$ | 5,37 |
| 51 | $2,3,5,6,\alpha4+7,\beta4+11$ | 6,6 | 102 | 1,2,3,5,6,7+10 | 6,16 |
|  | $\alpha^2+\beta^2=1$ |  | 103 | $1,2,3,5,6,4+7+\mu11$ | 6,8 |
| 53 | $2,3,5,6,11,7+\mu12$ | 5,26 | 104 | 1,2,3,5,6,4+7 | 6,25 |
| 54 | 2,3,5,6,7,11 | 6,6 | 105 | 1,2,3,5,6,4+7+10 | 6,26 |
| 55 | 2,3,5,6,11,12 | 5,26 | 132 | 1,4,7,10,11,12 | 4,23 |
| 59 | $1,2,3,4,11,7+\alpha12$ | 6,5 | 135 | 2,3,7,10,11,12 | 5,3 |
| 60 | 1,2,3,4,11,12 | 5,29 | 138 | 4,5,6,7,11,12 | 5,5 |
| 64 | $2,3,5,6,4+10,7+\mu(12-2\cdot11)$ | 5,31 | 142 | 1,5,6,7,11,12 | 5,7 |
| 65 | 2,3,5,6,7,4+10 | 6,12 | 146 | 2,3,4,7,11,12 | 5,6 |
| 66 | $2,3,5,6,4+10,2\cdot11-12$ | 5,31 | 150 | 1,2,3,7,11,12 | 5,8 |
| 68 | 2,3,5,6,4+10,1+7 | 6,12 | 157 | $1,2,3,7,4+10,-2\cdot11+12$ | 5,12 |
| 70 | $2,3,1+5,6,10,11-12$ | 5,32 | 160 | $1,2,3,10,\alpha11+12,7+\beta11$ | 5,4 |
| 73 | $2,3,5,6,10,7+\mu11+\beta12$ | 6,2 | 164 | $1,4,3+5,2-6,7,11-12$ | 5,16 |
| 74 | $2,3,5,6,10,7+\alpha12$ | 6,13 | 167 | $1,4,5,6,\alpha11+12,7+\beta11$ | 5,9 |

(continued)

| $r = 6$ | | |
|---|---|---|
| 178 | $2,3,5,6,\alpha 1 + 12, \beta 1 + 7$ | 5,17 |
| | $\alpha^2 + \beta^2 = 1$ | |
| 183 | $2,3,5,6,\alpha 11 + 12, 7 + \beta 11$ | 5,14 |
| | $\alpha^2 + \beta^2 \neq 0 \ \& \ (\alpha + 1)^2 + \beta^2 \neq 0$ | |
| 185 | $2,3,5,6,7,12$ | 5,14 |
| 187 | $2,3,5,6,7,11 - 12$ | 5,14 |
| 190 | $2,3,5,6,7 + \beta 10, \alpha 10 - 11 - 12$ | 5,14 |
| 194 | $1,2,3,4,\alpha 11 + 12, 7 + \beta 11$ | 5,15 |
| | $(\alpha + 1)^2 + \beta^2 \neq 0$ | |
| 196 | $1,2,3,4,7,11 - 12$ | 5,15 |
| 200 | $1,2,3,4,\alpha 10 - 11 + 12, 7 + \beta 10$ | 5,15 |
| | $\alpha^2 + \beta^2 = 1$ | |

# Appendix B

## B.1 Realizations of Lie Algebras on the Real Plane

Realizations of Lie algebras on the real plane [16] (Chap. 6) are given here (Table B.1).

The functions $1, x, \xi_1, \ldots, \xi_r$ are linearly independent. The functions $\eta_1, \ldots, \eta_r$ form a fundamental system of solutions for an $r$-order linear ordinary differential equation with constant coefficients $\eta^{(r)}(x) + c_1\eta^{(n-1)}(x) + \cdots + c_r\eta(x) = 0$.

**Table B.1** Realizations of Lie algebras on the real plane

| No. | Lie algebra basis |
|-----|-------------------|
| 1 | $\partial_x$ |
| 2 | $\partial_x, \partial_y$ |
| 3 | $\partial_x, y\partial_x$ |
| 4 | $\partial_x, x\partial_x + y\partial_y$ |
| 5 | $\partial_x, x\partial_x$ |
| 6 | $\partial_y, x\partial_y, \xi_1(x)\partial_y$ |
| 7 | $\partial_y, y\partial_y, \partial_x$ |
| 8 | $e^{-x}\partial_y, \partial_x, \partial_y$ |
| 9 | $\partial_y, \partial_x, x\partial_y$ |
| 10 | $\partial_y, \partial_x, x\partial_x + (x+y)\partial_y$ |
| 11 | $e^{-x}\partial_y, -xe^{-x}\partial_y, \partial_x$ |
| 12 | $\partial_x, \partial_y, x\partial_x + y\partial_y$ |
| 13 | $\partial_y, x\partial_y, y\partial_y$ |
| 14 | $\partial_x, \partial_y, x\partial_x + ay\partial_y, 0 < |a| \leq 1, a \neq 1$ |
| 15 | $e^{-x}\partial_y, e^{-ax}\partial_y, \partial_x, 0 < |a| \leq 1, a \neq 1$ |
| 16 | $\partial_x, \partial_y, (bx+y)\partial_x + (by-x)\partial_y, b \geq 0$ |
| 17 | $e^{-bx}\sin x\partial_y, e^{-bx}\cos x\partial_y, \partial_x, b \geq 0$ |

**Table B.1** (continued)

| No. | Lie algebra basis |
|---|---|
| 18 | $\partial_x, x\partial_x + y\partial_y, (x^2 - y^2)\partial_x + 2xy\partial_y$ |
| 19 | $\partial_x + \partial_y, x\partial_x + y\partial_y, x^2\partial_x + y^2\partial_y$ |
| 20 | $\partial_x, x\partial_x + \frac{1}{2}y\partial_y, x^2\partial_x + xy\partial_y$ |
| 21 | $\partial_x, x\partial_x, x^2\partial_x$ |
| 22 | $y\partial_x - x\partial_y, (1 + x^2 - y^2)\partial_x + 2xy\partial_y, 2xy\partial_x + (1 + y^2 - x^2)\partial_y$ |
| 23 | $\partial_y, x\partial_y, \xi_1(x)\partial_y, \xi_2(x)\partial_y$ |
| 24 | $\partial_x, x\partial_x, \partial_y, y\partial_y$ |
| 25 | $e^{-x}\partial_y, \partial_x, \partial_y, y\partial_y$ |
| 26 | $e^{-x}\partial_y, -xe^{-x}\partial_y, \partial_x, \partial_y$ |
| 27 | $e^{-x}\partial_y, e^{-ax}\partial_y, \partial_x, \partial_y, 0 < |a| \le 1, a \ne 1$ |
| 28 | $e^{-bx}\sin x\partial_y, e^{-bx}\cos x\partial_y, \partial_x, \partial_y, b \ge 0$ |
| 29 | $\partial_x, x\partial_x, y\partial_y, x^2\partial_x + xy\partial_y$ |
| 30 | $\partial_x, \partial_y, x\partial_x, x^2\partial_x$ |
| 31 | $\partial_y, -x\partial_y, \frac{1}{2}x^2\partial_y, \partial_x$ |
| 32 | $e^{-bx}\partial_y, e^{-x}\partial_y, -xe^{-x}\partial_y, \partial_x$ |
| 33 | $e^{-x}\partial_y, -x\partial_y, \partial_y, \partial_x$ |
| 34 | $e^{-x}\partial_y, -xe^{-x}\partial_y, \frac{1}{2}x^2e^{-x}\partial_y, \partial_x$ |
| 35 | $\partial_y, x\partial_y, \xi_1(x)\partial_y, y\partial_y$ |
| 36 | $e^{-ax}\partial_y, e^{-bx}\partial_y, e^{-x}\partial_y, \partial_x, -1 \le a < b < 1, ab \ne 0$ |
| 37 | $e^{-ax}\partial_y, e^{-bx}\sin x\partial_y, e^{-bx}\cos x\partial_y, \partial_x, a > 0$ |
| 38 | $\partial_x, \partial_y, x\partial_y, x\partial_x + (2y + x^2)\partial_y$ |
| 39 | $\partial_y, \partial_x, x\partial_y, (1 + b)x\partial_x + y\partial_y, |b| \le 1$ |
| 40 | $\partial_y, -x\partial_y, \partial_x, y\partial_y$ |
| 41 | $\partial_x, \partial_y, x\partial_x + y\partial_y, y\partial_x - x\partial_y$ |
| 42 | $\sin x\partial_y, \cos x\partial_y, y\partial_y, \partial_x$ |
| 43 | $\partial_x, \partial_y, x\partial_x - y\partial_y, y\partial_x, x\partial_y$ |
| 44 | $\partial_x, \partial_y, x\partial_x, y\partial_y, y\partial_x, x\partial_y$ |
| 45 | $\partial_x, \partial_y, x\partial_x + y\partial_y, y\partial_x - x\partial_y, (x^2 - y^2)\partial_x - 2xy\partial_y, 2xy\partial_x - (y^2 - x^2)\partial_y$ |
| 46 | $\partial_x, \partial_y, x\partial_x, y\partial_y, x^2\partial_x, y^2\partial_y$ |
| 47 | $\partial_x, \partial_y, x\partial_x, y\partial_y, y\partial_x, x\partial_y, x^2\partial_x + xy\partial_y, xy\partial_x + y^2\partial_y$ |
| 48 | $\partial_y, x\partial_y, \xi_1(x)\partial_y, \ldots, \xi_r(x)\partial_y, r \ge 3$ |
| 49 | $y\partial_y, \partial_y, x\partial_y, \xi_1(x)\partial_y, \ldots, \xi_r(x)\partial_y, r \ge 2$ |
| 50 | $\partial_x, \eta_1\partial_y, \ldots, \eta_r(x)\partial_y, r \ge 4$ |
| 51 | $\partial_x, y\partial_y, \eta_1\partial_y, \ldots, \eta_r(x)\partial_y, r \ge 3$ |
| 52 | $\partial_x, \partial_y, x\partial_x + cy\partial_y, x\partial_y, \ldots, x^r\partial_y, r \ge 2$ |
| 53 | $\partial_x, \partial_y, x\partial_y, \ldots, x^{r-1}\partial_y, x\partial_x + (ry + x^r)\partial_y, r \ge 3$ |
| 54 | $\partial_x, x\partial_x, y\partial_y, \partial_y, x\partial_y, \ldots, x^r\partial_y, r \ge 1$ |
| 55 | $\partial_x, \partial_y, 2x\partial_x + ry\partial_y, x^2\partial_x + rxy\partial_y, x\partial_y, x^2\partial_y, \ldots, x^r\partial_y, r \ge 1$ |
| 56 | $\partial_x, x\partial_x, y\partial_y, x^2\partial_x + rxy\partial_y, \partial_y, x\partial_y, x^2\partial_y, \ldots, x^r\partial_y, r \ge 0$ |

## B.2 Group Classification of Second-Order DODEs

Here $h(x) - h(x - \tau) = c$, $k = e^c$, $k_1 = kc$, $c_1 = \sin c$, $c_2 = \cos c$ where $c, c_3, k_2, k_3$ are arbitrary constants. The following invariants and functions are used in Table B.2.

$$I_1 = k^b y - [c_1 y'_\tau + (c_2 + bc_1)y_\tau],$$

**Table B.2** Group classification of second-order DODEs on the domain of real space

| No. | Lie algebra | Representation of second-order DODEs |
|---|---|---|
| 1 | $L_1^1$ $\partial_x$ | $y'' = f(y, y_\tau, y', y'_\tau)$ |
| 2 | $L_2^2$ $\partial_x, \partial_y$ | $y'' = f(y - y_\tau, y', y'_\tau)$ |
| 3 | $L_3^2$ $\partial_x, y\partial_x$ | $y'' = y'^3 f(y, y_\tau, \frac{1}{y'} - \frac{1}{y'_\tau})$ |
| 4 | $L_4^2$ $\partial_x, x\partial_x + y\partial_y$ | $y'' = \frac{1}{y} f(\frac{y_\tau}{y}, y', y'_\tau)$ |
| 5 | $L_5^2$ $\partial_x, x\partial_x$ | $y'' = y'^2 f(y, y_\tau, \frac{y'_\tau}{y'})$ |
| 6 | $L_6^3$ $\partial_y, x\partial_y, \xi_1(x)\partial_y$ | $y'' = (\xi_1' - \xi_1'^\tau)^{-1}\big(f\big(x, (\xi_1' - \xi_1'^\tau)(cy' - y + y_\tau)$ <br> $\quad - (\xi_1'c - \xi_1 + \xi_1^\tau)(y' - y'_\tau)\big) + \xi_1''(y' - y'_\tau)\big)$ |
| 7 | $L_7^3$ $\partial_y, y\partial_y, \partial_x$ | $y'' = y' f(\frac{y - y_\tau}{y'^2}, \frac{y'_\tau}{y'^2})$ |
| 8 | $L_8^3$ $e^{-x}\partial_y, \partial_x, \partial_y$ | $y'' = f(ky' - y'_\tau, k(y - y_\tau - y'_\tau) + y'_\tau) - y'$ |
| 9 | $L_9^3$ $\partial_y, \partial_x, x\partial_y$ | $y'' = f(y' - y'_\tau, cy' - y + y_\tau)$ |
| 10 | $L_{10}^3$ $\partial_y, \partial_x, x\partial_x + (x + y)\partial_y$ | $y'' = e^{-y} f(y' - y'_\tau, (y - y_\tau)e^{-y})$ |
| 11 | $L_{11}^3$ $e^{-x}\partial_y, -xe^{-x}\partial_y, \partial_x$ | $y'' = f\big(k(y + y') - (y_\tau + y'_\tau),$ <br> $\quad kc(y + y') - ky + y_\tau$ <br> $\quad - (2y' + y)\big)$ |
| 12 | $L_{12}^3$ $\partial_x, \partial_y, x\partial_x + y\partial_y$ | $y'' = (y - y_\tau)^{-1} f(y', y'_\tau)$ |
| 13 | $L_{13}^3$ $\partial_y, x\partial_y, y\partial_y$ | $y'' = (y' - y'_\tau) f\big(x, \frac{cy' - y + y_\tau}{(y' - y'_\tau)}\big)$ |
| 14 | $L_{14}^3$ $\partial_x, \partial_y, x\partial_x + ay\partial_y,$ <br> $0 < \|a\| \le 1, a \ne 1$ | $y'' = y'^{\frac{(a-2)}{(a-1)}} f\big(\frac{y'_\tau}{y'^2}, y'(y - y_\tau)^{\frac{(1-a)}{a}}\big)$ |
| 15 | $L_{15}^3$ $e^{-x}\partial_y, e^{-ax}\partial_y, \partial_x$ <br> $0 < \|a\| \le 1, a \ne 1$ | $y'' = f\big(k^a(y + y') - (y_\tau + y'_\tau), (k - k^a)(y + y')$ <br> $\quad - (1 - a)(ky - y_\tau)\big) - [(1 + a)y' + ay]$ |
| 16 | $L_{17}^3$ $e^{-bx}\sin x\partial_y, e^{-bx}\cos x\partial_y, \partial_x$ <br> $b \ge 0$ | $y'' = f(I_1, I_2) - (2by' + (b^2 + 1)y)$ |
| 17 | $L_{19}^3$ $\partial_x + \partial_y, x\partial_x + y\partial_y, x^2\partial_x + y^2\partial_y$ | $y'' = \frac{y'^{3/2}}{(x-y)}\big(f\big(y'(\frac{x - y_\tau}{y_\tau - y})^2,$ <br> $\quad \frac{(y_\tau - y)^2}{y'_\tau(x - y)^2}\big) - 2y'(y' + 1)\big)$ |
| 18 | $L_{20}^3$ $\partial_x, x\partial_x + \frac{1}{2}y\partial_y, x^2\partial_x + xy\partial_y$ | $y'' = y^{-3} f(\frac{y_\tau}{y}, y'y_\tau(\frac{y'_\tau}{y^2} - \frac{y_\tau}{y}))$ |
| 19 | $L_{24}^4$ $\partial_x, x\partial_x, \partial_y, y\partial_y$ | $y'' = \frac{y'^2}{(y - y_\tau)} f(\frac{y'_\tau}{y'})$ |
| 20 | $L_{25}^4$ $e^{-x}\partial_y, \partial_x, \partial_y, y\partial_y$ | $y'' = (ky' - y'_\tau) f\big(\frac{ky' - y'_\tau}{(k-1)y' - y + y_\tau}\big) - y'$ |
| 21 | $L_{26}^4$ $e^{-x}\partial_y, -xe^{-x}\partial_y, \partial_x, \partial_y$ | $y'' = \frac{f(I_3) + (k-1)y' - y + y_\tau}{(kc - k + 1)} - y'$ |
| 22 | $L_{27}^4$ $e^{-x}\partial_y, e^{-ax}\partial_y, \partial_x, \partial_y$ <br> $0 < \|a\| \le 1, a \ne 1$ | $y'' = \frac{1}{(k^a - k)}\big(f(I_4) + (a - 1)(ky' - y'_\tau)\big) - y'$ |

**Table B.2** (continued)

| No. | Lie algebra | Representation of second-order DODEs |
|-----|-------------|-------------------------------------|
| 23 | $L_{28}^4$  $e^{-bx}\sin x\partial_y, e^{-bx}\cos x\partial_y, \partial_x, \partial_y$ $b \geq 0$ | $y'' = \frac{f(I_5)-(b^2+1)[c_1 y_\tau'-(c_2+bc_1)(y-y_\tau)]}{[k^b-(bc_1+c_2)]} - 2by'$ |
| 24 | $L_{29}^4$  $\partial_x, x\partial_x, y\partial_y, x^2\partial_x + xy\partial_y$ | $y'' = f\left(\frac{y_\tau}{y}\right)\frac{y^2}{y}\left(\frac{y_\tau}{y} - \frac{y'}{y'}\right)^2$ |
| 25 | $L_{31}^4$  $\partial_y, -x\partial_y, \frac{1}{2}x^2\partial_y, \partial_x$ | $cy'' = y' - y_\tau' - f\left(2(y - y_\tau) - c(y' + y_\tau')\right)$ |
| 26 | $L_{32}^4$  $e^{-bx}\partial_y, e^{-x}\partial_y, -xe^{-x}\partial_y, \partial_x$ | $y'' = \frac{(f(I_6)-(b-1)^2[k(y+y')-(y_\tau-y_\tau')])}{(1-b)(k^b-k)} - [2y'+y]$ |
| 27 | $L_{33}^4$  $e^{-x}\partial_y, -x\partial_y, \partial_y, \partial_x$ | $y'' = \frac{f(I_7)+(y'-y_\tau')}{(k-1)}$ |
| 28 | $L_{34}^4$  $e^{-x}\partial_y, -xe^{-x}\partial_y, \frac{1}{2}x^2e^{-x}\partial_y, \partial_x$ | $y'' = \frac{k(y+y')-f(I_8)-(y_\tau+y_\tau')}{kc} - (2y'+y)$ |
| 29 | $L_{35}^4$  $\partial_y, x\partial_y, \xi_1(x)\partial_y, y\partial_y$ | $y'' = \frac{f(x)\left((\xi_1'-\xi_1'^\tau)(cy'-y+y_\tau)-(\xi_1'c-\xi_1+\xi_1^\tau)(y'-y'^\tau)\right)}{\xi_1'-\xi_1'^\tau}$ $+ \frac{\xi_1''(y'-y_\tau')}{\xi_1'-\xi_1'^\tau}$ |
| 30 | $L_{36}^4$  $e^{-ax}\partial_y, e^{-bx}\partial_y, e^{-x}\partial_y, \partial_x$ $-1 \leq a < b < 1, ab \neq 0$ | $y'' = \frac{(b-a)(b-1)\left(k^a(y+y')-(y_\tau+y_\tau')\right)-f(I_9)}{(k^b-kc)(b-1)}$ $- (a(y+y')+y')$ |
| 31 | $L_{37}^4$  $e^{-ax}\partial_y, e^{-bx}\sin x\partial_y, e^{-bx}\cos x\partial_y, \partial_x$ $a > 0$ | $y'' = \frac{\left(f(I_{10})((a-b)^2+1)[k^b c_1[ay+y']-[k^a y-y']]\right)}{[k^b(c_2+(b-a)c_1)-k^a]}$ $- (2b(ay+y')+a^2 y)$ |
| 32 | $L_{38}^4$  $\partial_x, \partial_y, x\partial_y, x\partial_x + (2y+x^2)\partial_y$ | $y'' = \ln\left((y'-y_\tau')^2 f\left(\frac{(y'-y_\tau')^2}{y-y_\tau}\right)\right)$ |
| 33 | $L_{39}^4$  $\partial_y, \partial_x, x\partial_y, (1+b)x\partial_x + y\partial_y$ $|b| \leq 1$ | $y'' = \left((y'-y_\tau')^{2b+1} f[(y-y_\tau)^b (y'-y_\tau')]\right)^{1/b}$ |
| 34 | $L_{40}^4$  $\partial_y, -x\partial_y, \partial_x, y\partial_y$ | $y'' = (y'-y_\tau')f\left(\frac{cy'-y+y_\tau}{y'-y_\tau'}\right)$ |
| 35 | $L_{42}^4$  $\sin x\partial_y, \cos x\partial_y, y\partial_y, \partial_x$ | $y'' = y\left(f(I_{11})(c_1 + \frac{y'}{y}(c_2 - \frac{y_\tau'}{y'})) - 1\right)$ |
| 36 | $L_{50}^{r+1}\partial_x, \eta_1(x)\partial_y, \ldots, \eta_r(x)\partial_y$ $r \geq 4$ | $\Phi_1(x, y, y_\tau, y', y_\tau', y'')$ |
| 37 | $L_{51}^{r+2}\partial_x, y\partial_y, \eta_1(x)\partial_y, \ldots, \eta_r(x)\partial_y$ $r \geq 3$ | $\Phi_2(x, y, y_\tau, y', y_\tau', y'')$ |
| 38 | $L_{54}^5$  $\partial_x, x\partial_x, y\partial_y, \partial_y, x\partial_y$ | $y'' = \frac{k_2(y'-y_\tau')^2}{y-y_\tau}$ |
| 39 | $L_{55}^5$  $\partial_x, \partial_y, 2x\partial_x + y\partial_y, x\partial_y, x^2\partial_x + xy\partial_y$ | $y'' = \frac{k_3}{(y-y_\tau)^3}$ |
| 40 | $L_{23}^4$  $\partial_y, x\partial_y, \xi_1(x)\partial_y, \xi_2(x)\partial_y$ | $y'' = \frac{I_{12}+\xi_1''(y'-y_\tau')}{\xi_1'-\xi_1'^\tau}$ |

$$I_2 = (c_2 - bc_1)[c_1 y_\tau' + (c_2 + bc_1)y_\tau] - [k^b c_1 y' + y_\tau],$$
$$I_3 = k_1(y_\tau - y - y' + y_\tau') + (k-1)(ky' - y_\tau'),$$
$$I_4 = (k^a - ak + a - 1)(ky' - y_\tau') - a(k^a - k)[(k-1)y' - y + y_\tau],$$
$$I_5 = [k^b(c_2 - bc_1) - 1][c_1 y_\tau' - (c_2 + bc_1)(y - y_\tau)]$$
$$+ [k^b - (c_2 + bc_1)][k^b c_1 y' - (y - y_\tau)],$$
$$I_6 = (k^b - bck + ck - k)\left(k(y + y') - (y_\tau - y_\tau')\right)$$
$$- (b-1)(k^b - k)\left(k_1(y + y') + ky - y_\tau\right),$$

$$I_7 = (k-1)(y - y_\tau - cy') + (k - c - 1)(y' - y'_\tau),$$

$$I_8 = c[k(y + y') - (y_\tau + y'_\tau)] - 2[k_1(y + y') + ky - y_\tau],$$

$$I_9 = [-k^{b+1}ac - k^{b+1}b + k^{b+1}bc + k^{b+1} + ack^2 - k^2c][k^a(y + y') - (y_\tau + y'_\tau)]$$
$$- (k^b - k_1)(b - 1)[(k - k^a)(y_\tau + y'_\tau) - (1 - a)k^a(ky - y_\tau)],$$

$$I_{10} = k^b(c_2 + (a - b)c_1 - k^a)(c_1(ay_\tau y'_\tau) - [c_2 + (b - a)c_1][k^a y - y_\tau])$$
$$- [k^b + k^a(c_1(a - b) - c_2)],$$

$$I_{11} = \frac{c_2 - \frac{y_\tau}{y} + c_1 \frac{y'}{y}}{c_1 + \frac{y'}{y}(c_2 - \frac{y'_\tau}{y'})}, \qquad I_{12} = \frac{f(x) + \zeta_1 \zeta_2}{\zeta_3 + \zeta_4},$$

where

$$\zeta_1 = \xi_1''(\xi_2'^\tau - \xi_2') + \xi_2''(\xi_1' - \xi_1'^\tau),$$

$$\zeta_2 = (\xi_1' - \xi_1'^\tau)(cy' - y + y_\tau) - (\xi_1'c - \xi_1 + \xi_1^\tau)(y' - y'_\tau),$$

$$\zeta_3 = \xi_2'^\tau(\xi_1'c - \xi_1' + \xi_1^\tau) - \xi_1'^\tau(\xi_2'c - \xi_2' + \xi_2^\tau),$$

$$\zeta_4 = \xi_2'(\xi_1 - \xi_1^\tau) - \xi_1'(\xi_2 - \xi_2^\tau).$$

The function $\Phi_1$ is

$$\Phi_1(x, y, y_\tau, y', y'_\tau, y'') = \phi(z_1, z_2),$$

where $y^{(0)} = y$, and

$$z_1 = y_\tau - \sum_{i=1}^{r} c_{i1} y^{(i-1)}, \quad z_2 = y'_\tau - \sum_{i=1}^{r} c_{i2} y^{(i-1)} \quad (r \geq 4).$$

The function $\phi(z_1, z_2)$ is such that

$$c_{j1}\phi_{z_1} + c_{j2}\phi_{z_2} = 0 \quad (j = 4, \ldots, r).$$

The function $\Phi_2$ is defined by the formula:

$$\Phi_2(x, y, y_\tau, y', y'_\tau, y'') = y_\tau - c_3 y'_\tau + \sum_{i=1}^{r}(c_3 c_{i2} - c_{i1})y^{(i-1)}, \quad r \geq 3,$$

where the constants $c_{j1}$ and $c_{j2}$ obey the equalities

$$c_3 c_{j2} - c_{j1} = 0 \quad (j = 4, \ldots, r + 1).$$

The functions $1, x, \xi_1(x), \xi_2(x), \xi_3(x)$ are linearly independent. The functions $\eta_1(x), \eta_2(x), \eta_3(x), \ldots, \eta_r(x)$ form a fundamental system of solutions of an $r$-order ordinary differential equation with constant coefficients,

$$\eta^{(r)} + c_r \eta^{(r-1)} + \cdots + c_2 \eta' + c_1 \eta = 0.$$

# Index

Y.N. Grigoriev et al., *Symmetries of Integro-Differential Equations*,
Lecture Notes in Physics 806,
DOI 10.1007/978-90-481-3797-8, © Springer Science+Business Media B.V. 2010